Multicultural and Diversity Strategies for the Fire Service

Herbert Z. Wong

Aaron T. Olson

PEARSON
Prentice
Hall

Pearson Education, Upper Saddle River, New Jersey

Library of Congress Control Number: 2008921106

Publisher: Julie Levin Alexander
Publisher's Assistant: Regina Bruno
Senior Acquisitions Editor: Stephen Smith
Associate Editor: Monica Moosang
Development: TripleSSS Press Media Development, Inc.
Editorial Assistant: Patricia Linard
Director of Marketing: Karen Allman
Executive Marketing Manger: Katrin Beacom
Marketing Specialist: Michael Sirinides
Marketing Assistant: Lauren Castellano
Managing Production Editor: Patrick Walsh
Production Liaison: Julie Li
Production Editor: Lisa Garboski, bookworks
Media Product Manger: John Jordan
Media Project Manager: Stephen J. Hartner
Manufacturing Manager: Ilene Sanford
Manufacturing Buyer: Pat Brown
Senior Designer Coordinator: Christoper Weigand
Cover Designer: Christoper Weigand
Cover Photo: Ray Kemp
Director, Image Research Center: Melinda Reo
Manager, Rights and Permissions: Zina Arabia
Manager, Visual Research: Beth Brenzel
Manager, Cover Visual Research and Permissions: Karen Sanatar
Image Permission Coordinator: Vicki Menanteaux
Composition: Aptara, Inc.
Printing and Binding: Edwards Brothers
Cover Printer: Phoenix Color Corporation

Pearson Education Ltd.
Pearson Education Singapore Pte. Ltd.
Pearson Education Canada, Ltd.
Pearson Education—Japan

Pearson Education Australia Pty. Limited
Pearson Education North Asia Ltd.
Pearson Educación de Mexico, S.A. de C.V.
Pearson Education Malaysia Pte. Ltd.

10 9 8 7 6 5 4 3 2 1
ISBN 13: 978-0-13-238807-8
ISBN 10:   0-13-238807-3

*To the fire service officers, who, in their day-to-day work,
demonstrate the professionalism that is required in
our multicultural society, and to the
memory of Glenn Michael Perry*

Contents

APPENDIXES

Foreword

The United States is one of the most diverse countries in the world. As our cities, counties, and states continue to become more multicultural, skills and competencies with cultural differences need to become the norm of human behavior. Enhanced awareness of the value of our multicultural environment needs to be a common focus among our emergency service workers.

Because the fire service in the United States is a microcosm of American society, it is becoming more multicultural in its workforce makeup. Progressive and responsive fire departments recognize the value of educating and training their personnel to be culturally competent, value cultural differences, and acquire cross-cultural skills. They realize this is an ongoing process that requires being open to new paradigms because of changing demographics. Changing multicultural communities require a fire service that is dynamic, efficient, and progressive in meeting the needs of the diverse people they serve. Failure to remain acutely aware of the richness of the varied cultures we serve and failure to meet that richness with a robust approach to embracing diversity in all we do are failures of our basic tenet to protect and serve.

Multicultural and Diversity Strategies for the Fire Service is a useful tool enabling the fire services to meet such challenges. It will prove helpful as a resource in fire science college courses or recruit and in-service fire service academies in all jurisdictions. Moreover, its contents should be useful to a variety of first-responder practitioners in emergency medical services, as well. The major sections of *Multicultural and Diversity Strategies for the Fire Service* effectively address the key multicultural needs of fire services. The practical contents provide critical information and insight that will improve firefighter performance and professionalism. The subject matter presented on cultural-specific information is on the leading edge. As a complete learning system, *Multicultural and Diversity Strategies for the Fire Service* is partnered with appropriate learning supplements such as an Instructor's Manual, chapter quizzes, a student study guide, and PowerPoint presentations for each chapter. This systematic educational approach will aid fire service in training awareness and application of appropriate strategies to meet the service desires of our diverse communities and organizations.

The authors are themselves an advanced and exceptional multicultural writing team. Both have trained fire service personnel in multicultural and diversity topics, areas in which the fire service has great needs but limited expertise. Both are coauthors of the fourth edition of *Multicultural Law Enforcement: Strategies for Peacekeeping in a Diverse Society* and have expanded their expertise to the fire service. Dr. Herbert Z. Wong, Ph.D., is an organizational psychologist, research director, and college professor who has worked within city, county, state, and federal levels of government agencies in training fire service officers. Aaron T. Olson, M.Ed., is a retired

Oregon State Police patrol supervisor who spent the majority of his career working side-by-side with firefighters on calls for assistance; he is now a college professor and organizational consultant. Herb and Aaron have compiled what I predict will become a classic text in multicultural and diversity strategies for the fire service. I am confident in recommending this text as a solid foundation for developing multicultural awareness and key diversity strategies at all levels of the fire service. I encourage all who use this text to take what you learn from the authors and put into action the strategies, information, and tools of this exceptional work for the betterment of yourself, your agency, and the culturally diverse community you serve.

Kevin S. Brame
Deputy Fire Chief
City of North Las Vegas, Nevada

Preface

This first edition of *Multicultural and Diversity Strategies for the Fire Service* is a textbook for all fire service agencies and academies, colleges, and universities. It is also designed to assist all levels of fire service representatives in understanding the pervasive influences of culture, race, and ethnicity in the workplace and in multicultural communities. The text focuses on the cross-cultural contact that fire service personnel have with citizens and coworkers from diverse backgrounds. We include material on the following: immigrants, undocumented immigrants, women, gays, and lesbians in the fire service; and demographic data using the most current population estimates and projections available to date.

Both coauthors have worked closely and extensively with firefighters. The first author (Wong) has helped fire agencies to recruit and develop strategies for selection and retention issues involving minority and women candidates. He has worked extensively to train fire service personnel in cultural awareness and workplace diversity topics, as well. The second author (Olson) has worked side-by-side with firefighters to deal with emergencies such as car fires, medical assistance, accidental deaths, and other issues in his role as an Oregon State Police officer. In his training and consulting roles, he also teaches specific topics to fire service personnel to include multicultural and diversity strategies. Both of us have sought the information and input from fire service personnel to write this important textbook to be useful to firefighters.

Throughout these pages, we stress the need for awareness, understanding of cultural differences, and respect toward those of different backgrounds. We encourage all representatives of the fire service to examine preconceived notions they might hold of particular groups. We outline for all fire service personnel why they should build awareness and promote cultural understanding and tolerance within their agencies.

An increasing number of leaders in the fire service and their employees have accepted the premise that greater multicultural competency must be a key objective of all management and professional development. We know this firsthand as educators and professional trainers for fire departments because fire service personnel are hungry for and responsive to multicultural training and education that is relevant to their profession. This need is driven by changing demographics in our communities and workforce.

In a nation and world in which news is reported immediately, the public is exposed almost daily to multicultural contact between public safety agencies and citizens. Also, community members have become increasingly sophisticated and watchful with how members of multicultural groups are treated by public servants. Professional fire service agencies and other public safety entities should expect public scrutiny and the legal accountability that accompanies it.

Multicultural and Diversity Strategies for the Fire Service provides practical information and guidelines for fire service executives, managers, supervisors, firefighters, instructors, and the college student. It is the only fire service textbook of its kind that offers multicultural and diversity strategies for this highly respected profession. With multicultural knowledge, competency, and tolerance, those who are employed in the fire service will enhance their image while demonstrating greater professionalism with the changing multicultural workforce and community.

Herbert Z. Wong, Ph.D.

Aaron T. Olson, M.Ed.

Acknowledgments

This first edition has benefited from expert content contributions by the following people and organizations:

Marilyn Loden, President, Loden Associates, Inc., Tiburon, California

Char Miller, Professor of History, Trinity College, San Antonio, Texas

George Thompson, Ph.D., President, Verbal Judo Institute, Auburn, New York

David Benton, Civilian Policy Advisor to the Commandant, U.S. Coast Guard, Washington, DC

John Brothers, Executive Director, Quincy Asian Resources, Inc., Quincy, Massachusetts

Patricia DeRosa, President, ChangeWorks Consulting, Randolph, Massachusetts

Lubna Ismail, President, Connecting Cultures, Washington, DC

Lourdes Rodriguez-Nogues, Ed.D., President, Rasi Associates, Boston, Massachusetts

Victoria Santos, President, Santos & Associates, Newark, California

A very special thank-you goes to Stephen Smith, Senior Acquisitions Editor, for his invaluable insights, perseverance, and dedication to this project, and to the members of the Brady team at Prentice Hall for their assistance and support.

We owe special thanks to the following individuals for their helpful contributions:

Marc Crain, Deputy Chief, Clackamas County Fire District #1—Training Division, Clackamas, Oregon

Lieutenant Scott Steiner, Tualatin Valley Fire and Rescue, Aloha, Oregon

Kim Ah-Low, Georgia

Joe Canton, Ph.D., California

Chung H. Chuong, California

Wilbur Herrington, Massachusetts

Jim Kahue, Hawaii

Dinh Van Nguyen, California

Paula Parnegian, Massachusetts

We would also like to thank the following reviewers:

John P. Alexander
Adjunct Instructor
Connecticut Fire Academy
Enfield, Connecticut

Miles N. Allen, CAI
Winston, Georgia

Craig Bryan, Captain
City of Forest Park Fire Department
Division of Training
Forest Park, Ohio

Robert L. Darr, FF/Paramedic
North Pulaski Fire-Rescue
Jacksonville, Arkansas

Lynne Dees, M.F.A., L.P., Assistant Professor
Emergency Medicine Education
University of Texas Southwestern Medical Center at Dallas
Dallas, Texas

David W. Goldblum, Ex-Chief
Stewart Manor Fire Department
Stewart Manor, New York

Ted Huffman, Commander
Fire Training Academy Cuyahoga
Community College
Parma, Ohio

Matthew Marcarelli, Lieutenant
City of New Haven Fire Department
Northford, Connecticut

Jerry A. Nulliner, Division Chief
Fishers Fire Department
Fishers, Indiana

J.T. O'Neal
Deputy Fire Marshal
Henderson, Nevada

Ken Riddle, Deputy Fire Chief
Las Vegas Fire & Rescue
Las Vegas, Nevada

Michael Ryan, M.S.
Office of Emergency Management Fairfax
County, Virginia

Douglas P. Skinner, NREMT-P
Training Officer
Loudoun County Fire Rescue
Leesburg, Virginia

Thomas Y. Smith
NFPA Member, GA, FLA & NPQS
Instructor
Fire Science Program Director
West Georgia Technical College
LaGrange, Georgia

Kevin S. Walker, JD MBA
Eastern Oregon University
Spokane, Washington

About the Authors

Herbert Z. Wong, Ph.D., an organizational psychologist, provides cultural awareness and diversity training to fire service and other public safety officers on local, state, and federal levels nationwide. Dr. Wong is professor of psychology and research director with the Graduate School of Professional Psychology at John F. Kennedy University. He is president of Herbert Z. Wong & Associates, a management consulting firm to over 300 businesses, universities, government agencies, and health care corporations, specializing in multicultural management and workforce diversity. In 1990, Dr. Wong cofounded and  was president of the National Diversity Conference, which became the Society for Human Resource Management's Workplace Diversity Conference. He developed and provided the national Training-of-Trainers programs for the seven-part Valuing Diversity film series used in over 4,000 organizations worldwide. For 14 years, Dr. Wong served as the executive director of a comprehensive Community Mental Health Center (CMHC) in San Francisco, which provided crisis intervention, inpatient, partial hospitalization, outpatient, and prevention services in 22 languages. In addition, Dr. Wong has written books, extensive reports, and publications on workforce diversity, cultural competency, and multicultural leadership issues.

Aaron T. Olson, M.Ed., is an organization and training consultant, specializing in staff development and multicultural training workshops for the fire service (regular recruit and supervisor in-service academies), medical, 9-1-1, law enforcement, corrections, and transportation professionals. He is a criminal justice instructor at Portland Community College, Portland, Oregon, where he teaches Courts, Criminal Investigations, Cultural Diversity, and Evidence. He designed the first accredited cultural diversity course and curriculum for the college's criminal justice program in 2001. He is a retired Oregon State Police patrol sergeant and shift supervisor with 26 years of police experience in communications, recruiting, and patrol assignments. He has instructed at Oregon's Department of Public Safety Standards and Training, teaching recruit, supervision, mid-management, and executive management courses. He established and still delivers public safety workshops to immigrants and refugees at the Immigrant Refugee Community Organization (IRCO) and has written instructor manuals and textbooks dealing with workforce diversity, leadership, communications, and train-the-trainer instructor courses.

The Changing Fire Department: A Microcosm of Society

1 CHAPTER

Key Terms

Overview

The gender, racial, ethnic, and lifestyle composition of fire departments is changing in the United States. In this chapter we examine the increasingly pluralistic workforce and provide examples of racism and cultural insensitivity within fire departments (Figure 1.1). We present suggestions for defusing racially and culturally rooted conflicts and address issues related to women, men, lesbians, and gay men in fire departments. The chapter ends with recommendations for all employees who work within a diverse workforce and particularly emphasizes the role of the chief executive.

Objectives

After completing this chapter, participants should be able to:

+ Distinguish the differences from the alarm to the firehouse.
+ Describe the demographics of society.
+ Discuss the implications of the changing workforce.
+ Describe the dimensions of diversity.

FIGURE 1.1 ◆ The Statue of Liberty in the United States is the symbol of freedom, immigration, and the nucleus for the microcosm of diversity in America.

- Discuss ethnic and racial issues.
- Discuss women and gender issues.
- Discuss gay and lesbian issues.
- Describe the role of leadership: react or respond?
- Explain the influencing factors for the changing fire department.
- Explain why fire departments need to bridge cultural and racial gaps within their organizations.

LEARNING TASKS

fire agency

Use synonymously as fire department and firehouse.

Emergency responders recognize the importance of staying current with changes to the U.S. DOT's National Standard Curriculum and their local EMS system. The same holds true for being knowledgeable and competent on the impact of cultural group dynamics to an organization. Fire departments are no exception and are a microcosm of society. Fire agencies are expected to adapt to the changes in society. In this text, the terms *firehouse*, *fire department*, and ***fire agency*** are used interchangeably. The uses of these different terms reflect the transitions and changes that are happening within the fire service today. Talk with those at your agency and find out how they are embracing change. You should be able to:

- Find out the demographics in your state, county, and city.
- Find out the dimensions of diversity in your county and city fire departments. Every state, county, and city has varying demographics. Fire departments are continuing to reflect the communities they serve in their employment of more diverse groups. These groups include women, gays, lesbians, nonwhites, different ethnic groups, and older new hires.
- Become familiar with your local fire department's policies and procedures in valuing and embracing a diverse workforce.

A BRIEF EXPLANATION ON PERCEPTION

Perception, also referred to as observation, is what we see with our eyes or sight. In more detail, perception also includes the awareness of our environment. *Perception* is a Latin word that dates back to the fourteenth century. We can appreciate people using the word *perception* because it is how they see things; and it is their commentary on the subject, topic, or person they are being asked about.

FIRE DEPARTMENTS ARE NOT EXEMPT

Racism and **discrimination** are two words that most people don't use when they describe fire departments. These words do align themselves to law enforcement because of documented cases of police brutality and racial profiling for decades (Figure 1.2). This is a perception that continues to stick with police departments. Too many national and regional incidents since the Rodney King beating have resurfaced to prevent people from dismissing this perception. Unfortunately, the behavior of racism and discrimination has been documented as pervasive in fire departments too, as shown in Chicago, Dearborn, and San Francisco.

Chicago Fire Department   The Chicago Fire Department has received decades of attention. In 1973, a federal court decree ordered Chicago's fire department to hire and promote more minorities because of its discriminatory hiring and promotion practices. In 1997 a video surfaced showing several firefighters drinking and using racial slurs at a firehouse party. The discipline for this misconduct netted seven dismissals and 21 suspensions, but an arbitrator later reversed this decision, citing the

racism

Total rejection of others by reason of race, color, or, sometimes more broadly, culture.

discrimination

The denial of equal treatment to groups because of their racial, ethnic, gender, religious, or other form of cultural identity.

FIGURE 1.2   ◆   Community members hold protest signs regarding the adverse treatment of African American Tawana Brawley during a demonstration in New York City.

city had brought the charges too late. On February 2, 2004, a Chicago firefighter transmitted racial slurs over his radio and was later given a 90-day suspension and transfer. This radio transmission was not an isolated incident because the department's internal affairs division was investigating six other separate radio transmissions in which racial slurs were also used (*USA Today,* 2004).

Dearborn Fire Department   In November 2005, a Jewish firefighter with the Dearborn, Michigan, Fire Department filed in Wayne County Circuit Court that the city of Dearborn had violated his state civil rights. Tom Begres, age 36, alleged that his coworkers used **anti-Semitic** slurs against him and that he had been denied career advancement because of his religion. Information of a firefighter using a city computer to purchase Nazi paraphernalia and also the use of an offensive Japanese epithet were mentioned by Begres's attorney (Warikoo, 2005).

San Francisco Fire Department   In July 2005, the city of San Francisco, California, agreed to a $400,000 settlement of a lawsuit brought by a female firefighter who indicated on-the-job drinking at the fire department had contributed to sexual harassment against her. Kristen Odlaug, 41, sued the department claiming that alcohol was prevalent at one firehouse in the Richmond District and that executive management tolerated on-duty drinking, which influenced the sexual harassment against her by coworkers. The allegations that several firefighters were drinking on duty was sustained, and the department's second-in-command, Deputy Chief Fred Sanchez, was asked to and did retire in April 2005 (Derbeken, 2005).

anti-Semitism
Latent or overt hostility toward Jews, often expressed through social, economic, institutional, religious, cultural, or political discrimination and through acts of individual or group violence.

INTRODUCTION

"The 21st century will be the century in which we (America) redefine ourselves as the first country in world history that is literally made up of every part of the world," said Kenneth Prewitt, director of the U.S. Census Bureau, 1998–2001 (Prewitt, 2001).

multiculturalism
The existence within one society of diverse groups that maintain their unique cultural identify while accepting and participating in the larger society's legal and political system.

Multiculturalism and diversity are at the very center of America and best describe the demographics of our nation accurately. The word **multiculturalism** does not refer to a movement or political force, nor is it an anti-American term. A multicultural community is one that is composed of many different ethnic and racial groups. The United States, compared to virtually all other nations, has experienced unparalleled growth in its multicultural population. Reactions to these changes range from appreciation and even celebration of diversity to an absolute intolerance of differences.

The United States has always been a magnet for people from nearly every corner of the earth, and, consequently, U.S. demographics continue to undergo constant change. In efforts to be both proactive and responsive to diverse communities, firefighters and groups from many backgrounds around the country are working to connect more closely with each other. Leaders from the fire service and the community have realized that they both benefit when each group seeks mutual assistance and understanding. The firefighter occupation requires a certain level of comfort and professionalism in interacting with people from all backgrounds, whether working with community members to build trust or dealing with calls, alarms, and coworkers.

Through increased awareness, knowledge, and skills, the fire service as a profession can increase its **cultural competence**. Developing cultural competence is a process that evolves over time, requiring that individuals and organizations place a high value on the following (adapted from Cross, Bazron, Dennis, and Isaacs, 1989):

- Developing a set of principles, attitudes, policies, and structures that will enable all individuals in the organization to work effectively and equitably across all cultures
- Developing the capacity to acquire and apply cross-cultural knowledge
- Developing the capacity to respond to and communicate effectively within the cultural contexts that the organization serves

The strategies used to approach and build rapport with one's own cultural group may very well result in unsuspected difficulties with another group. The acts of approaching, communicating, questioning, assisting, and establishing trust with members of culturally diverse groups require special knowledge and skills. Acquiring knowledge and skills that lead to sensitivity does not mean preferential treatment of any one group; rather, it contributes to improved communication with members of all groups.

Individuals must seek a balance between downplaying and denying the differences of others, and distorting the role of culture, race, and ethnicity. In an effort simply to "respect all humans equally," we may inadvertently diminish the influence of culture or ethnicity, including the role it has played historically in our society.

THE MELTING POT MYTH AND THE MOSAIC

Multiculturalism, also referred to as cultural pluralism, violates what some consider being the "American way of life." However, from the time our country was founded, we have never been a homogeneous society. The indigenous peoples of America (the ancestors of the American Indians) were here long before Christopher Columbus "discovered" them. There is even strong evidence that the first Africans who set foot in this country came as free people, 200 years before the slave trade from Africa began (Rawlins, 1992). Furthermore, the majority of people in America can claim to be the children, grandchildren, or great-grandchildren of people who have migrated here. Americans did not originate from a common stock. Until fairly recently, America has been referred to as a *melting pot,* a term depicting an image of people coming together and forming a unified culture. One of the earliest usages of the term was in the early 1900s, when a famous American playwright, Israel Zangwill, referring to the mass migration from Europe, said, "America is God's crucible, the great Melting-Pot where all the races of Europe are melting and re-forming. Germans and Frenchmen, Irishmen and Englishmen, Jews and Russians—into the Crucible with you all! God is making the American!" (Zangwill, 1908).

This first use of the term *melting pot* was not designed to incorporate anyone except Europeans. Did the melting pot ever exist, then, in the United States? No, it never did. Yet people still refer to the belief, which is not much more than a romantic myth about the "good old days." African Americans, brought forcibly to this country between 1619 and 1850, were never part of the early descriptions of the melting pot. Likewise, Native American people were not considered for the melting pot. It is not coincidental that these groups were nonwhite and therefore not "meltable." Furthermore, throughout our past, great efforts have been made to prevent any additional diversity. Most notable in this regard was the Chinese Exclusion Act in 1882, which denied Chinese laborers the right to enter America. Early in the twentieth century organized labor formed the Japanese and Korean Exclusion League "to protest the

cultural competence

A developmental process that evolves over an extended period of time; both individuals and organizations are at various levels of awareness, knowledge, and skills along a continuum of cultural competence. Culturally competent organizations place a high value on the following: (1) developing a set of principles, attitudes, policies, and structures that will enable all individuals in the organization to work effectively and equitably across all cultures; and (2) developing the capacity to acquire and apply cross-cultural knowledge and respond to and communicate effectively within the cultural contexts that the organization serves.

influx of 'Coolie' labor and in fear of threat to the living standards of American workingmen" (Kennedy, 1986, p. 72). Immigration was discouraged or prevented if it did not add strength to what already existed as the European-descended majority of the population (Handlin, 1975).

Even at the peak of immigration (late 1800s), New York City exemplified how different **immigrant** groups stayed separate from each other, with little of the "blending" that people often imagine took place (Miller, 2006). Three-fourths of New York City's population consisted of first- or second-generation immigrants (including Europeans and Asians); 80 percent did not speak English, and there were 100 foreign-language newspapers in circulation. The new arrivals were not accepted by those who had already settled, and newcomers found comfort in an alien society by choosing to remain in ethnic enclaves with people who shared their culture and life experiences.

The first generation of every immigrant and refugee group seeing the United States as the land of hope and opportunity has always experienced obstacles in **acculturation** (i.e., integration) into the new society. In many cases, people resisted Americanization and kept to themselves. Italians, the Irish, eastern European Jews, the Portuguese, Germans, and virtually all other groups tended to remain apart when they first came. Most previously settled immigrants were distrustful and disdainful of each newcomer group. "Mainstreaming" began to occur only with children of immigrants (although some people within certain immigrant groups tried to assimilate quickly). For the most part, however, society did not permit a quick shedding of previous cultural identity. History has never supported the metaphor of the melting pot, especially with regard to the first and second generations of most groups of newcomers. Despite the reality of past multicultural disharmony and tension in the United States, the notion of the melting pot prevailed.

The terms *mosaic* and *tapestry* more accurately and idealistically portray a view of diversity in America. They describe a society in which all colors and backgrounds contribute their parts to form society as a whole, but one in which groups are not required to lose their characteristics in order to "melt" together. The idea of a mosaic portrays a society in which all races and ethnic groups are displayed in a form that is attractive because of the very elements of which it is made. Each group is seen as separate and distinct in contributing its own color, shape, and design to the whole, resulting in an enriched society.

REACTIONS TO DIVERSITY: PAST AND PRESENT

Accepting **diversity** has always been a difficult proposition for most Americans (Miller, 2006). Typical criticisms of immigrants, now and historically, include "They hold on to their cultures," "They don't learn our language," "Their customs and behavior are strange," and "They form cliques." Many newcomers, in fact, have historically resisted Americanization, keeping to ethnic enclaves. They were not usually accepted by mainstream society.

Are the reactions to newcomers today so different from people's reactions to earlier waves of immigrants? Let's look at reactions to the Irish, who by the middle of the nineteenth century constituted the largest group of immigrants in the United States, making up almost 45 percent of the foreign-born population. Approximately 4.25 million people left Ireland, mainly because of the potato famine. Many of these immigrants had come from rural areas but ended up in cities on the East Coast. Most were illiterate; some spoke only Gaelic (Kennedy, 1986). Their reception in America

immigrant

Any individual who moves from one country, place, or locality to another. An alien admitted to the United States as a lawful permanent resident.

acculturation

The process of becoming familiar with and comfortable in another culture. The ability to function within a different culture or environment, while retaining one's own cultural identity.

diversity

The term used to describe a vast range of cultural differences that have become factors needing attention in living and working together. Often applied to the organizational and training interventions in an organization that seek to deal with the interface of people who are different from each other. Diversity has come to include race, ethnicity, gender, disability, and sexual orientation.

was anything but welcoming, exemplified by the plethora of signs stating, "Jobs available, no Irish need apply." Kennedy further writes:

> The Irish . . . endure[d] the scorn and discrimination later to be inflicted, to some degree at least, on each successive wave of immigrants by already settled "Americans." In speech and in dress, they seemed foreign; they were poor and unskilled and they were arriving in overwhelming numbers. . . . The Irish found many doors closed to them, both socially and economically. When their earnings were not enough . . . their wives and daughters obtained employment as servants (1986, p. 18).

If this account were rewritten without specific references to time and cultural group, it would be reasonable to assume it describes contemporary reactions to newcomers. One could take this quotation and substitute Jew, Italian, or Polish at various points in history. Today, it could be used in reference to Cubans, Somalis, Afghans, Mexicans, Haitians, Serbs, or Ethiopians. If we compare immigration today with that during earlier periods in U.S. history, we find similarities as well as significant differences. In the past few decades, we have received people from cultures more dramatically different than those from Western Europe. For example, many of our "new Americans" from parts of Asia or Africa bring values and languages not commonly associated with or related to mainstream American values and language. Middle Easterners bring customs unknown to many U.S.-born Americans. (For cultural specifics, refer to Chapter 5.) Many **refugees** bring scars of political persecution or war trauma, the nature of which the majority of Americans cannot even fathom. The relatively mild experiences of those who came as voluntary migrants do not compare with the tragedies of many of the more recent refugees. True, desperate economic conditions compelled many early European immigrants to leave their countries (and thus their leaving was not entirely voluntary). However, their experiences do not parallel, for example, war-torn eastern European refugees who came to the United States in the 1990s.

Disparaging comments were once made toward the very people whose descendants would, in later years, constitute much of mainstream America. Many fourth- and fifth-generation immigrants have forgotten their history (Miller, 2006) and are intolerant of the "foreign ways" of emerging immigrant groups. Every new group seems to be met with some suspicion and, in many cases, hostility. Adjustment to a new society is and has always been a long and painful process, and the first-generation immigrant group suffers, whether Irish, Jewish, Polish, Afghani, Laotian, Filipino, or Russian. It must also be remembered that many groups did not come to the United States of their own free will but rather were victims of a political or economic system that forced them abruptly to cut their roots and escape their homelands. Although grateful for their welcome to this country, such newcomers did not want to be uprooted. Many new Americans did not have any part in the creation of events that led to the flight from their countries.

refugee
A person who flees for safety and seeks asylum in another country. In addition to those persecuted for political, religious, and racial reasons, "economic refugees" flee conditions of poverty for better opportunities elsewhere.

THE ALARM AND THEN THE FIREHOUSE

Firefighters respond to all types of calls for assistance. Working together is essential in dealing with medical emergencies, motor vehicle accidents, water rescue emergencies, fire suppression, training, and all the other encompassing operations. Unless one is a firefighter or works closely with firefighters, it is difficult to understand or grasp what they do. Firefighters are unique in many aspects. One of the authors (Olson, 2007), a

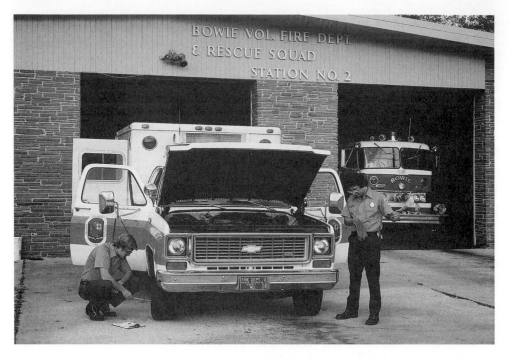

FIGURE 1.3 ◆ Fire stations, also known as firehouses, serve as the workstation and second home for firefighters pulling their shift through America.

retired Oregon State Police patrol supervisor who worked on calls for assistance with firefighters and now is a fire service academy and in-service trainer, and a criminal justice college instructor, shares the following observations based on his personal and professional experiences with firefighters:

• One specific aspect that separates firefighters from all professions, including law enforcement and corrections, is that firefighters are together usually for 24-hour shifts.
• People working a regular job can usually put on a false appearance when interacting with coworkers or citizens. They can take their breaks and lunch hour to escape from stress. In-between calls, firefighters are together in the firehouse (Figure 1.3).
• Nonfirefighters get to go home after working a day at the office. The firehouse is the firefighter's home, and it becomes a combination of work and home. Coworkers become a second family and, inherent with any family relationship, are more open with each other in communication and interpersonal behavior.
• As first responders, firefighters constantly face the risk of injury and death. The investigation of a firefighter killed in the line of duty ranks as one of the worst calls a police officer has to investigate.
• Combined with the dangers of the job and the homelike setting of the firehouse, it is understandable why firefighters need to be culturally competent, be responsive, and take care of each other.

THE DEMOGRAPHICS OF SOCIETY

THE DEFINITION OF DEMOGRAPHICS

Demographics are the characteristics of a population. Demographics are inclusive and classified by age, race, ethnicity, national origin, gender, occupation, religion, political affiliation, sexual orientation, marital status, and more. The study of demographics is used for but not limited to market research, sociological analysis, political campaigns, needs assessments, and organization evaluations. Effective leaders and organizations pay close attention to demographics at the global, national, state, and local levels. This information allows leaders and organizations to recognize who their customers and constituents are. Intelligent leaders and organizations use the information derived from demographics to meet the needs of their stakeholders. We cannot overstate the importance of demographics and will revisit its value in further discussions of leadership issues in this chapter and again in Chapter 7.

THE OVERLAP OF RACE, CULTURE, AND ETHNICITY

Before entering into a discussion of demographics of various minority and immigrant groups, we must mention how, in this twenty-first century, demographic estimates and projections are likely to fall short of counting the true mix of people in the United States. In Chapter 5 of this book, we discuss characteristics and fire service–related issues of African Americans Asian and Pacific Americans, Latino and Hispanic Americans, Arab Americans, and other Middle Eastern groups, and Native Americans. This categorization is merely for the sake of convenience; an individual may belong to two or more groups. For example, a black Latino, such as a person from the Dominican Republic or Brazil, may identify himself or herself as both black and Latino. **Race** and ethnic background (e.g., in the case of a black Latino) are not necessarily mutually exclusive. Hispanic is considered an **ethnicity**, not a race. Therefore people of Latino descent can count themselves as part of any race. An individual who in the 1990 census counted himself or herself as black and now chooses both black and white is considered one person with two races in the 2000 census ("Multiracial Data," 2000). Public safety officials need to be aware of the overlap between race and ethnicity and that many individuals consider themselves to be multiracial. "Every day, in every corner of America, we are redrawing the color lines and are redefining what race really means. It's not just a matter of black and white anymore; the nuances of brown and yellow and red mean more—and less than ever" ("Redefining Race in America," 2000).

The United States, a heterogeneous society, is an amalgam of races, cultures, and **ethnic groups**. The first photograph in the "Redefining Race in America" article just mentioned shows a child whose ethnicity is Nigerian, Irish, African American, Native American, Russian Jewish, and Polish Jewish (from parents and grandparents). Her U.S. census category may be simply "other." When we interpret population statistics, we have to understand that the face of America is changing. In 1860 there were only three census categories: black, white, and quadroon (a person who has one black grandparent or the child of a mulatto and a white). In the 2000 census, there were 63 possible options for marking racial identity, or twice that if people responded to whether or not they were of Hispanic ethnicity. As the 2000 census director, Kenneth Prewitt, wrote, the concept of classification by race is human-made and endlessly complex.

> What is extraordinary is that the nation moved suddenly, and with only minimal public understanding of the consequences, from a limited and relatively closed racial taxonomy to one that has no limits. In the future, racial categories will no doubt become more numerous, and why not? What grounds does the government have to declare "enough is enough"? When there were only three or even four or five categories, maybe "enough is enough" was plausible. But how can we decide, as a nation, that what we allow for on the

race

A group of persons of (or regarded as of) common ancestry. Physical characteristics are often used to identify people of different races. These characteristics should not be used to identify ethnic groups, which can cross racial lines.

ethnicity

Refers to the background of a group with unique language, ancestral, often religious, and physical characteristics. Broadly characterizes a religious, racial, national, or cultural group.

ethnic group

Group of people who conceive of themselves, and who are regarded by others, as alike because of their common ancestry, language, and physical characterisics.

TABLE 1.1 ◆ Difference in Population by Race and Hispanic Origin: 1990 to 2000 (rounded to the nearest hundred thousand)				
Ethnicity/Race	*1990*	*2000*	*Difference*	*Percent Difference*
Hispanic origin	22.4	35.3	13.0	58.00%
White, non-Hispanic	188.1	194.6	6.4	3.40%
Black, non-Hispanic	29.2	34	4.7	16.20%
Asian	6.6	10.1	3.5	52.40%
American Indian and Alaska Native	1.8	2.1	0.3	15.30%
Native Hawaiian and Other Pacific Islander	0.3	0.3	0	8.50%
Other Race	0.3	0.5	0.2	87.80%
Total	248.7	281.4	32.7	13.20%

Source: U.S. Census Bureau, 2001.

Note: The 2000 data do not include persons who described themselves as being of more than one of the above racial categories, as the 1990 data did not include this option.

census form of today—63 racial groups or 126 racial/ethnic ones—is the "right" number? It can't be, nor can any other number be "right." There is no political or scientifically defensible limit (Prewitt, 2001).

CHANGING POPULATION

Changes in population characteristics between 1990 and 2000 have been dramatic. Table 1.1 and Figure 1.4 graphically present the relative sizes of various ethnic populations.

According to population projections, by 2050, the non-Hispanic white population will decline to 53 percent from 69 percent in 2000. In 2050, 25 percent of the population will be of Hispanic origin; 15 percent will be black; 9 percent Asian; and 1 percent American Indian, Eskimo and Aleut, and Pacific Islander. The non-Hispanic white population will be the slowest-growing race group (U.S. Census Bureau, 2002)

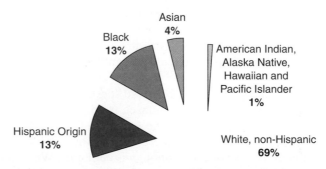

FIGURE 1.4 ◆ Resident Population by Race and Hispanic Origin: 2000.
Source: U.S. Census Bureau, *Statistical Abstract of the United States: 2002.*

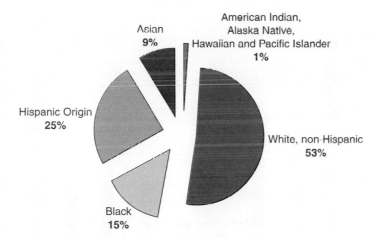

FIGURE 1.5 ◆ U.S. Population Projection by Race and Hispanic Origin Status: 2050. (It should be noted that the total percentage value is 103 resulting from rounding off percentages compared to 100.)
Source: U.S. Census Bureau, *Statistical Abstract of the United States: 2002.*

(Figure 1.5). This demographic shift has already occurred in some large cities across the country and in a number of areas in California, where the minority has become the majority. This change has had a huge impact on many institutions in society, not the least of which is public safety for call for service.

Immigrants Immigration is not a new phenomenon in the United States (Figure 1.6). Virtually every citizen, except for indigenous peoples of America, can claim to be a descendant of someone who migrated (whether voluntarily or not) from another country. Immigration levels per decade reached their highest absolute numbers ever at the end of the last century, when the number of immigrants surpassed 9 million

FIGURE 1.6 ◆ A citizenship swearing in ceremony for immigrants at Monticello, Virginia.

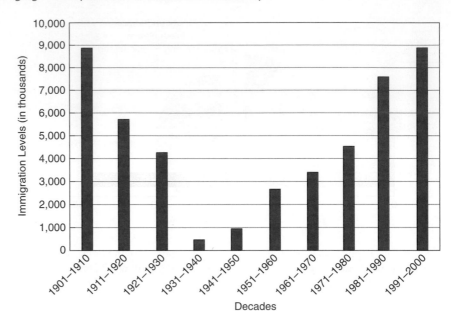

FIGURE 1.7 ◆ Emigration to America in the Twentieth Century.
Source: U.S. Census Bureau, *Statistical Abstract of the United States: 2002.*

from 1991 to 2000 (see Figure 1.7). In addition, immigrants from 1980 to the present have come from many more parts of the world than those who arrived at the turn of the twentieth century. The U.S. Census Bureau reported in March 2002 that approximately 32.5 million U.S. residents (11.5 percent of the population) had been born in other countries.

Among the foreign born in 2002, 52.2 percent were from Latin America, 25.5 percent from Asia, 14.0 percent from Europe, and the remaining 8.3 percent from other regions of the world. The foreign-born population from Central America, including Mexico, made up more than two-thirds of the foreign born from Latin America and more than one-third of the total foreign born (see Figure 1.8). In terms of geographic

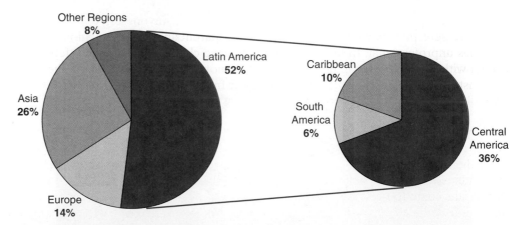

FIGURE 1.8 ◆ Emigration to America by World Region, 2002.
Source: U.S. Census Bureau, Current Population Reports, February 2003: "The Foreign Born Population in the United States: March 2002."

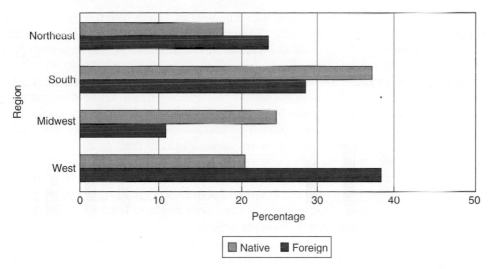

FIGURE 1.9 ◆ Distribution of Native and Foreign-Born Populations by Regional Percentages of Each Population.
Source: U.S. Census Bureau, Current Population Reports, February 2003. "The Foreign Born Population in the United States: March 2002."

distribution within the country, 11 percent of the foreign-born population lived in the Midwest, 23 percent lived in the Northeast, 28 percent in the South, and 38 percent in the West (see Figure 1.9) (U.S. Census, 2002).

Even though most Americans, with the exception of the indigenous peoples, have been immigrants at some time in their lineage, anti-immigrant sentiment is common. Especially in a time of recession, many people perceive that immigrants are taking jobs away from "real Americans" (forgetting that legal immigrants are also "real" Americans). However, the issues surrounding immigration are not as clear-cut as they may at first appear. Despite the problems that are inevitably created when large groups of people have to be absorbed into a society, some immigrant groups stimulate the economy, revitalize neighborhoods, and eventually become fully participatory and loyal American citizens. Nevertheless, if a firefighter has an anti-immigrant bias, negative attitudes may surface when that firefighter interacts with immigrants, especially under stressful circumstances. When firefighters or anyone for that matter is under pressure, most negative attitudes become apparent and communication usually becomes unprofessional. Indeed, some citizens have claimed that public safety personnel with whom they have been in contact do not attempt to understand them or demonstrate little patience in communicating or finding a translator. (See Chapter 4 for a discussion of communication issues and first responders.)

Undocumented Immigrants   There are two major groups of undocumented immigrants: those who cross U.S. borders without having been "inspected" and those who enter the country with legal documents as temporary residents but then violate their legal admission status by extending their stay. Initially, Mexicans and other Latin Americans come to most people's minds when they hear the terms *illegal alien* and *undocumented worker*. Additionally, however, people from the Dominican Republic enter through Puerto Rico; because Puerto Ricans are U.S. citizens, they are considered legal. Therefore, officers may be in contact with "Puerto Ricans" who are actually from the Dominican Republic and have therefore come to the United States under an illegal pretext. (There is also the smuggling of Asians into the United States, which includes women brought in for the sex trade.) People from other parts of the world may come to the United States on a tourist visa and then decide to remain permanently (e.g., Canadians).

alien

Any person who is not a citizen or national of the country in which he or she lives.

Some undocumented **aliens** come to the United States hoping to remain legally by proving that they escaped their homeland because of political repression, claiming that if they were to return, they would face persecution or death (i.e., they are seeking asylum). People who are often deported as undocumented arrivals are those who come as "economic refugees" (i.e., their economic status in their home country may be desperate). Undocumented aliens generally have few occupational skills and are willing to take menial jobs that many American citizens will not accept. They fill economic gaps in various regions where low-wage labor is needed. Further, outer appearances are not an accurate guide to who does and does not have legal status. Both legal and illegal immigrants may live in the same neighborhoods.

The former Immigration and Naturalization Service (INS) estimated that, in January 2000, there were over 7 million undocumented people living in the United States. In its 2003 report on the subject, the former INS, now U.S. Citizenship and Immigration Services (USCIS), places the growth of this population at 350,000 annually. This figure is 75,000 per year higher than was estimated before the 2000 census, primarily because of improved means of counting this hard-to-track population. In May 2006, there were an estimated 12 million undocumented people living in the United States (Ferraro, 2006).

Illegal immigrants lack documents that would enable them to obtain legal residence in the United States. The societal consequences are far-reaching. Public safety (fire services, emergency medical services, and law enforcement) personnel, politicians, and social service providers, among others, have had to deal with the many concerns related to housing, education, safety, employment, spousal violence, and health care regarding this illegal immigrant problem. Again, this is due to the increasing number of undocumented immigrants that contributes to calls for assistance.

THE DEFINITION OF CULTURE

culture

A way of life developed and communicated by a group of people, consciously or unconsciously, to subsequent generations. It consists of ideas, habits, attitudes, customs, and traditions that help to create standards for a group of people to coexist, making a group of people unique. In its most basic sense, culture is a set of patterns for survival and success that a particular group of people has developed.

Although there are many facets of culture, the term **culture** is defined as beliefs, values, patterns of thinking, behavior, and everyday customs that have been passed on from generation to generation. Culture is learned rather than inherited and is manifested in largely unconscious and subtle behavior. With this definition in mind, consider that most children have acquired a general cultural orientation by the time they are five or six years old. For this reason, it is difficult to change behavior to accommodate a new culture. Many layers of cultural behavior and beliefs are subconscious. Additionally, many people assume that what they take for granted is also taken for granted by all people, and they do not even recognize their own culturally engrained behavior. Anthropologist Edward T. Hall (1959) said, "Culture hides much more than it reveals and, strangely enough, what it hides, it hides most effectively from its own participants." In other words, people are blind to their own deeply embedded cultural behavior.

To further understand the hidden nature of culture, picture an iceberg (Ruhly, 1976). The visible part of the iceberg is the tip, which typically constitutes only about 10 percent of the mass. Like most of culture's influences, the remainder of the iceberg is submerged beneath the surface. This means that there will be a natural tendency for the firefighter to interpret behavior and motivations from coworkers and citizens from the firefighter's cultural point of view. This tendency is due largely to an inability to understand behavior from alternative perspectives and because of the inclination toward **ethnocentrism** (i.e., an attitude of seeing and judging all other cultures from the perspective of one's own culture). In other words, an ethnocentric person would say that there is only one way of being "normal" and that is the way of his or her own culture. This being the case can lead to prejudice.

THE DEFINITION OF PREJUDICE

Prejudice is a judgment or opinion formed before facts are known, usually involving negative or unfavorable thoughts about groups of people. Discrimination is action based on prejudiced thought. It is not possible to force people to abandon their own prejudices in the firefighter workplace or when working in the community. Because prejudice is thought, it is private and does not violate any law. However, because it is private, a person may not be aware when his or her judgments and decisions are based on prejudice. In fire departments, the expression of prejudice as bias discrimination and racism is illegal and can have severe consequences. All firefighters must consider the implications of prejudice in their day-to-day work as it relates to their actions toward the public and their coworkers.

Prejudice is encouraged by stereotyping, which is a shorthand way of thinking about people who are different. The **stereotypes** that form the basis of a person's prejudice can be so fixed that they can easily justify their own racism, sexism, or other bias and make such statements as, "I am not prejudiced, but let me tell you about those _____ I had to deal with today." Coffey, Eldefonson, and Hartinger (1982) discuss the relationship between selective memory and prejudice:

> A prejudiced person will almost certainly claim to have sufficient cause for his or her views, telling of bitter experiences with refugees, Koreans, Catholics, Jews, Blacks, Mexicans, and Puerto Ricans, or Indians. But in most cases, it is evident that these "facts" are both scanty and strained. Such a person typically resorts to a selective sorting of his or her own memories, mixes them up with hearsay, and then over generalizes. No one can possibly know all refugees, Koreans, Catholics, and so on. (p. 8)

These examples provided by Coffey, Eldefonson, and Hartinger do not stop with their short list. Whites, women, men, and every imaginable dimension of diversity have been targeted by a prejudicial person. Prejudice is faulty information about individuals and groups that usually leads to discrimination.

PEER RELATIONSHIPS AND PREJUDICE

Expressions of prejudice in fire departments may go unchallenged because of the need to conform to or fit into the group. Firefighters do not make themselves popular by questioning peers or challenging their attitudes. It takes a leader to voice an objection or to avoid going along with group norms. Studies have shown that peer behavior in groups reinforces acts of racial bias. For example, when someone in a group makes ethnic slurs, others in the group may begin to express the same hostile attitudes more freely. This behavior is particularly relevant in fire departments, given the nature of the subculture and strong influence of peer pressure. Thus fire department leaders must not be ambiguous when directing their subordinates to control their expressions of prejudice, even among peers. Furthermore, according to some social scientists, the strong condemnation of any behavior of prejudice can at times affect a person's feelings.

ethnocentrism

Using the culture of one's own group as a standard for the judgment of others, or thinking of it as superior to other cultures that are merely different.

prejudice

The inclination to take a stand for one side (as in a conflict) or to cast a group of people in favorable or unfavorable light, usually without just grounds or sufficient information.

stereotype

To believe or feel that people and groups are considered to typify or conform to a pattern or manner, lacking any individuality. Thus a person may categorize behavior of a total group on the basis of limited experience with one or a few representatives of that group. Negative stereotyping classifies many people in a group by slurs, innuendoes, names, or slang expressions that depreciate the group as a whole and the individuals in it.

To use pressure from authorities or peers to people who are prejudiced from acting on those biases can, in the long run, waken the prejudice itself, especially if the prejudice is not virulent. People conform; that is, people will behave differently, even if they still hold the same prejudicial thoughts. Even if they are still prejudiced, they will be reticent to show it. National authorities have become much more vocal about dealing directly with racism and prejudice in fire departments as an institution, especially in light of the quantity of allegations of racial, ethnic, and sexual discrimination across the country.

A process of socialization takes place when change has been mandated by top management and a person is forced to adopt a new standard of behavior. When a mistake is made and the expression of prejudice occurs, a fire department will pay the price (in adverse media attention, lawsuits, citizen complaints, human relations commissions' involvements, or dismissal of the chief or other personnel). Government officials' public expressions are subject to a great deal of scrutiny. Alison Berry-Wilkinson, a lawyer and expert on harassment issues, cited the case of a prosecutor who was publicly reprimanded for a hallway comment to another lawyer during a murder trial: "I don't believe either of those chili-eating bastards." The court stated: "Lawyers, especially . . . public officials, [must] avoid statements as well as deeds . . . indicating that their actions are motivated to any extent by racial prejudice" (*People v. Sharpe,* 789 p.2d 659 [1989], Colorado, in Berry-Wilkinson, 1993, p. 2d). Berry-Wilkinson's concluding statement following the reporting of this case reads: "What once may have been acceptable is now definitely not and may bring discipline and monetary sanctions. While public employees may be free to think whatever they like, they are not free to say whatever they think. A public employee's right to free speech is not absolute." (p. 2d).

When firefighters are not in control of their prejudices (in either their speech or their behavior), the negative publicity affects the reputation of all firefighters (career or volunteer). Unfortunately, the adverse attention reinforces some negative stereotypes that firefighters are **racists** or **bigots**. Yet, because of publicized instances of discrimination, firefighters have become increasingly aware of correct and incorrect behavior toward ethnic and female minorities. They also know they are not powerless in confronting expressions of prejudice. No one has to accept sweeping stereotypes (e.g., "You can't trust a woman," "All white males are sexists and racists," "_____ was hired or promoted because of his or her minority status," and so on). To eliminate the manifestation of prejudice, people have to begin to interrupt biased and discriminatory behavior at all levels. Firefighters have to be willing to remind peers that ethnic slurs and offensive language, as well as differential treatment of certain groups of people, are neither ethical nor professional. Firefighters need to change the aspect of their culture that discourages speaking out against acts or speech motivated by prejudice. A firefighter or civilian employee who does nothing in the presence of racist or other discriminatory behavior by his or her peers becomes a silent accomplice.

racist
One with a closed mind toward accepting one or more groups different from one's own origin in race or color.

bigot
A person who steadfastly holds to bias and prejudice, convinced of the truth of his or her own opinion and intolerant of the opinions of others.

THE CHANGING WORKFORCE

As microcosms of their communities, fire departments increasingly include among their personnel more women, ethnic and racial minorities, and gays and lesbians (Figure 1.10). Although such groups are far from achieving parity in most agencies in the United States, advances have been made. In many regions of the country, the workforce of today's fire departments differs greatly from those of the past; the profound shift in demographics has resulted in notable changes in the fire service.

FIGURE 1.10 ◆ This Emergency Medical Technician team positions a supine patient on a long spine board, reflecting the growing diversity in this honorable profession.

FIRE SERVICE OPERATIONS AND SERVICES

According to the U.S. Fire Administration (USFA) and National Fire Protection Association (NFPA) Profile Report, as of December 2003, there were an estimated 30,542 fire departments and 51,650 firehouses in the United States. The totals included 1,096,900 firefighters (296,850 career and 800,050 volunteers). Career firefighters include paid full-time (career) uniformed firefighters regardless of assignments (e.g., suppression, prevention/inspection, and administrative). Career firefighters included here work for public municipal fire departments; they do not include career firefighters who work for the state or federal government or in government or in private fire brigades. Volunteer firefighters include any active part-time or on-call firefighters. Active volunteers are defined as being involved in firefighting. Seventy-four percent of career firefighters are in communities that protect a population of 25,000 or more. Ninety-three percent of the volunteer firefighters are in departments that protect a population of 25,000 or less, and over 50 percent are located in small, rural departments that protect a population of 2,500 or less. Twelve percent of all fire departments are career or mostly career and protect 61 percent of the U.S. population. Eighty-eight percent of fire departments are volunteer or mostly volunteer and protect 39 percent of the U.S. population. The total also included firefighters by age group: 16 to 19 (3.7 percent), 20 to 29 (20.8 percent), 30 to 39 (30.3 percent), 40 to 49 (26.9 percent), and 50 and over (18.4 percent). Table 1.2 shows the 2003 response breakdown for fire departments across the United States.

A DANGEROUS AND SELFLESS PROFESSION

It would be irresponsible to comment that firefighting is not a dangerous profession. It is not only a dangerous profession but also a selfless service (Figure 1.11). The men and women who make up the ranks of all fire departments put their lives on the line every time they work their shift. As the fire service workforce becomes

TABLE 1.2 ◆ U.S. Fire Departments 2003 Response Breakdown

Incident Response	Number	% Change from 2002
Fires	1,584,500	−6.1
Medical aid	13,631,500	+5.6
False alarms	2,189,500	+3.5
Mutual aid/assistance	987,000	+11.1
Hazmat	349,500	−3.2
Other hazardous (Arcing wires, bomb removal, etc.)	660,500	+9.4
All other (smoke scares, lockouts, etc.)	3,003,500	+9.5
Total	22,406,000	+5.2

Source: U.S. Fire Administration and National Fire Protection Association Profile Report, 2005.

more diversified, the dangers are still present, as indicated in the following information from 2003:

- ◆ A total of 105 firefighters were fatally injured while on duty. Of these, 25 were career, 58 were volunteer, and 22 were nonmunicipal (those not employed by local, public fire departments).
- ◆ A total of 78,750 firefighters were injured in the line of duty. Of the injuries, 38,045 occurred on the fire ground.
- ◆ There were an estimated 15,900 collisions involving fire department emergency vehicles while responding to or returning from incidents.

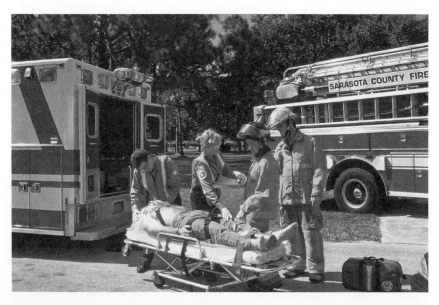

FIGURE 1.11 ◆ Firefighters and all EMS personnel have earned the respect and reputation of being selfless first responders at calls for assistance, as shown here in the patient care of this accident victim.

BECOMING MORE DIVERSIFIED

According to a U.S. Department of Labor Bureau of Labor Statistics report, as of 2004, the career firefighter female and minority ranks were women, 5.1 percent; black or African American, 8.4 percent; Hispanic or Latino, 8.6 percent; and Asian, 1.3 percent. When combined, this totals 23.4 percent, compared to white male career firefighters who had a 76.6 percent representation. In 1983, the U.S. Department of Labor Bureau of Labor Statistics had reported women at 1 percent, blacks at 7.3 percent, and Hispanic origin at 3.8 percent. Most observers will comment there has been a measurable increase in the minority employment status of firefighters since 1983, but critics will cite the percentages still do not match the demographics of the United States. This is particularly true when looking at the small amount of career women firefighters employed with fire departments. Table 1.3 shows the delineation of career firefighters for women, black, and Hispanic origin from 1983 to 2003.

TABLE 1.3 ◆ Delineation of Career Firefighters for Women, Black, and Hispanic Origin from 1983 to 2003

Year	Total	Women	Black	Hispanic Origin
1983	170,000	1,700 (1.0%)	12,400 (7.3%)	6,500 (3.8%)
1984	168,000	200 (0.1%)	8,400 (5.0%)	4,700 (2.8%)
1985	186,000	1,500 (0.8%)	11,300 (6.1%)	6,100 (3.3%)
1986	205,000	3,900 (1.9%)	16,000 (7.8%)	9,000 (4.4%)
1987	204,000	1,800 (0.9%)	15,700 (7.7%)	10,800 (5.3%)
1988	195,000	4,100 (2.1%)	16,800 (8.6%)	12,500 (6.4%)
1989	188,000	6,600 (3.5%)	22,600 (12.0%)	8,300 (4.4%)
1990	205,000	4,900 (2.4%)	22,600 (11.0%)	9,000 (4.4%)
1991	200,000	2,800 (1.4%)	17,800 (8.9%)	9,400 (4.7%)
1992	190,000	4,000 (2.1%)	12,000 (6.3%)	7,000 (3.7%)
1993	188,000	6,200 (3.3%)	14,100 (7.5%)	9,400 (5.0%)
1994	195,000	4,000 (2.1%)	18,000 (9.2%)	11,000 (5.6%)
1995	237,000	6,000 (2.5%)	36,000 (15.2%)	11,000 (4.6%)
1996	217,000	4,000 (1.8%)	30,000 (13.8%)	11,000 (5.1%)
1997	218,000	7,000 (3.2%)	27,000 (12.4%)	10,000 (4.6%)
1998	228,000	5,000 (2.2%)	27,000 (11.8%)	9,000 (3.9%)
1999	223,000	4,000 (1.8%)	25,000 (11.2%)	12,000 (5.4%)
2000	233,000	7,000 (3.0%)	21,000 (9.0%)	12,000 (5.2%)
2001	250,000	7,000 (2.8%)	32,000 (12.8%)	21,000 (8.4%)
2002	248,000	8,000 (3.3%)	24,000 (9.6%)	23,000 (9.3%)
2003	258,000	9,000 (3.4%)	21,000 (8.1%)	16,000 (6.2%)
Annual Average 1999–2003	242,400	7,000 (2.8%)	24,600 (10.1%)	16,800 (6.9%)

Source: 2003 U.S. Department of Labor, Bureau of Labor Statistics, Household Data Survey.

Consistent with police departments throughout the United States, there are more minorities and women career firefighters in the urban and major metropolitan fire departments compared to smaller and rural fire departments. The reader needs to know that the occupations of firefighter and police officer have traditionally been professions dominated by white males. However, due to changing demographics, federal and state equal employment opportunity laws, and the attraction to these professions, both fire service and law enforcement organizations are slowly and gradually becoming more diversified.

DIMENSIONS OF DIVERSITY

To make sense of the different groups in our workplace and society, we need to have functional categories and terms. The primary and secondary dimensions of diversity, discussed by organizational diversity consultant Marilyn Loden (2006), provide an outline of familiar information on this subject. We define a dimension as an element or component that forms a personality or entity. The information in the dimensions of diversity is not new but is helpful in providing a social, functional construction of individual and group characteristics for understanding the diversity of people in the workforce and our society. The benefits of this awareness and using these differences as sources of strength are improved interpersonal relationships and improved citizen contacts.

PRIMARY DIMENSIONS OF DIVERSITY

A primary dimension is a social constructed core characteristic a person is born with that remains with the individual in all stages of life. People have a minimum of six dimensions (Loden, 2006). Some view religion as a primary dimension because of strong personal beliefs. Also, there is debate whether sexual orientation is either genetic or a choice. The six primary dimensions are:

1. Age
2. Ethnicity
3. Gender
4. Mental/physical abilities and characteristics
5. Race
6. Sexual orientation

Most people are aware of the meaning of these components; however, for the sake of clarity, the following terms are included in sexual orientation: *heterosexual, homosexual* (*lesbian* or *gay*), *bisexual,* and *transgender*. All of these primary dimensions are characteristics that have contributed to being advantaged or disadvantaged in the workforce or society. In addition, victims of hate bias crimes have been targeted because of their ethnicity, gender, disability status, race, or sexual orientation.

LODEN'S SECONDARY DIMENSIONS OF DIVERSITY AND COMPONENTS

A secondary dimension is a social constructed characteristic a person acquires as the result of a choice that person makes or a choice someone else makes for him or her (Loden, 2006). All of the characteristics of secondary dimensions are components that contribute to the study of demographics. The secondary dimensions of diversity would include but are not limited to:

1. Communication style
2. Education
3. Family status
4. Military experience
5. Organizational role and level
6. Religion
7. First language
8. Geographic location
9. Income
10. Work experience
11. Work style
12. Others

Fire service personnel are influenced by these secondary dimensions in their personal and professional lives. They have had positive and negative experiences with coworkers and supervisors who may have a disposition toward the secondary dimensions of diversity similar to or different from themselves. Firefighters need to be cognizant of these dimensions with their coworkers and leaders with their subordinates. Tensions between supervisors and coworkers are often caused by the differences in secondary dimensions.

Figure 1.12 shows how the primary and secondary dimensions of diversity influence people in the workforce and society.

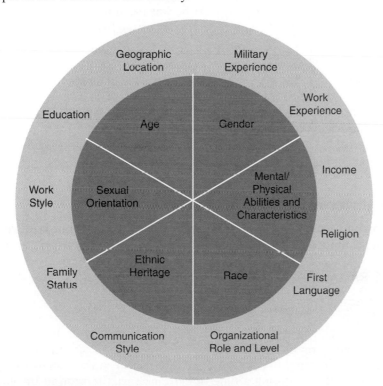

FIGURE 1.12 ◆ Dimensions of Diversity.
Source: Loden, 1996, p. 16. Reproduced with permission of Marilyn Loden.

While each dimension adds a layer of complexity, it is the dynamic interaction among all the dimensions of diversity that influences one's self image, values, opportunities, and expectations. Together, the primary and secondary dimensions give definition and meaning to our lives by contributing to a synergistic, integrated whole—the diverse person (Loden, 1996, p. 16).

ETHNIC AND RACIAL ISSUES

Along with all other professions in the history of the United States, fire departments have experienced racism, segregation, and discrimination. Even before Los Angeles, California, hired its first black firefighter, volunteer and paid minority firefighters provided their dedicated services to their communities throughout this country. The following information from the Los Angeles Fire Department Historical Archives, Los Angeles, California, provides a snapshot of history in how the Los Angeles Fire Department was undergoing cultural changes but was still harnessed by racism and segregation.

George W. Bright

George W. Bright, hired October 2, 1897, was the first black member of the Los Angeles Fire Department. He was appointed by the fire commission as a call man and assigned to Engine Co. No. 6. Less than a month later on November 1, 1897, Bright was promoted to a full-time hose man and assigned to Engine Co. No. 3. On January 31, 1900, he was promoted to driver third class and assigned to Chemical Engine Co. No. 1. George Bright lived next door at 125 Belmont Avenue.

On August 1, 1902, George Bright was promoted to lieutenant. In those days chief officers made the promotions. However, before the commission would certify his promotion, Bright, being the first colored to express desires for such advancement, was required to go to the Second Baptist Church and obtain an endorsement from his minister and congregation.

The department, to avoid Bright from commanding white firemen, gathered up all the colored and Mexican American firemen and formed the city's first all-black fire company (Los Angeles Fire Department, 1999).

The Los Angeles Fire Department is not alone. Racism has occurred and continues in fire departments regardless of size or region. This is documented all too often in personnel complaints, civil rights investigations, lawsuits, and court decisions.

DEFUSING RACIALLY AND CULTURALLY ROOTED CONFLICTS

Racism and sexism exist within our fire departments; firefighters are not immune to social ills (Figure 1.13). One of the greatest challenges for firefighters is dealing with their own racism and sexism. The first step in addressing the problem is for fire department personnel, on all levels, to admit that racism and sexism exist rather than denying it. The next step is to combat and defuse racism and sexism by using the U.S. Army model developed during a time of extreme racial tension in the early 1970s (Shipler, 1992). Obviously, no model of training will bring guaranteed success and alleviate all acts of prejudice and racism. However, professional groups can build on

FIGURE 1.13 ◆ An African American firefighter in full uniform, standing next to his fire engine.

each other's attempts, especially when these have proven to be successful. We realize that firefighters are not soldiers, but some military approaches are adaptable to fire departments and law enforcement and still have application today. According to researcher David Shipler, the basic framework for combating and defusing racism (we agree on sexism too) in the military has been:

- **Command commitment:** The person at the top sets the tone all the way down to the bottom. Performance reports document any bigoted or discriminatory behavior. A record of racial slurs and discriminatory acts can derail a military career.
- **Training of advisers:** Military personnel are trained at the Defense Equal Opportunity Management Institute in Florida as equal opportunity advisers. These advisers are assigned to military units with direct access to commanders. They conduct local courses to train all members of the unit on race relations.
- **Complaints and monitoring:** The advisers provide one channel for specific complaints of racial and gender discrimination, but they also drop in on units unannounced and sound out the troops on their attitudes. Surveys are conducted and informal discussions are held to lessen racial tensions.

Sondra Thiederman (1991), a cultural diversity consultant and author, provides nine tips that can help organizational managers and leaders identify and resolve conflicts that arise because of cultural (not only racial) differences in the workplace. She says that the following guidelines are applicable no matter what cultures, races, religions, or lifestyles are involved:

1. Give each party the opportunity to voice his or her concerns without interruption.
2. Attempt to obtain agreement on what the problem is by asking questions of each party to find out specifically what is upsetting each person.
3. During this process, stay in control and keep employees on the subject of the central issue.
4. Establish whether the issue is indeed rooted in cultural differences by determining:
   a. Whether the parties are from different cultures or subcultures.
   b. Whether the key issue represents an important value in each person's culture.
   c. How each person is expected to behave in his or her culture as it pertains to this issue.
   d. Whether the issue is emotionally charged for one or both of the parties.
   e. Whether similar conflicts arise repeatedly and in different contexts.

5. Summarize the cultural (racial, religious, or lifestyle) differences that you uncover.
6. State the negative outcomes that will result if the situation is not resolved (be specific).
7. State the positive outcomes that will result if the situation is resolved (be specific).
8. Negotiate terms by allowing those involved to come up with the solutions.
9. Provide positive reinforcement as soon as the situation improves.

Thiederman's approach is based on conflict resolution and crisis intervention techniques. Fire department leadership must encourage the use of conflict resolution techniques by firefighters of all backgrounds as a way of handling issues prior to their becoming flash points. With professionalism and patience, the use of conflict resolution techniques to reduce racial and ethnic problems will work within and outside of the firehouse.

WOMEN IN FIRE DEPARTMENTS

Historically, women have always been part of the general workforce in American society, although usually in jobs that fulfilled traditional female employment roles, such as nurses, secretaries, schoolteachers, waitresses, and flight attendants. The first major movement of women into the general workforce occurred during World War II. With men off to war, women entered the workforce in large numbers and successfully occupied many nontraditional employment roles. After the war, 30 percent of all women continued working outside the home ("History of Women in Workforce," 1991, p. 112). According to the 2000 U.S. Bureau of Labor Statistics, 60 percent of women and 75 percent of men were in the United States labor force. This is a significant increase compared to after World War II in 1945.

In 1972, the passage of the Equal Employment Opportunity (EEO) Act applied to state and local governments the provisions of Title VII of the Civil Rights Act of 1964. The EEO Act prohibited employment discrimination on the basis of race, color, religion, sex, or national origin. Selection procedures, criteria, and standards were changed or eliminated and/or made "job related." This law played an important role in opening up police and fire departments to women (Figure 1.14). Adoption of affirmative action policies, now illegal in many states, along with court orders and injunctions, also played a role in bringing more women into fire departments.

Considering how long organized fire departments have existed in the United States, women entered relatively late into firefighter positions within them. EEO was a major catalyst in the first generation of women firefighters getting hired over 30 years ago. Even though this was a milestone, the number of women in fire departments remains small and is increasing at a slow rate. According to a U.S. Department of Labor Bureau of Labor Statistics report, as of 2004, only 5.1 percent of career firefighters were female. This is an increase of 2.1 percent compared to its 2000 report of employed female firefighters. However, this number is significantly less when compared to women police officers and corrections officers. See Table 1.4.

In 2005, the 1,088 fire departments that employed women firefighters had 14 women as department chiefs and 703 women in leadership positions (Women in Fire Service, 2005). In looking at these numbers, why are the percentages of women firefighters considerably less than women in law enforcement, corrections, and other

FIGURE 1.14 ◆ Two EMTs in uniform and displaying their protective gear.

TABLE 1.4 ◆ Women in Law Enforcement Labor Force				
Police Supervisor/ Manager	*Detective and Criminal Investigator*	*Police and Sheriff Patrol Officer*	*Bailiffs, Correctional Officers, and Jailers*	*Firefighter*
21.2%	20.2%	13.3%	28.4%	5.1%

Source: 2004 U.S. Bureau of Labor Report, Women in the Labor Force.

professions? Terry Golway, an author and historian, provides some insight and offers the following explanation:

> While Americans have long admired their firefighters, not until September 11 did many of us fully appreciate the job's heroism and dangers. Popular culture and the media have made many of us experts in police work (or so we think). Firefighters, however, were nearly invisible in paperback novels, films, and primetime television. Then, suddenly, they became international symbols of sacrifice, courage, and dedication to duty. They achieved a status seldom granted to mere mortals: They have become models for action figures, available in toy departments and stores near you. Less celebrated are controversies that also have their roots in the history and traditions of the fire service. The American firehouse has been an outpost of masculinity like few others

in contemporary life. Long after women police officers, soldiers, and sportswriters have become commonplace, women firefighters remain a rarity. New York offers the most explicit example, with no more than 30 women among its more than 11,000 firefighters. It is no coincidence that one of the many books celebrating firefighters after September 11 used the word *Brotherhood* in the title. Similarly, fire departments remain overwhelmingly white even in cities that are majority–minority, or close to it. The Fire Department of New York is 90 percent white; the departments of Boston, Philadelphia, and Chicago between 70 and 75 percent white. Those statistics obviously indicate that fire departments have been slow to accept African-Americans, Latinos, and Asians. They also are reminders of the American fire service's guildlike traditions, which may be out of favor in the twenty-first century but nevertheless offer the service a cohesion and sense of family crucial in times of peril and tragedy. Firefighters often have a combat soldier's view of the larger world (and of their superiors). They trust each other, and only each other. And they take dim view of those they consider outsiders. The public, however, seems willing to grant the fire service a pass on its struggles with workplace diversity (Golway, 2005).

GENDER ISSUES

Research on gender issues confronting women in fire departments focus on discrimination and sexual harassment, role barriers, the "brotherhood," differential treatment, and career versus family.

Discrimination and Sexual Harassment Although discrimination and sexual harassment exist in both private and public sectors, they have been problems documented in fire departments—an occupation that is overwhelmingly male (Figure 1.15). The

FIGURE 1.15 ◆ A happy toddler being held by an African American female EMT.

predominantly male makeup and stereotypical macho image of fire departments have lead to some problems of sexual harassment in the workplace. Harassment on the basis of sex is a violation of Section 703 of Title VII of the Civil Rights Act (29 CFR Section 1604.11 [a] [1]) and is defined as unwelcome or unsolicited sexual advances, requests for sexual favors, and other verbal or physical conduct of a sexual nature when:

- ◆ Submission of such conduct is made either explicitly or implicitly a term or condition of an individual's employment; or
- ◆ Submission to, or rejection of, such conduct by an individual is used as the basis for employment decisions affecting such individual; or
- ◆ Such conduct has the purpose or effect of unreasonably interfering with an individual's work performance or creating an intimidating, hostile, or offensive working environment.

Empirical data on fire departments across the nation disclose that sexual harassment has been a problem. Approximately 85 percent of women firefighters have experienced some form of sexual harassment at work or in their volunteer departments, but not all women report this harassment through their chain of command (FEMA, 1999a). There are several reasons why women firefighters who have been harassed choose not to report it or make a formal complaint. In the 1999 FEMA publication titled *Many Women Strong: A Handbook for Women Firefighters,* five reasons are listed by women firefighters who were sexually harassed:

1. I didn't want to be labeled a troublemaker, and I didn't feel a positive outcome was possible.
2. Much of the harassment occurred when I was on probation and felt I could not speak out. . . . I did attempt to speak to my officers; all of them shrugged off my appeals for help.
3. I didn't want to be "singled out" even more.
4. I want to keep my job. It's clear that those who seek legal recourse can't come back to work.
5. I didn't want to get someone suspended or fired. I just wanted it to stop. (1999b)

Fire department chiefs and leadership must institute a zero-tolerance sexual harassment policy and send that message throughout the department. Sexual harassment training and its prevention must be conducted. The training should also include reporting procedures and the emphasis on eliminating a hostile working environment.

When discrimination or harassment takes place, the results can be devastating in terms of the involved employees' careers, the internal environment of the organization, and the department's public image. The importance of training all fire department employees on the issues of sexual harassment cannot be stressed enough. With policies and procedures as well as training, sexual harassment is expected to decrease in the fire department workforce as the once male-dominated occupation makes its transition to mixed-gender, multiethnic, and multilifestyle organizations. Discrimination and sexual harassment training should deal not only with legal and liability issues but also with deep-seated attitudes about differences based on sex. Fortunately, progressive fire departments have been instituting these changes, and these responsive fire departments are growing in number throughout the United States.

Role Barriers   Barriers based on gender are diminishing, both in the general population and within fire departments. For example, ideas about protection differ by gender—who protects whom? In American society, women may protect children, but it has been more socially acceptable and traditional for men to protect women. In the act of protecting, the protectors become dominant and the protected become subordinate. Although this gender-role perception has not completely broken down in fire departments, law enforcement, and corrections, it is subsiding due to the increasing number of women and young males in the workforce ranks. Many traditional males initially had difficulty with the transition as women came into the dangerous, male-dominated occupations that men felt required "male" strength and abilities. The result has been described as a clash between cultures—the once male-dominated workforce versus the new one in which women are integral parts of the organizational environment. The traditional male firefighter, socially conditioned to protect women, felt he had the added responsibility of protecting the women firefighters with whom he worked. These feelings, attitudes, and perceptions can make men and women in firefighter positions uncomfortable with each other. Women sometimes feel patronized, overprotected, or merely tolerated rather than appreciated and respected for their work. Again, these attitudes and perceptions are diminishing as many in the new generation of male firefighters are more willing to accept women in firefighting.

The Brotherhood   Women who are accepted into the "brotherhood" of fire departments have generally had to become "one of the guys" (Figure 1.16). (Refer to Chapter 4 for more information on how language used in the brotherhood excludes women.) However, a woman who tries to act like one of the guys comes across as disingenuous and phony. On the other hand, if a woman is too feminine or not sufficiently aggressive, men will not take her seriously. Women are confronted with a dilemma: They must be aggressive enough to do the job but feminine enough to be acceptable to male peers. To succeed, women have to stay within narrow bands of acceptable behavior and exhibit only certain traditionally masculine and feminine qualities. Walking this fine line is difficult. These same challenges have been experienced by women in the military, law enforcement, and corrections. This phenomenon is not unique to these professions. An article on a woman ironworker reported that today's female ironworkers are still pio-

FIGURE 1.16 ◆ Two firefighter emergency medical technicians checking out their rescue equipment on a fire truck.

neers. No matter how skilled she becomes, she has to prove herself over and over again. "'What it is is attitude,' [one woman ironworker] says. 'I know that I'm on male-dominated home turf'" ("Male Dominated Occupation," 2000, p. A3).

Differential Treatment Many women in fire departments have indicated they are treated differently by supervisors that are men and they are frequently held back from promotions or special assignments. A 2002 Census Bureau report indicated that in private corporations "Women hold nearly half the executive and managerial jobs in the United States, up from only about a third in 1983" ("Women Land More Top Jobs," 2003, p. A16). This is far from the case in fire departments.

Firefighter executives must determine whether their female firefighters are receiving opportunities for assignments and training that will provide the groundwork and preparation for their eventual promotion. They need to determine whether female firefighters are applying for promotions in numbers proportionate to their representation in the department. If not, perhaps women firefighters need encouragement from their supervisors. It is also possible that the promotion process disproportionately screens out female firefighters. Research shows that the more subjective the process is, the less likely women are to be promoted. The use of assessment center, "hands-on" testing is said to offer some safeguards against the potential for the perception of bias against women. Utilizing structured interviews and selecting interview board members who represent different races and both sexes can also minimize this risk. Many departments are now training interview board members on interviewing techniques. The Institute for Women in Trades, Technology and Science has developed a well-received half-day training session for supervisors, "Creating a Supportive Work Environment," to address issues of integration and retention of women.

Career versus Family Many women in fire departments are faced with another dilemma—trying to raise a family and have a successful career, two goals that are difficult to combine. Women, regardless of whether they are married with children or single custodial parents who had children when they entered the profession, frequently find that they have difficulty balancing their commitments to family and work. Child care for a single parent working shift work is a challenge, especially a 24-hour shift because practically no child-care resources are available other than family or friends from 6 P.M. to 6 A.M. Unfortunately, the majority of fire departments do not provide child-care support for their personnel. This is one area in which improvement is needed to be creative and resourceful to correct an inherent problem.

GAY AND LESBIAN ISSUES

Sexual orientation includes homosexual, heterosexual, bisexual, and transgender (Figure 1.17). For the purposes of this discussion, the terms *lesbian, homosexual, gay, bisexual,* and *transgender* will be used to describe the sexual orientation of underrepresented groups in fire departments. These terms are defined as follows: gay, a male homosexual; homosexual, characterized by sexual attraction to those of the same sex as oneself; lesbian, a homosexual woman; and bisexual, characterized by sexual attraction to both men and women. The term **transgender** covers a range of people, including heterosexual cross-dressers, homosexual drag queens, and transsexuals who believe

transgender

The term *transgender* covers a range of people, including heterosexual cross-dressers, homosexual drag queens, and transsexuals, who believe they were born in the wrong body. This term includes those who consider themselves to be both male and female, or intersexed, as well as those who take hormones to complete their gender identity without a sex change.

FIGURE 1.17 ◆ People march in the New York City Gay/Lesbian Pride Day.

they were born in the wrong body. There are also those who consider themselves to be both male and female, or intersexed, and those who take hormones and believe that is enough to complete their gender identity without a sex change.

Gay and lesbian firefighters are similar to heterosexual firefighters in that they desire to conform to the norms of the fire department they work for and to prove their worth as members of that organization. As any other group, gays and lesbians seldom engage in behaviors that would challenge those norms or shock or offend fellow firefighters. Unfortunately, there are too many incidents in which gays and lesbians have been subjected to discrimination and harassment by fellow firefighters. Discrimination and harassment are immoral, unethical, and illegal. As of June 2003, 14 states have passed laws prohibiting discrimination based on sexual orientation. If your state, county, or city government does not have antidiscrimination laws protecting gays and lesbians in the workplace, your fire department should adopt a policy that does. The policy should establish that:

1. Sexual orientation is not a hindrance in hiring, retention, or promotion.
2. Hiring is based solely on merit as long as the individual meets objective standards of employment.
3. Hiring is done on the basis of the identical job-related standards and criteria for all individuals.

Fire department managers and supervisors must routinely check to ensure that this policy is being carried out as intended.

TRAINING ON GAY, LESBIAN, AND TRANSGENDER ISSUES

Cultural awareness programs that train fire department personnel on diversity within communities and in the workforce must also educate employees on gay, lesbian, and transgender issues. The training should address and show the falsehoods of stereotypes and myths. It must cover legal rights, including a discussion of statutes and departmental policies on nondiscrimination and the penalties for violating them. These penalties included liability for acts of harassment and discrimination. Often, involving openly gay or lesbian firefighters (from other departments, if necessary) in these training programs provides the best outcome. Ideally, this training will enable employees to know the gay or lesbian firefighters they work with as human beings, reduce personal prejudices and false assumptions, and thus change negative behavior. This type of training furthers the ideal of respect for all people. A secondary benefit of this training is the decreased likelihood of personnel complaints and lawsuits by gay or lesbian firefighters or community members against your department.

LEADERSHIP: REACT OR RESPOND?

THE FIRE EXECUTIVE

The fire executive or chief should follow specific guidelines to meet the challenge of providing fire services to a multicultural and multiracial community. In this section and throughout this text, the terms *fire chief*, **fire executive**, and *fire service executive* are used interchangeably to reflect local, regional, and national venue differences when referring to fire service leaders (Figure 1.18). As emphasized previously, he or she must first effectively manage the diversity within his or her own organization. Progressive fire chiefs and fire service executives are aware that before employees can be asked to value diversity in the community, it must be clear that diversity within the organization is valued. Managing diversity in the fire department workplace is therefore of high priority.

fire executive

Current term in the fire service for fire chief and other fire service senior officers.

FIGURE 1.18 ◆ In demonstrating leadership, firefighters, emergency medical technicians, and citizens participate in a critical incident stress debriefing.

Executive leadership and team building are crucial to managing a diverse workforce and establishing good minority–community relations. The fire chief must take the lead in this endeavor by:

- Demonstrating commitment
- Developing strategic, implementation, and transition management plans
- Managing organizational change
- Developing fire service–community partnerships
- Providing new leadership models

Demonstrate Commitment   The organization must adopt and implement policies that demonstrate a commitment to providing fire services to a diverse society. These policies must be developed with input from all levels of the organization and community. Valuing diversity and treating all persons with respect must be the imperative first from the fire chief. His or her personal leadership and commitment are the keystones to implementing policies and awareness training within the organization and to building bridges to the community successfully. One of the first steps is the development of a "macro" mission statement for the organization that elaborates the philosophy, values, vision, and goals of the department to foster good relationships with a diverse workforce and community. All existing and new policies and practices of the department must be evaluated to see how they may affect women, members of diverse ethnic and racial groups, gays, lesbians, bisexuals, and transgender employees on the fire department. Recruitment, hiring, and promotional practices must be reviewed to ensure that there are no institutional barriers to different groups in the fire department. The fire chief stresses, via mission and values statements, that the department will not tolerate discrimination or abuse against protected classes within the community or within the department itself. The policy statements should also include references to discrimination or bias based on physical disability or age.

The executive must use every opportunity to speak out publicly on the value of diversity and to make certain that people inside and outside the organization know that upholding those ideals is a high priority. He or she must actively promote policies and programs designed to improve community relations and use marketing skills to sell these programs, both internally and externally. Internal marketing is accomplished by involving senior management and the firefighter labor association in the development of the policies and action plans. The fire chief can use this opportunity to gain support for the policies by demonstrating the value to the fire department's effectiveness and to firefighter safety of having community support. External marketing is accomplished by involving representatives of community-based organizations in the process.

Fire chiefs must institute policies that develop positive attitudes toward a multicultural workplace and community even as early as the selection process. Throughout the hiring process, candidates for employment must be carefully screened. The questions and processes can help determine candidates' attitudes and beliefs and, at the same time, make them aware of the fire department's strong commitment to a multicultural workforce.

Develop Strategic, Implementation, and Transition Management Plans   Textbooks and courses that teach strategic, implementation, and transition management planning are available to fire department leaders. The techniques, although not difficult, are quite involved and are not the focus of this book. Such techniques and methodologies are planning tools, providing the road map that the organization uses to implement programs and to guide the fire department through change. An essential component is the action plan that identifies specific goals and objectives. Action plans include budgets and timetables and establish accountability of who is to accomplish what by when. Multiple action plans involving the improvement of fire service community relations in a diverse society would be necessary to cover such varied components as policy and procedure changes; affirmative action recruitment (where legal), hiring, and promotions; cultural awareness training; and community involvement.

Manage Organizational Change   The department leadership is responsible for managing change processes and action plans (Figure 1.19). This is an integral part of implementation and transition management, as discussed previously. The fire chief must ensure that any new policies, procedures, and training result in increased employee responsiveness and awareness of the diversity in the community and within the organization's workforce. He or she must require that management staff continually monitor progress on all programs and strategies to improve fire service–community relations. Additionally, the chief must ensure that all employees are committed to those ideals. Managers and supervisors need to ensure application of these established philosophies and policies of the department, and they must lead by example. When intentional deviation from the system is discovered, retraining and discipline should be quick and effective. Firefighters must be rewarded and recognized for their ability to work with and within a multicultural community. The reward systems for employees, especially first- and second-line supervisors, would recognize those who foster positive relations with individuals of different gender, ethnicity, race, or sexual orientation both within and outside the organization. As we have illustrated, the fire chief, management staff, and supervisors are role models and must set the tone for the sort of behavior and actions they expect of employees.

Develop Fire Department–Community Partnerships   Progressive fire departments have adopted community-based fire service as one response strategy to meet the needs and challenges of a pluralistic workforce and society. The establishment of

FIGURE 1.19   ◆   Two firefighters demonstrate their individual leadership and communication skills in providing a scene assessment.

community partnerships is a very important aspect of meeting the challenges. For example, a cultural awareness training component will not be as effective if fire department–community partnerships are not developed, utilized, and maintained. The fire chief establishes and maintains ongoing communications with all segments of the community.

Provide New Leadership Models   In the past all methods or models of management and organizational behavior were based on implicit assumptions of a homogeneous, white male workforce. Even best sellers such as *The One-Minute Manager* and *In Search of Excellence,* which continue to be useful management tools, are based on that traditional assumption. Managers must learn to value diversity and overcome personal and organizational barriers to effective leadership, such as stereotypes, myths, unwritten rules, and codes. New models of leadership must be incorporated into fire departments to manage the multicultural and multiracial workforce.

Jamieson and O'Mara (1991) address the topic of motivating and working with a diverse workforce, explaining that the leader must move beyond traditional management styles and approaches. They indicate that the modern manager must move from the traditional one-size-fits-all management style to a "flex-management" model (Figure 1.20). They describe flex management as not just another program or quick fix but one that is "based on the need to individualize the way we manage, accommodating differences and providing choices wherever possible" (p. 31). The flex-management model they envision involves three components:

- **Policies:** published rules that guide the organization
- **Systems:** human resources tools, processes, and procedures
- **Practices:** day-to-day activities

The model is based on four strategies: matching people to jobs, managing and rewarding performance, informing and involving people, and supporting lifestyle and

FIGURE 1.20 ◆ Two firefighters arrive at the scene of a call for assistance with the tools, skills, and flexibility to handle the call.

life needs. Five key management skills are required of the modern manager to use this model successfully:

1. Empowering others
2. Valuing diversity
3. Communicating responsibly
4. Developing others
5. Working for change

Good leaders not only acknowledge their own ethnocentrism but also understand the cultural values and biases of the people with whom they work. Consequently, such leaders can empower, value, and communicate with all employees more effectively. Developing others involves mentoring and coaching skills, important tools for modern managers. To communicate, responsible leaders must understand the diverse workforce from a social and cultural context and flexibly utilize a variety of verbal and nonverbal communication strategies with employees. Modern leaders are also familiar with conflict mediation in cross-cultural disputes.

Jamieson and O'Mara (1991) explain that to establish a flex-management model, managers must follow a six-step plan of action that includes:

1. Defining the organization's diversity
2. Understanding the organization's workforce values and needs
3. Describing the desired future state
4. Analyzing the present state
5. Planning and managing transitions
6. Evaluating results

They contend that to be leaders in the new workforce, most managers will have to "unlearn practices rooted in old mindsets, change the way their organization operates, shift organizational culture, revamp policies, create new structures, and redesign human resource systems" (p. 25).

The vocabulary of the future involves leading employees rather than simply managing them. Hammond and Kleiner (1992) wrote about the distinction between the two:

> One of the first things companies must look at in multicultural environments is the leadership vs. management issue. Leadership, in contrast to management, deals with values, ethics, perspective, vision, creativity and common humanity. Leadership is a step beyond management; it is at the heart of any unit in any organization. Leadership lies with those who believe in the mission and through action, attitude, and attention pass this on to those who have to sustain the mission and accomplish the individual tasks. People want to be led, not managed; and the more diverse the working population becomes the more leadership is needed.

Hammond and Kleiner explain that in a multicultural and multiracial society and workforce, "the genius of leadership" includes:

1. Learning about and understanding the needs of the diverse people you want to serve—not boss, not control—but serve.
2. Creating and articulating a corporate mission and vision that your workers can get excited about, participate in, and be a part of.
3. Behaving in a manner that shows respect to and value for all individual workers and their unique contributions to the whole. Those you can't value, you can't lead (p. 13).

This approach to leadership was echoed in the introduction to *Transcultural Leadership—Empowering the Diverse Workforce:*

> Transcultural Leadership addresses a new global reality. Today productivity must come from the collaboration of culturally diverse women and men. It insists that leaders change organizational culture to empower and develop people. This demands that employees be selected, evaluated, and promoted on the basis of *performance and competency,* regardless of sex, race, religion, or place of origin. Beyond that, leaders must learn the skills that enable men and women of all backgrounds to work together effectively (Simons, Vazquez, and Harris, 1993).

If management is to build positive relationships and show respect for a pluralistic workforce, it needs to be aware of differences, treat all employees fairly, and lead. The differing needs and values of a diverse workforce require flexibility by organizations and their leaders. Modern leaders of organizations recognize not only that different employees have different needs but also that these needs change over time. The fire chief's goal is to bridge cultural and racial gaps within his or her organization. This means continual mentoring and support programs for all of its firefighters and leaders.

SUMMARY

Firefighters who traditionally worked in predominantly white male workforces must learn to work with increasing numbers of women, gays, lesbians, blacks, Hispanics, Asians, and others within our diverse society. In this chapter we have suggested that to be effective in this new environment, firefighters must have a working knowledge of conflict resolution techniques to reduce racial and ethnic problems.

The chapter focused on concerns and issues of members of underrepresented ethnically and racially diverse groups as well as women, gays, and lesbians in fire departments. The importance of support and mentoring programs for women and diverse groups was stressed. Such programs help them make transitions into organizations, cope with stress, and meet their workplace challenges more effectively.

In the chapter we provided suggestions for fire chiefs whose jurisdictions are pluralistic and whose workforces are diverse. Fire chiefs must be committed to setting an organizational tone that does not permit bigoted or discriminatory acts and must act swiftly against those who violate these policies. They must monitor and deal quickly with complaints both within their workforce and from the public they serve.

■■

Discussion Questions and Issues

1. ***Defusing Racially and Culturally Rooted Conflicts***. What training does the fire department academy in your region provide on defusing racially and culturally rooted conflicts? What training of this type does your local fire department provide to firefighters? What community (public and private) agencies are available as referrals or for mediation of such conflicts? Discuss what training should be provided to firefighters to defuse, mediate, and resolve racially and culturally rooted conflicts. Discuss what approaches a fire department should utilize.

2. ***Women in Fire Departments***. How many women firefighters are there in your local fire department? How many of those women are in supervisory or management positions? Are any of the women assigned to nontraditional roles? Have there been incidents of sexual harassment of women employees? If so, how were the cases resolved? Has the fire department implemented any programs to increase the employment of women, such as flextime, child care, mentoring, awareness training, or career development? Has the agency been innovative in the recruitment efforts for women applicants? Discuss your findings in a group setting.

3. ***Diversity in Fire Departments***. Comment on the diversity in your local fire department. What is the breakdown in your fire department's hierarchy? For example, who holds supervisory or management positions? Have there been reported acts of discrimination against people of diverse backgrounds? Has the fire department implemented any programs to increase the employment of minorities? Discuss your findings in a group setting.

Website Resources

Visit these websites for additional information related to the content of Chapter 1:

African American Firefighter Museum: http://www.aaffmuseum.org
It is a website resource for the heritage of African American firefighters.

Bureau of Labor Statistics: http://www.bls.gov
It is a website resource of the federal government in the field of labor economics and statistics.

International Association of Black Professional Fire Fighters: http://www.iabpff.org
It is a website resource on black professional firefighters.

Institute for Women in Trades, Technology & Science: http://www.iwitts.com
This website features fact sheets, news articles, and publications on women in nontraditional careers.

National Association of Hispanic Firefighters: http://www.nahf.org
This website provides resource information on Hispanic firefighters.

National Fire Protection Association: http://nfpa.org
This website is a resource on fire and building safety standards.

U.S. Census Bureau: http://www.census.gov
This website provides comprehensive information about changing demographics.

U.S. Fire Administration: http://www.usfa.fema.gov
This website provides training and educational opportunities for the fire service and allied organizations. It also features fire statistics and other invaluable information related to fire service.

Women in the Fire Service, Inc. (WFS): http://www.wfsi.org
This website provides networking, data, and communications for fire service women.

References

"America Is Accepting Gays Easier." (2000, June 17). *Contra Costa Times,* p. A27.

Berry-Wilkinson, Alison. (1993). "Be Careful What You Say When. . . ." *Labor Beat,* 5(1), 16.

Blanchard, Kenneth H., and Spencer Johnson. (1983). *The One-Minute Manager.* New York: Berkley.

Coffey, Alan, Edward Eldefonson, and Walter Hartinger. (1982). *Human Relations: Law Enforcement in a Changing Community*, 3rd ed. Englewood Cliffs, NJ: Prentice-Hall.

Cross, T., Bazron, B. J., Dennis, K. W., and Isaacs, M. R. (1989). *Toward a Culturally Competent System of Care. Volume 1: Monograph on Effective Services for Minority Children Who Are Severely Emotionally Disturbed.* Washington, DC: CASSP Technical Assistance Center, Georgetown University Child Development Center.

Derbeken, Jaxon Van. (2005). "San Francisco Fire Department Settles Sexual Harassment Suit." *San Francisco Chronicle*, July 25, 2005. Retrieved on January 13, 2008 from http://www.sfgate.com/cgi-bin/article.cgi? f=/c/a/2005/07/25/BAGMVDSS8Q1.DTL &hw=san+francisco+fire+department+ settles+sexual+harassment+suit&sn=004 &sc=528

Federal Emergency Management Administration. (FEMA). (1999a, September). *Many Faces, One Purpose: A Manager's Handbook on Women in Firefighting.* Washington, DC.

Federal Emergency Management Administration. FEMA. (1999b, November). *Many Women Strong: A Handbook for Women Firefighters*. Washington DC.

Ferraro, Thomas. (2006). "Immigrants Set for Massive Rallies." *ABC News*, May 1, 2006. Retrieved June 24, 2006, from http:// abcnews. go.com/US/wireStory?id=1908521

Golway, Terry. (2005). "Firefighters." *American Heritage Magazine*, November/December 2005, Volume 56, Issue 6. Retrieved on January 14, 2008 from http://www. americanheritage.com/people/articles/ web/20051212-firefighter-fireman-fire-fighting-fire-benjamin-franklin-fdny.shtml

Hall, Edward T. (1959). *The Silent Language.* Greenwich, CT.: Fawcett.

Hammond, Teresa, and B. Kleiner (1992). "Man aging Multicultural Work Environments." *Equal Opportunities International*, 11 (2).

Handlin, Oscar. (1975). *Out of Many: A Study Guide to Cultural Pluralism in the United States.* Anti-Defamation League of B'nai B'rith, published through Brown & Williamson Tobacco Corporation.

Harris, P. R. (1994). *High Performance Leadership: HRD Strategies for the New Work Culture.* Amherst, MA.: Human Resource Development Press.

"History of Women in Workforce." (1991, September 19). *Business Week*, p. 112.

Jamieson, David, and Julie O'Mara. (1991). *Managing Workforce 2000: Gaining the Diversity Advantage.* San Franciso: Jossey-Bass.

Kennedy, John F. (1986). *A Nation of Immigrants.* New York: Harper & Row.

Loden, Marilyn. (1996). *Implementing Diversity.* New York: McGraw-Hill.

Loden, Marilyn. (2006). Organizational diversity consultant with Loden Associates, Inc., Tiburon, California, personal communication, July 2006.

Los Angeles Fire Department Historical Archives. (1999). *History of the Black Firemen.* Retrieved February 22, 2007, from http://www.lafire.com/black_ff/black .htm

"Male Dominated Occupation." (2000, October 11). *Contra Costa Times*, p. A3.

Miller, Char. (2006). Professor of history, Trinity College, San Antonio, Texas, personal communication, July 2006.

"Multiracial Data in Census Adds Categories and Controversies." (2000, December 10). *San Francisco Chronicle,* p. A14.

Naisbit, John, and Patricia Aburdene. (1986). *Re-inventing the Corporation.* New York: Warner Books, p. 243.

National Fire Protection Association. (2005). *Firefighters and Fire Departments (U.S.).* Retrieved on 1-14-2008 from http://www. nfpa.org/displayContent.asp?categoryID= 955&itemID=23688

National Fire Protection Association. (2006a). *Fire Department Calls.* Retrieved on 1-14-2008 from http://www.nfpa.org/ displayContent.asp?categoryID= 955&itemID=23850

National Fire Protection Association. (2006b). *Firefighter Injuries by Type of Duty.* Retrieved on 1-14-2008 from http://www. nfpa.org/displayContent.asp?categoryID= 955&itemID=23466

National Fire Protection Association. (2007). *Firefighter Deaths by Type of Duty.* Retrieved on 1-14-2008 from http://www.

nfpa.org/displayContent.asp?categoryID=955&itemID=23471

"Obstacles Hinder Minority Women." (2003, August 7). *Contra Costa Times,* p. C1.

Olson, Aaron T. (2007). Professor of criminal justice at Portland Community College, organizational consultant and retired Oregon State Police patrol supervisor, Portland, Oregon, personal communication, February 2007.

Peters, Thomas J., and Robert H. Waterman. (1984). *In Search of Excellence.* New York: Warner Books.

Prewitt, Kenneth. (2001, fall). "Beyond Census 2000: As a Nation, We Are the World." *The Carnegie Reporter,* 1(3): 1.

Rawlins, Gary H. (1992, October 8). "Africans Came 200 Years Earlier." *USA Today,* p. 2a.

"Redefining Race in America." (2000, September 18). *Newsweek,* p. 38.

"Report Recommends Using Military Model to Defuse Racism in Police Departments." (1992, May 26). *New York Times,* p. A17.

Ruhly, Sharon. (1976). *Orientations to International Communication: Modules in Speech Communication.* Chicago: Science Research Associates.

Shipler, David. (1992, May 26). "Report Recommends Using Military Model to Defuse Racism in Police Departments." *New York Times,* p. A17.

Simons, George, Carmen Vazquez, and Philip Harris. (1993). *Transcultural Leadership—Empowering the Diverse Workforce.* Houston, TX: Gulf Publishing.

Thiederman, Sondra. (1991). *Bridging Cultural Barriers for Corporate Success.* San Francisco: Jossey-Bass.

USA Today. (2004). "Slurs Broadcast over Chicago Fire Radios." Retrieved on January 14, 2008 from http://www.usatoday.com/news/nation/2004-03-10-racism-chicago-fd_x-htm

U.S. Department of Labor Bureau of Labor Statistics. (2000). *Current Population Survey.* Retrieved on January 23, 2006 from http://www.bls.gov/data/home.htm.

U.S. Census Bureau. (2002). "The Foreign Born Population in the United States," *The U.S. Census Bureau Current Population Survey,* March 2002.

U.S. Census Bureau. (2003a). *Current Population Survey:* "The Foreign Born Population in the United States: March 2002." Washington, D.C.: U.S. Government Printing Office.

U.S. Department of Labor Bureau of Labor Statistics. (2003b). *Household Data Survey.* Retrieved on January 19, 2006 from http://stats.bls.gove/data/home.htm.

U.S. Department of Labor Bureau of Labor Statistics. (2004a). *Household Data Annual Average and Survey.* Retrieved on January 19, 2006 from http://stats.bls.gov/cps

U.S. Department of Labor Bureau of Labor Statistics. (2004b). *Women in Law Enforcement Labor.* Retrieved on January 19, 2006 from http://stats.bls.gov/cps

Warikoo, Niraj. (2005). "Jewish Firefighter Sues Dearborn." *Detroit Free Press,* December 22, 2005. Retrieved on January 13, 2006 from http://nl.newsbank.com/nl-search/we/Archives?s_site=freep&f_site=freep&f_sitename+Detroit+Free+Press&p_theme=gannett&p_product=FP&p_action=search&p_field_basc-0=&p_text_base-0=jewish+firefighter+sues+dearborn&Search=Search&p_perpage=10&p_maxdocs=200&p_query-name=700&s_search_type=keyword&p_sort=_rank_%3AD&p_field_date-0–YMD_date&p_params_date-0=date%3AB%2CE&p_text_date-0=

"Women Land more Top Jobs, But Not Top Pay." (2002, March 25). *Contra Costa Times,* p. A32.

Zangwill, Israel. (1908). *The Melting Pot: Drama in Four Acts.* New York: Macmillan.

CHAPTER 2 # Recruitment, Retention, and Promotion in Fire Departments

Key Terms

affirmative action, p. 63
applicant screening, p. 59
glass ceiling, p. 62

paradigm shift, p. 66
recruiting incentives, p. 57
recruitment strategies, p. 52

role identification, p. 65
role modeling, p. 65

Overview

In this chapter we discuss recruitment trends (Figure 2.1) and the probability that recruitment of women and minorities will be an ongoing challenge for fire agencies. A brief summary of the demographics of women and minorities in the fire service is provided, including examples of successful recruitment efforts in state, county, and local agencies across the country. We discuss reasons for recruitment difficulties and offer strategies for success. The retention and promotion of minorities and women are addressed in the final section of the chapter.

Objectives

After completing this chapter, participants should be able to:

- Understand the demographic makeup of women and minorities in the fire service.
- Discuss the ongoing challenges of recruitment trends with respect to women and minorities in fire service agencies.
- Explain recruitment difficulties and strategies for recruitment success.
- Understand the importance of retention and promotion of minorities and women in fire service careers.
- Identify promotional policies and practices in fire service agencies that would demonstrate the valuing of differences in our workplaces and communities.

FIGURE 2.1 ◆ Recruitment, retention, and promotion are central issues for a diverse workforce serving diverse communities including fire service.

LEARNING TASKS

Knowing the needs and gaps in the fire service for the ethnic, racial, and cultural groups in one's service area is critical toward understanding the multicultural challenges of your area. Talk with your agency and find out how it is involved in recruiting, hiring, and retaining a diverse workforce. Be able to:

- Find out the recruitment activities and strategies that are highlighted for your area.
- Identify the top six actions within your local area's recruitment efforts for your county and city fire departments.

PERCEPTIONS

The following quotes draw attention to the recruitment crisis for fire service occupations and illustrate the importance of recruitment, hiring, retention, and promotion of women and minorities (Figure 2.2).

> For us, the challenge is not to accept where the fire service is and where this union is in terms of diversity because doing so would be a failure of leadership of the highest order. We must also admit that a lack of diversity in the fire service is not new. It's a challenge that has existed since the beginning of our proud profession and too little has been done about it (General President Harold Schaitberger, International Association of Fire Fighters [IAFF], 2006).

> Fire Chief Doug Holton said that adding diversity to the city's firehouses was among his goals when he took over the department two years ago. He had served in a Milwaukee fire department that had less than 20 percent minority employees when more than half the city was composed of minority members. "You have to look at the history of fire service in this country and realize that 25 years ago, you did not have Hispanic or females," Holton said. "Are things getting better? Yes. Are we where we want to be? No" (Brown, 2005).

FIGURE 2.2 ◆ Fire chiefs and senior leadership provide the key to effective fire service and successful implementation of multicultural and diversity strategies in the workplace.

As highlighted by Fire Chief Ed Wilson of the Portland, Oregon, Fire Department, "It would be naive to think with 750 employees, representing four generations with different values, working in a 24-hour setting, that we're not going to have problems. . . . What we're trying to do here is just change the culture" (Bernstein, 2003).

One of the greatest challenges facing fire service organizations today is the successful recruitment and retention of highly qualified employees. Community safety can be compromised when substantial experience and training is lost through staff turnover and vacancy. It is imperative, then, to recruit, select, and retain the kind of personnel who will bring to the department and to the community a strong commitment to and talent for the job. Fire service departments throughout the United States are facing the challenge of diversifying their workforces and professionally serving their multicultural communities (Coleman, 2003; Fulbright, 2004). As the demographics of districts, towns, and cities change, the makeup of their public-safety officers should reflect such changes. However, currently, nationwide, the firefighting profession is dominated by white men (Fulbright, 2004) (Figure 2.3).

According to Bergman (2004), "The nation's Hispanic and Asian populations would triple over the next half century. Non-Hispanic whites would represent about one-half of the total population by 2050." From 2000 to 2050, the non-Hispanic white population would increase from 195.7 million to 210.3 million, an increase of 14.6 million or 7 percent. This group is projected to actually lose population in the 2040s and would comprise just 50.1 percent of the total population in 2050, compared with 69.4 percent in 2000. Nearly 67 million people of Latino/Hispanic origin (who may be of any race) would be added to the nation's population between 2000 and 2050. Their numbers are projected to grow from 35.6 million to 102.6 million, an increase of 188 percent. Their share of the nation's population would nearly double, from 12.6 percent to 24.4 percent.

The Asian and Pacific American population is projected to grow 213 percent, from 10.7 million to 33.4 million. Its share of the nation's population would double,

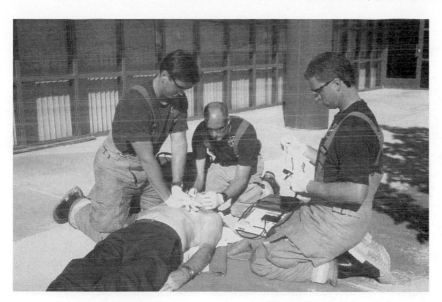

FIGURE 2.3 ◆ Currently, nationwide, the firefighting profession is dominated by white men.

from 3.8 percent to 8 percent. The African American population is projected to rise from 35.8 million to 61.4 million in 2050, an increase of about 26 million or 71 percent. That would raise its share of the country's population from 12.7 percent to 14.6 percent. These are the demographic realities of the future. These are the populations and communities from which the firefighters of the future are to be recruited and hired.

INTRODUCTION

Our society is becoming more and more diverse. In California, demographers predict that, by the 2010 census, no single racial or ethnic group will be in the majority. That trend is occurring in some major cities in other states as well. Some demographers say that in the future every group will be a minority. To recruit and retain a representative staff and provide effective services, fire service executives must have a clear understanding of their community and their own workforce. The focus upon the recruitment and retention of qualified employees, and especially of women, blacks, Asians, Hispanics, and other diverse groups, has become a concern and a priority of fire departments nationwide. Many agencies are having difficulty finding qualified applicants, resulting in a general recruitment crisis, and it appears that this is the case regardless of the economic condition of state or local governments. The recruitment pool of eligible and qualified candidates is diminishing. The crisis is multidimensional and is discussed in detail in this chapter.

RECRUITMENT OF A DIVERSE WORKFORCE

The recruitment of women and members of diverse groups has been a concern and a priority of fire service agencies nationwide for a few decades. This is not a new issue in the history of the fire service in the United States, as indicated in the prior quotes. In the 1970s and even up to today, many fire departments in cities and local municipalities

FIGURE 2.4 ◆ IAFF leadership and organizing support of issues are critical to their successful outcomes in contract areas and in other initiatives such as diversity and multicultural strategies for recruitment.

have had to be brought into the courts and made to enter into consent decrees to address the lack of diversity in the fire service. In consulting with senior management of fire departments by one of the authors (Wong), the firefighter leadership tended to view consent decrees as only moderately useful. Current leaders with the fire service in the United States, including the leadership of the IAFF, are keenly aware of and are taking steps to remedy and address the issue of a diverse workforce within the fire service (Schaitberger, 2006) (Figure 2.4).

PROFILE OF PERSONNEL AND THEIR CAPABILITIES

The U.S. Fire Administration, FEMA, and the National Fire Protection Association (NFPA) performed the "Needs Assessment of the U.S. Fire Service," a cooperative study authorized by U.S. Public Law 106-398, in 2002. One part of the report profiles the demographic makeup of fire service personnel in the United States. According to the FEMA and NFPA report, as of December 2002, is the following profile:

- **Total number of firefighters in the United States:** There are just over a million active firefighters in the United States, of which slightly over three-fourths are volunteer firefighters. Nearly half the volunteers serve in communities with a population of less than 2,500. Table 2.1 (from the 2002 report) summarizes the number of career and voluntary firefighters by size of community in the United States.

- **Communities with a population of less than 2,500:** Communities with a population of less than 2,500 (21 percent of fire departments in the United States), nearly all of which are entirely or mostly volunteer departments, deliver an average of four or fewer volunteer firefighters to a midday house fire (Figure 2.5). Because these departments average only one career firefighter per department, it is likely that most of these departments often fail to deliver the minimum of four firefighters needed to initiate an interior attack safely on such a fire due to the inability to access the volunteer firefighters immediately. Recruitment in these small communities is a problem involving difficulties in mobilizing volunteer personnel in general, as well as representation of women and minorities as volunteers.

TABLE 2.1 ◆ Number of Career, Volunteer, and Total Firefighters by Size of Community

Population Protected	Career Firefighters	Volunteer Firefighters	Total Firefighters
1,000,000 or more	32,700	150	32,850
500,000 to 999,999	28,400	4,900	33,300
250,000 to 499,999	26,600	4,250	30,850
100,000 to 249,999	39,750	8,550	48,300
50,000 to 99,999	37,750	11,000	48,750
25,000 to 49,999	40,000	29,300	69,300
10,000 to 24,999	38,850	86,050	124,900
5,000 to 9,999	12,200	112,300	124,500
2,500 to 4,999	5,050	157,600	162,650
Under 2,500	4,800	408,750	413,550
Total	266,100	822,850	1,088,950

- **Communities with a population of at least 50,000:**   An estimated 48,750 firefighters serve in fire departments that protect communities with a population of at least 50,000 and have fewer than four career firefighters assigned to first-due engine companies (Figure 2.6). It is likely that, for many of these departments, the first-arriving complement of firefighters often falls short of the minimum of four firefighters needed to initiate an interior attack safely on a structure fire, thereby requiring the first-arriving firefighters to wait until the rest of the first-alarm responders arrive. For communities with a population of at least 50,000, recruitment issues involve getting sufficient personnel in general and recruiting minority and women in particular.

- **Volunteer firefighters lack proper training:**   An estimated 413,550 firefighters, most of them volunteers serving in communities with a population of less than 2,500, are involved in structural firefighting but lack formal training in those duties. For volunteer firefighters,

FIGURE 2.5 ◆ Small community fire stations staffed mostly with volunteers constitute 21 percent of the fire departments in the United States, as illustrated by the Branchville Volunteer Fire Company and Rescue Squad.

FIGURE 2.6 ◆ An estimated 73,000 firefighters serve in fire departments that protect communities with a population of at least 50,000 and have fewer than four career firefighters assigned to first-due engine companies.

recruitment issues involve getting sufficient volunteers, involving minorities and women as volunteers, and obtaining training for the volunteers recruited (Figure 2.7).

The FEMA and NFPA (2002) needs assessment report clearly specifies the shortage of career and volunteer firefighter personnel and the inadequacy of resources and training to fully deploy firefighting personnel in times of community need. The needs assessment study, however, did not ask any questions with regard to the ethnic/racial demographics or the gender makeup of fire department personnel, although such information may have important bearing on the effectiveness of fire service departments, recruitment, and outreach efforts in many communities today.

The following information with regard to the ethnic/racial and gender demographics of fire service departments was obtained from the noted sources, primarily from ethnic/racial and gender-specific fire service associations and networks.

FIGURE 2.7 ◆ For larger fire departments with multiple stations, equipment, and personnel, recruitment issues involve getting sufficient volunteers, involving minorities and women as volunteers, and obtaining training for the volunteers recruited.

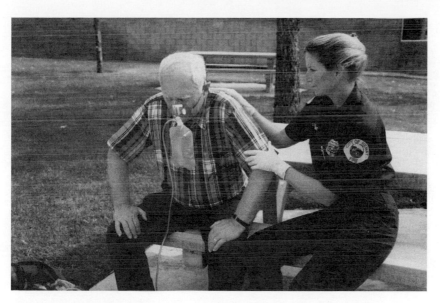

FIGURE 2.8 ◆ Female firefighters would represent about 2.3 percent of the total career firefighters in the United States today.

- **Women firefighters and officers:**   The total number of women firefighters and officers included in the Women in Fire Service (WFS) current database is 5,857. Adding a conservative 5 percent to account for women not included in these statistics gives a current estimate of some 6,155 women active in career-level structural fire suppression in the United States (WFS, 2005). Female firefighters would represent about 2.3 percent of the total career firefighters in the United States today (Figure 2.8). WFS (2005) delineated the following number of women in promoted career positions in the United States today:

> Engineers: 221
> Fire marshals: 2
> Sergeants: 6
> Lieutenants: 202
> Captains: 193
> Chiefs (other than top-level chief): 79
> Chiefs of department: 14

According to WFS, as of January 2006, there were at least 29 career-level or combination (i.e., with some career and some volunteer personnel) fire agencies in the United States whose top-level chief was a woman.

According to a U.S. Department of Labor Bureau of Labor Statistics report, as of 2004, women career firefighters was 5.1 percent, compared to male career firefighters at 94.9 percent, and compared to white male career firefighters at 76.6 percent. The numbers provided by the WFS (i.e., 2.3 percent women firefighters) and the Department of Labor Bureau of Labor Statistics (i.e., 5.1 percent women firefighters) seemed to differ with regard to the percentage of women in career firefighting positions.

- **Racial and ethnic minority firefighters and officers:**   Davis (2006) estimates that there are approximately 20,000 career and volunteer firefighters who are Latino/Hispanic Americans (approximately 2 percent of all firefighters in the United States). According to a U.S. Department of Labor Bureau of Labor Statistics report, as of 2004, the career firefighter minority ranks were black or African American, 8.4 percent; Hispanic or Latino, 8.6 percent; and Asian, 1.3 percent

FIGURE 2.9 ◆ Demographic data delineating the ethnic and racial, as well as the gender, makeup of fire service personnel within both state and local municipalities are needed to provide a comprehensive picture of the firefighter workforce in the United States.

(compared to white male career firefighters who had a 76.6 percent representation). The numbers for the Latino/Hispanic American group as provided by the National Association of Hispanic Firefighters (NAHF) seem to be dramatically different from those provided by the Department of Labor Bureau of Labor Statistics.

Accurate and current demographic data delineating the ethnic and racial, as well as the gender, makeup of fire service personnel within both state and local municipalities are needed to provide a comprehensive picture of the firefighter workforce in the United States (Figure 2.9).

CHALLENGES AND DIFFICULTIES

As noted by Shannon (2003), at least 65 percent of our nation's cities and towns do not have enough fire stations to achieve the widely recognized response-time guidelines. Those guidelines recommend that first-call companies in "built upon" areas of the city be located to ensure travel distances within one and one-half miles. That guidance is consistent with the requirements of NFPA standards that firefighters respond within four minutes, 90 percent of the time. However, arriving on scene in time is not enough if one arrives without the necessary resources and personnel to make a difference (Figure 2.10).

The FEMA and NPFA (2002) needs assessment survey found fire departments protecting communities of at least 1 million citizens had at least four career firefighters assigned to engines. But the availability of career firefighters breaks down into much smaller numbers in less populated communities: Only 60 percent of departments protecting communities of 250,000 to 1 million had four career firefighters assigned to engines. In departments serving populations of 100,000 to 250,000 people, only 44 percent could make that claim. And in communities with populations between 10,000 and 100,000, just 20 to 26 percent of departments offered the necessary coverage.

FIGURE 2.10 ◆ Fire service faces many workplace challenges and accountabilities to the local communities served.

According to FEMA and NFPA (2002), most smaller communities protected by an all volunteer or mostly volunteer fire department responded with four or more firefighters to a midday house fire; but for some, the total response was limited to only two firefighters on the scene. As noted by Shannon (2003), what remains unclear and unmeasured is how long it took to assemble those firefighters. When fewer than four firefighters are on scene, first responders face a cruel choice between initiating an interior attack without proper personnel to secure their own safety during high-risk operations, or delaying the interior fire attack until additional personnel arrive. Obviously, the latter increases the danger to occupants and overall damage to the property. Both NFPA, *Standard on Fire Service Occupational Safety and Health*, and federal OSHA regulations require a minimum of two firefighters to back up an initial team of at least two firefighters working in a hazardous environment, referred to as the "two-in-and-two-out rule."

Closing these gaps requires the recruitment, hiring, and retention of more firefighters (Figure 2.11). There are no shortcuts. Just to staff the number of fire stations required to meet response-time guidelines, NFPA estimates 25,000 to 35,000 more career firefighters are needed; and to address the staffing of existing departments so that firefighters safely and effectively mount an interior attack on a fire, another 50,000 career officers are needed (Shannon, 2003).

For a period of time during the late 1990s and into 2000, the United States experienced the lowest unemployment rate in 30 years (3.9 percent) because of a robust economy. There were plenty of jobs, but the number of applicants for fire service positions was substantially lower than in the past. One of the authors (Wong) provided recruitment strategies in both the public and private arenas and found that with an

FIGURE 2.11 ◆ Recruitment and job fairs are competitive events for employers recruiting job seekers.

overabundance of jobs (technical, service, sales, etc.) to choose from, civil service positions often suffered. Many jobs in private industry, especially high-tech jobs, come with high salaries, stock options, signing bonuses, company cars, and year-end bonuses. Public-sector employers have difficulty competing with such incentives (Figure 2.12). Another factor in recruitment difficulties entailed court decisions (and pretrial settlements) against such companies as Coca-Cola, Texaco, and Wall Street's Smith Barney in race bias lawsuits. U.S. firms scrambled to recruit and promote talented

FIGURE 2.12 ◆ Many jobs in private industry, especially high-tech jobs, come with high salaries, stock options, signing bonuses, company cars, and year-end bonuses; and public-sector employers have difficulty competing with such incentives.

women and minorities. Many large companies, aware that the populations within their recruitment areas were diversifying, knew that their employee demographics had to match those of the world outside or they, too, would be subject to lawsuits and/or criticism. This is particularly true of companies that sell products or services to the public. Their recruitment efforts, therefore, focused on women and members of diverse groups. The wide-open market also made for transitory employees, causing retention problems for employers, especially of trained employees. During this period, smaller fire service and public agencies lost their employees as larger agencies lured experienced officers away with promises of benefits. The effort was compared to the National Football League (NFL) draft, in which everyone was competing for qualified candidates.

Compounding the fire service recruitment problem was the negative impact of highly publicized court cases and scandals involving the fire service in major cities—Chicago, Dearborn, and San Francisco, as mentioned in Chapter 1. These court cases, it is suggested, "tarnished the image" of the fire service work. At the same time, departments were not fully staffed because many employees who had been hired in the fire service expansion wave of the 1960s and 1970s were retiring.

Other reasons that young people are not applying for careers in the fire service are related to the traditional organizational structure of most agencies. It has been suggested that the independent-minded youth of today are less tolerant of the rigid hierarchy to which most fire service agencies still adhere (Blair, 2005). In addition, most young people do not like being micromanaged.

Serious budget shortfalls in most states, which started in 2002 and extended into 2006, forced cities and counties to reduce fire service personnel through layoffs, attrition, and curtailing department size. Some fire service agencies benefited because unemployed people turned to civil service jobs, including fire departments; however, most fire service agencies did not, as their budgets were also reduced. Whenever there is a recession, it creates a problem because cities and counties must lay off or terminate those employees who were hired most recently. Women and people of diverse ethnic and racial groups, therefore, are usually the first to lose their jobs, thereby canceling any gains made in terms of a multicultural workforce, including women.

Cultural Diversity at Work, a newsletter addressing multicultural workplace issues, presented a list of the most frequent causes of the failure to attract and retain high-level minority and female employees. The following is a brief synopsis of the implicated causes most pertinent to the fire service (Micari, 1993):

1. **Senior management is not sending the "diversity message" down the lines.**   Senior management does not always demonstrate commitment in the form of value statements and policies emphasizing the importance of a diverse workforce.
2. **Informal networking channels are closed to outsiders.**   Women and minorities often experience discomfort within the traditional all-white, all-male informal career networks, including extracurricular activities. For example, some women firefighters are uncomfortable participating in firefighter association activities that involve recreational gambling, fishing, sports activities, or the roughhousing that can take place at meetings.
3. **In-house recruiters are looking in the wrong places.**   Recruiters must use different methods and resources than those they have traditionally used to find diverse candidates.
4. **Differences in life experience are not taken into account.**   Some applicants, both women and minorities, will have had life experiences that differ from those of the traditional job candidate. For example, the latter has typically had some college experience, with a high grade point average, and is often single (and thus does not have the many responsibilities associated with marriage and children). On the other hand, many minority job candidates have had to work

through school and are married and may have children. Therefore, their grade point averages may have suffered.

5. **Negative judgments are made based on personality or communications differences.** Although not everybody's style reflects cultural, racial, or gender characteristics, there are distinct differences in communication style and personality between women and minorities and traditional white male job applicants. Communication style differences are especially apparent when English is the applicant's second language. Generally speaking, women are often not as outwardly assertive as men in communication. Sometimes these factors can affect the outcome of a preemployment interview and become barriers to a job offer.

6. **The candidate is not introduced to people who are like him or her.** It is important that the candidate or new hire meet people of the same gender, race, or ethnicity within the agency. A mentor or support group may be crucial to a successful transition into the organization.

7. **Organizations are not able or willing to take the time to do a thorough search.** Recruitment specialists indicate that searches for qualified minority and women candidates take four to six weeks longer than others. These searches take commitment, resources, and time.

8. **Early identification is missing from the recruitment program.** Programs that move students into the proper fields of study early on and that provide the education required for the job are essential. Examples of these programs include internships, scholarships, and fire service programs (e.g., volunteer programs) that bring future job candidates into contact with the organization.

According to Blair (2005), two key issues were identified as impediments in today's recruitment of firefighters:

1. **Identification of recruiting problems.** The problems most frequently reported were a decreasing number of qualified applicants; the inability to offer competitive compensation; and the difficulty in recruiting women and Asians, blacks, Hispanics, and other minority candidates.

2. **Identification and specification of appropriate and unbiased testing on which candidates are likely to pass or fail** as predictors of future success in job performance.

To address the latter, departments should perform a statistical analysis of the selection processes and tests they use by gender and ethnic or racial group to determine whether women or minorities are being disproportionately eliminated by these tests. The department should then evaluate whether those screening devices contain outdated or biased questions or questions that are no longer job related. If the tests contain questions that have an adverse impact on targeted groups, then these obstacles must be studied further to determine whether they can be removed.

RECRUITMENT STRATEGIES

recruitment strategies
Approaches used by organizations to attract and bring in job applicants and candidates to enable a diverse workforce.

To build a diverse workforce, **recruitment strategies** used in the past will no longer be sufficient and will not provide agencies with high-quality applicants. Runnett (1999) summarizes the following key components involved in the Charlottesville, Virginia, Fire Department's recruitment strategies:

1. The posting period for the job of firefighters in most municipalities tends to be very short. Generally, applications for firefighters are accepted for a brief period, usually two weeks once per year and only in some cases on a biannual basis. It is extremely difficult to keep any candidate focused on the process with such a short posting period whereby candidates must find the listing in the newspaper. As part of the department's efforts, fire service had to cast a wider net to recruit the best candidates. With the era of the 1,000 plus prospective candidates being over, fire service agencies need to allow applications to be received at any time during the year or have a notification waiting list, and send out reminder letters prior to posting the job.

Fire service agencies need to be more proactive and "applicant friendly" in their approaches to the job application process.

2. The written test is a comprehensive competitive examination, with the highest scores continuing in the hiring process. If fire service agencies are to achieve greater diversity, they must recruit candidates that will perform well on this type of test. Recruiting candidates that do well in the comprehensive examination is one of not only the greater challenges but also the greater opportunities to have an impact on the organization's diversity. It is important that the department recruit at local colleges and universities, including predominantly minority institutions. College students are accustomed to taking comprehensive examinations and tend to excel in this area.

3. Recruiting college students has its challenges. Family pressure for the college student to succeed is one such challenge. The family of a prospective fire service candidate most likely has incurred the ever-increasing expense of educating the candidate and is expecting him to reap the financial rewards of his hard work to obtain a college degree. One finding by Runnett (1999) revealed that many cultures still consider the fire service a "blue-collar" occupation, not necessarily fitting for a college graduate. As such, fire service must develop a comprehensive marketing strategy to demonstrate the need for higher education to meet the challenges of an ever-changing occupation in fire service. This marketing plan must be multidimensional, focused on attracting prospective candidates during their grade school years as well as college. It needs to incorporate advertising, mentoring programs with local schools and colleges, community outreach efforts, and a career-focused website to enhance the fire service image in the community.

4. Competition with other public agencies and the private sector is another factor. Many organizations are trying to recruit the best people. During periods of a strong economy, the threat of layoff and organizational rightsizing is diminished in the private sector. In addition, the traditional stability of the municipal sector is threatened as many government agencies are finding that they too must rightsize to be competitive. In most cases, the private sector has the flexibility to offer attractive and competitive pay and benefit packages. The public sector is locked into a pay and benefit system that is cumbersome and difficult to modify. This reality, along with a rigid work schedule and the danger of working in the fire service, can make recruiting good people a formidable challenge. It is very difficult to recruit highly skilled candidates if the compensation and benefit package is not competitive with the existing marketplace. Fire service compensation and benefit packages must be competitive with those of other public agencies and the private sector.

5. Recruiting from the military offers the fire service many advantages. The military lifestyle lends itself well to the hierarchical fire service departments. According to recent military publications, the majority of the people in the services today have obtained some higher education preparing them for the fire service entry-level testing. All the armed services have a mandatory physical fitness program that all but assures these candidates will meet the physical test requirements of the entry-level program. At least two significant obstacles must be overcome to recruit military personnel. Fire service benefit packages must be competitive with those of other public agencies and the private sector. The fire service should consider offering to assist with relocation costs. One possible scenario would require a contract between an employee and the agency, whereby the agency will ensure (a) relocation costs in exchange for a five- to ten-year employment commitment, (b) developmental opportunities, (c) succession planning to develop high-potential employees, and (d) sensitivity to issues of cultural and racial diversity and gender.

Agencies with the most success in recruiting women and minorities have had specific goals, objectives, and timetables in place; these policies must be established at the top level of the organization. At the same time, management must commit to not lowering standards for the sake of numbers and deadlines. The philosophy and procedures required are discussed in the following sections.

COMMITMENT

Recruiting minority and women applicants, especially in highly competitive labor markets, requires commitment and effort. Fire service executives must communicate that commitment to their recruiting staff and devote the resources necessary to achieve recruitment goals. This genuine commitment must be demonstrated both inside and outside of the organization. Internally, chief executives should develop policies and procedures that emphasize the importance of a diverse workforce. Affirmative action and/or programs that target certain applicants (where legal) will not work in a vacuum. Chief executives must integrate the values that promote diversity and affirmative action into every aspect of the agency, from its mission statement to its day-to-day service training. Externally, fire service executives should publicly delineate the specific hiring and promotion goals of the department to the community through both formal (e.g., media) and informal methods (e.g., community safety messages, networking with organizations representing the diverse groups) (Figure 2.13). While chief executives promote the philosophy, policies, and procedures, committed staff who are sensitive to the needs for affirmative hiring and promotions carry them out. Executives should also build partnerships with personnel officials so that decisions clearly reflect the hiring goals of the department. It is recommended that fire service executives audit the personnel selection process to ensure that neither the sequencing of the testing stages nor the length of the selection process is hindering the objective of hiring women and minorities (Blair, 2005).

Although fire executives may be genuine in their efforts to champion diversity and affirmative action hiring, care must be taken that their policies and procedures do not violate Title VII of the Civil Rights Act of 1964. Also, agencies in states that have enacted laws prohibiting affirmative action hiring programs and the targeting of "protected classes" (e.g., California's Proposition 209, passed in 1996, which is discussed later in this chapter) must adhere to these regulations. A knowledgeable personnel department or legal staff should review policies and procedures prior to implementation.

FIGURE 2.13 ◆ Firefighters' union, city officials, and other executives discuss some of the firefighters' human resource issues.

PLANNING

As noted in the WFS (2005) steps for recruitment, when fire executives talk about firefighter recruitment, they usually mean the effort that is spent, shortly before an application period opens, to get candidates to apply for job openings. Recruitment is the way a fire department attracts new members. Fire chiefs who wish to diversify their department's workforce or attract specific groups of people (college graduates, licensed paramedics) have learned to plan and to target those groups in the recruiting effort.

A productive recruitment drive is just part of what it takes to increase the number of women and minorities within a fire department. For recruitment to be really effective, managers must establish a positive climate within the department before encouraging women and minorities to become firefighters. Fire departments must also begin to recognize, and take advantage of, the recruitment impact of most of their public activities. Expanding the concept of recruitment in these two directions will make the recruitment drive itself more productive and will increase the likelihood that the women who are recruited will actually become firefighters.

A strategic marketing plan would include the action steps that commit the objectives, goals, budget, accountability, and timetables for the recruitment campaign to paper. Demographic data should form one of the foundations for the plan, which must also take into account the current political, social, and economic conditions of the department and the community. To avoid losing qualified applicants to other agencies or to private industry, the plan should provide for fast-tracking the best candidates through the testing and screening processes. Fast-tracking is an aggressive recruitment process wherein a preliminary, qualifying check quickly takes place concerning the driving and background history of the applicant before any other screening process occurs. Applicants who do not meet established standards are immediately notified. The remaining applicants are then interviewed by a trained ranking officer or civilian holding the position for which they are applying. Applicants earning a passing score on the personal interview immediately receive a letter advising them that they passed, thereby giving them tangible evidence that the agency is seriously considering hiring them. The letter also informs them that they must pass additional qualifying steps, which must be commenced immediately.

In today's labor market, the best-qualified candidates are not willing to wait months for the selection process to run its course. Public safety organizations must develop valid selection approaches that are also timely. Otherwise, by the time the selection process has run its course, the best-qualified candidates may have found other jobs.

The strategic recruitment plan should include an advertising campaign that targets:

- Community-based human service agencies
- Community-based and regional ethnic, cultural, and racial advocacy and networking agencies
- Community-based job training and job referral programs and agencies
- Colleges and universities
- Military bases and reserve units
- Churches, temples, synagogues, and other places of worship
- Gymnasiums, fitness and martial arts studios, athletic clubs, and the like

Participation of women and minority officers within the department in recruitment efforts is crucial (see Recruiting Incentives), as is the involvement of groups and organizations that represent the target groups. If there are high-profile minorities and women, such as athletes and business executives, within the community, they should be enlisted to promote fire service work as a career through media releases and endorsements on fliers and brochures.

In summary, the key steps in planning include:

- ◆ Making an assessment of the department's current recruitment practices
- ◆ Establishing a positive climate for a diverse workforce
- ◆ Developing a strategic marketing plan
- ◆ Utilizing a strategic recruitment plan

RESOURCES

Adequate resources, including money, personnel, and equipment, must be made available to the recruitment effort. Financial constraints challenge almost every organization's recruitment campaign. The size or financial circumstances of an agency may necessitate less expensive—and perhaps more innovative—approaches. For example, many small fire service jurisdictions can combine to implement regional testing. One large county on the West Coast successfully formed a consortium of agencies and implemented regional testing for fire service candidates. To participate, each agency pays into an account based on the population of its jurisdiction. Alternatively, each agency can pay according to how many applicants it hired from the list. The pooled money is then used for recruitment advertising (e.g., billboards, radio, television, newspapers) and the initial testing processes (e.g., reading, writing, and physical tests, including proctors). The eligibility list is then provided to each of the participating agencies, which continue the screening process for applicants in whom they have an interest. Fire service agencies should not see other agencies as competitors with respect to recruiting. By combining their efforts, they may be able to:

- ◆ Save money (consolidate resources)
- ◆ Develop a larger pool of applicants
- ◆ Become more competitive with private industry and other public agencies
- ◆ Test more often
- ◆ Reduce the time it takes from application to hire

In terms of recruiting a diverse workforce, the second benefit listed—developing a larger pool of applicants—is central to reaching beyond the traditional applicant pool. Fire service agencies, in taking advantage of the Internet's global coverage, can post position openings in the effort to recruit individuals with a wide variety of backgrounds and skills. Agencies should also create their own websites describing their departments and offering information on recruitment and on how to obtain applications. Department websites have been found to be valuable recruiting tools when used to highlight the work of women and minorities within an agency.

SELECTION AND TRAINING OF RECRUITERS

A recruiter is an ambassador for the department and must be selected carefully. Fire service recruiters should reflect the diversity within the community and include women. Full-time recruiters are a luxury most often found only in large agencies. The benefit of a full-time recruiter program is that usually the employees in this assignment have received some training in marketing techniques. They have no other responsibilities or assignments and can, therefore, focus on what they do and do it well. They develop the contacts, resources, and skills to be effective. Whether full-time, part-time, or assigned on an as-needed basis, however, the following criteria should be considered when selecting recruiters:

- ◆ Commitment to the goal of recruiting
- ◆ Belief in a philosophy that values diversity
- ◆ Ability to work well in a community environment
- ◆ Knowledge of the ethnic, cultural, racial, and gender networks and agencies
- ◆ Belief in and ability to market a product: fire service as a career
- ◆ Comfort with people of all backgrounds and ability to communicate this comfort
- ◆ Ability to discuss the importance of entire community representation in fire service work and the advantages to the department without sounding patronizing

Recruiters must be given resources (e.g., budget and equipment) and must have established guidelines. They must be highly trained with respect to their role, market research methods, public relations, and cultural awareness. They also need to understand, appreciate, and be dedicated to organizational values and ethics. They must be aware and in control of any biases they might have toward individuals or groups of people who might be different from themselves.

RECRUITING INCENTIVES

It is recommended that fire service executives consider using financial and other **recruiting incentives** to recruit women and minorities where such programs are lawful (monetary incentive programs may be adversely affected by Fair Labor Standards Act considerations). Financial and other incentive programs are especially useful to agencies that cannot afford the luxury of full-time recruiters. They are used to encourage officers to recruit bilingual whites, women, and ethnically and racially diverse candidates, informally, while on or off duty. One possible program would give officers overtime credit for each person they recruit in those categories who is able to make the eligibility list, additional credit if the same applicant is hired, and further credit for each stage the new officer passes until the probation period ends. Certain departments offer officers (not assigned to recruiting) compensatory time for recruiting candidates who make the eligibility list. Department members can also receive additional compensatory time for recruiting candidates who are bilingual or from protected classes. The department defines a member of a protected class as African American, Asian/Pacific, Latino/Hispanic, Native American, or female. Encouraging all department members to be involved in the recruitment effort, including the promotion of fire service as a career, is usually effective. To be competitive in recruiting employees, especially minorities and women, agencies must offer incentives just as corporations do.

recruiting incentives
Financial and other incentives to recruit women and minorities where such programs are lawful (monetary incentive programs may be adversely affected by Fair Labor Standards Act considerations).

COMMUNITY INVOLVEMENT

A pluralistic community must have some involvement early in the recruitment effort of candidates for the fire service. Representatives from different ethnic and racial backgrounds should be involved in initial meetings to plan a recruitment campaign. They can assist in determining the best marketing methods for the groups they represent and by personally contacting potential candidates. They should be provided with recruitment information (e.g., brochures and posters) that they can disseminate at religious institutions, civic and social organizations, schools, and cultural events. Community-based safety meetings also offer opportunities for officers to put messages out regarding agency recruiting. Community leaders representing the diversity of the community should also be involved in the selection process, including sitting on oral boards for applicants. Many progressive agencies have encouraged their officers to join community-based organizations, in which they interact with community members and are able to involve the group in recruitment efforts for the department, as is seen in the following example:

> Dozens of suburban police chiefs, fire chiefs and municipal leaders will convene Wednesday in Arlington Heights to discuss the importance of attracting minorities to local police and fire departments. . . . "We see from the 2000 census that the suburban population is changing," said the Rev. Clyde H. Brooks, president of the human rights conference. "In some cases, though, the makeup of our police and fire departments isn't changing. That can cause barriers to form between these departments and the public they serve." Brooks said he hopes the communities participating in Wednesday's conference will leave at the end of the day with a renewed commitment to diversification and with new ideas about how to achieve that goal. Brooks, the first black member of Mount Prospect's board of fire and police commissioners, will be one of the speakers at the conference; he will talk about how to go about recruiting minorities for police and fire positions. Other talks will focus on topics such as diversity-related policy making and the "dos and don'ts" of achieving a diverse workforce (Arado, 2002 p. C1).

SELECTION PROCESSES

Prior to initiating any selection processes, fire service agencies must assess the satisfaction level of current employees and the workplace environment of the department.

SATISFACTION LEVEL OF EMPLOYEES

The first step before outreach recruitment can take place is for the department to look inward. Are any members experiencing emotional pain or suffering because of their race, ethnicity, nationality, gender, or sexual orientation? A department seeking to hire applicants from these groups cannot have internal problems, either real or perceived, related to racism, discrimination, or hostility toward female or homosexual officers. A department with a high turnover rate or a reputation for not promoting women, minorities, or homosexuals will also deter good people from applying. The department must resolve any internal problems before meaningful recruitment can occur. To determine the nature and extent of any such problems, fire service agencies can perform an assessment of all their employees through anonymous surveys about their work environment. There should be a review of policies and procedures (especially those related to sexual harassment) and an examination of statistical information, such as the number of officers leaving the department and their reasons for doing so, and which employees are promoted. The goal is not only to evaluate the workplace environment for women and minorities but also to determine what steps need to be taken to dissolve barriers confronting them. See Appendix A for a sample survey that can be used for this purpose.

Supervisors and managers must talk with all members of their workforce on a regular basis to find out whether any issues are disturbing them. They must then demonstrate that they are taking steps to alleviate the source of discomfort, whether this involves modifying practices or simply discussing behavior with other employees.

The training program for new recruits should also be reviewed and evaluated to ensure that new firefighters are not being arbitrarily eliminated or subjected to prejudice or discrimination as is evident in the following example with the Los Angeles Fire Department:

> The case of a Westchester firefighter who found dog food in his dinner is typical of the hazing, harassment, discrimination and retaliation that goes on within the Los Angeles Fire Department, according to an audit released by City Controller Laura Chick on Thursday. The audit also examined academy graduations from 1998 to 2004 and found that Fire Chief William Bamattre frequently allowed failing minority recruits to become rookie firefighters in an apparent attempt to bolster the department's diversity, Chick said. . . . The audit surveyed about 25 percent of the fire department's minorities, women and rookies, and found that a majority had willingly or unwillingly participated in hazing. In addition, 80 percent of the women surveyed knew of or had been the victim of sexual harassment at work. Nearly 90 percent of blacks and 40 percent of Hispanics reported discrimination, and 38 percent said they feared retaliation if they reported any such hostile activity (Hewitt, 2006).

By the time a recruit has reached this stage of training, much has been invested in the new firefighter; every effort should be made to see that he or she completes the program successfully. Negligent retention, however, is a liability to an organization. When it is well documented that a trainee is not suitable for retention, release from employment is usually the best recourse regardless of race, ethnicity, lifestyle, or gender.

Role models and mentoring programs should be established to give recruits and junior officers the opportunity to receive support and important information from senior officers of the same race, ethnicity, gender, or sexual orientation. However, many successful programs include role models of different backgrounds than the recruits.

APPLICANT SCREENING

Fire service agencies must conduct **applicant screening** to assess applicants along a range of dimensions that include, but are not limited to, the following:

applicant screening
The methods, processes, and procedures used by fire service agencies to assess applicants along a range of dimensions to perform fire service roles, tasks, and functions.

- Basic qualifications such as education, requisite licenses, and citizenship
- Intelligence and problem-solving capacity
- Psychological fitness
- Physical fitness and ability
- Current and past illegal drug use
- Character as revealed by criminal record, driving record, work history, military record, credit history, and reputation
- Aptitude and ability to serve others
- Racial, ethnic, gender, sexual orientation, and cultural biases

The last dimension, testing for biases, deserves particular attention. An agency whose hiring procedures screen for unacceptable biases demonstrates to the community that it seeks public servants who will carry out their duties with fairness, integrity, diligence, and impartiality—firefighters who will respect the civil rights and dignity of the people with whom they serve and work. Such screening should include not only the use of interviews to determine attitudes and bias but also careful review of the candidate by personnel staff. The review should consider the applicant's own statements about racial issues, as well as information from interviews that provides clues to how the applicant feels about and treats members of other racial, ethnic, gender, and sexual orientation groups. These interviews would include questions on:

- How the applicant has interacted with other groups
- What people of diverse groups say about the applicant
- Whether the applicant has ever experienced conflict or tension with members of diverse groups or individuals and how he or she handled these experiences

Because of the emphasis on community service, fire service recruiters must also seek applicants who demonstrate the mentality and ability to serve others (not just fight fires). Recruiters therefore are looking for candidates who are adaptable, analytical, communicative, compassionate, courageous (both physically and morally), culturally sensitive, decisive, disciplined, ethical, goal-oriented, incorruptible, mature, responsible, and self-motivated. Agencies also expect firefighters to have good interpersonal and communication skills, as well as sales and marketing abilities, so applicants should be screened for these attributes and for their desire for continued learning and their ability to work in a rapidly evolving environment.

The search for recruits should focus on those who demonstrate an ability to take ownership of problems and work with others toward solutions. These potential officers must remain open to people who disagree or offer different opinions. Fire service agencies should make their goals and expectations clear to candidates for employment, recognizing that some will not match the department's vision, mission, and values and therefore should be screened out at the beginning of the process.

It is in the best interest of fire service agencies, before implementing tests, to complete a job analysis. Although a time-consuming process, the final result is a clear description of the job for which applicants are applying and being screened. Utilizing the job analysis data, tests and job performance criteria can be developed and become part of the screening process. Applicants, when provided a copy of the job analysis, can decide in advance whether they fit the criteria.

The Federal Emergency Management Administration (FEMA) has produced two guides to assist fire service agencies seeking to recruit and retain more women in fire service positions (FEMA, 1999a, 1999b). The resulting publication, *Many Faces, One Purpose: A Manager's Handbook on Women in Firefighting* (1999a), provides assistance to federal, state, and local fire service agencies on examining their policies and procedures to identify and remove obstacles to hiring and retaining women at all levels of the organization. The guide also provides a list of resources for agencies to use when they plan or implement changes to their current policies and procedures. The second FEMA publication, *Many Women Strong: A Handbook for Women Firefighters* (1999b), provides information and assistance to women who are fire service job-applicant candidates and current firefighters on the job.

EXAMPLES OF SUCCESSFUL RECRUITING PROGRAMS

Successful recruitment programs vary by community. If recruitment efforts result in a large enough pool of qualified applicants, the pool will contain individuals of all backgrounds. The following agencies have had success in recruiting female and minority applicants and might be contacted to determine the strategies they used:

- **Minneapolis Fire Department:** Minneapolis had its first female fire executive in 2004, and 17 percent of its 380 uniformed firefighters are women.

FIGURE 2.14 ◆ Against great odds, after more than 20 years of struggling with the biases and prejudices of the Firefighter Union, Fire Station Three is the first all-female firefighting company in San Diego.

- **San Diego Fire-Rescue Department:** Women account for about 8 percent of the 880 uniformed firefighters assigned to its station houses, compared with the national average of 2.5 percent. The San Diego Fire-Rescue Department, which has a female assistant chief, is considered one of the best departments for women to work in, according to Women in the Fire Service (WFS), an advocacy group based in Madison, Wisconsin (Kershaw, 2006) (Figure 2.14).

- **Montgomery County's Fire and Rescue Services:** Montgomery County Council members and minority leaders criticized the numbers of minority and women candidates in the department; and the county executive ordered county administrators to toss out the written examination the county has used for years and to reexamine the way it recruits minorities and makes hiring decisions. The County Council approved a $200,000 special appropriation to bolster recruitment efforts. "We're very pleased with the diversity of this new recruit class," said David Weaver, a County Council spokesman. "The county executive made it very clear that the lack of diversity in the last class was totally unacceptable, and he challenged the fire department and the personnel office to go back and find a way to ensure that future classes better reflected the diversity of our community, and they've achieved that." This county in Maryland reestablished a minority recruiter position that had been cut from the payroll in recent years because of budget cuts and reinterviewed more than 100 applicants who had been rejected in the last round of job applications. The fire department has come under criticism in the past for having an exclusively white command staff and in recent years has substantially increased minority recruitment (Snyder, 2004).

- **Chattanooga Fire Department:** David Brooks, Jr., assistant fire marshal and president of the local chapter of the International Association of Black Professional Firefighters, said the local department is progressive "in the city, the state and the region. . . . As far as minorities, at least a third of the academy is African-American" (O'Neal, 2004). All firefighters train together, and their ability is judged by the same standards and assessment testing for promotion, which is uniform. Firefighters must learn not only firefighting but also emergency medical care and extrication, hazardous materials handling, and the needs of homeland security. A major opportunity for diversity in the ranks came about in 1998 when the department instituted assessment-based testing for promotion, rather than relying on seniority.

- **St. Louis Fire Department:** In 1974, the U.S. Justice Department sued St. Louis in federal court, charging racial discrimination. At the time, less than 10 percent of St. Louis firefighters were African American. City officials agreed to begin hiring one black firefighter for every white firefighter until the racial makeup of the department reflected the city as a whole. The St. Louis Fire Department still follows that agreement because it hasn't reached that goal. As of June 2002, the department is 42 percent black, while the city is 51 percent African American (Little and Levins, 2002).

RETENTION AND PROMOTION OF A DIVERSE WORKFORCE

THE GLASS CEILING

Selecting recruiting officers who reflect the gender, racial, ethnic, and sexual orientation demographics of the community is one important challenge of fire service agencies. Retention and promotion are equally important. Retention of any employee is usually the result of good work on the part of the employee and a positive environment wherein all employees are treated with dignity and respect. A high rate of retention is most likely in organizations that meet the basic needs of employees and offer reasonable opportunities for career development. In fact, once an agency earns a reputation for fairness, talented men and women of all ethnicities and races will seek out that agency and remain longer.

glass ceiling

An invisible but often perceived barrier that prevents some ethnic or racial groups and women from becoming promoted or hired.

The lack of promotions of protected classes and women to supervisor and command ranks has been cited as a severe problem in the fire service for at least three decades by scholars and fire service researchers. Authors and advocates for the promotion of women have used the term **glass ceiling** to describe an unacknowledged barrier that inhibits those officers from reaching ranks above middle levels and higher. The Glass Ceiling Commission, a federal bipartisan group studying diversity in the workplace (1994–1995), discovered that the glass ceiling has not been broken to any significant extent in most organizations, including fire service agencies.

Within organizations, women and others of diverse backgrounds are frustrated when promotional opportunities seem more available to white males than to them. The disenchantment that often accompanies frustration frequently leads to low productivity and morale, early burnout, and resignation because opportunities appear better elsewhere. Lack of attention to equal opportunity promoting practices at some fire service agencies has resulted in court-ordered promotions. These have a negative impact on a department's operations and relationships, both internally and externally, and often lead to distrust and dissatisfaction.

Although women are making gains, they still constitute only a small proportion of fire service supervisors and managers (WFS, 2005). Fire service executives and city or county managers cannot afford to minimize the consequences of poor retention and inadequate promotional opportunities for women and diverse groups within their organizations.

Chief executives must determine whether female officers are applying for promotions in numbers that are proportional to their numbers in the department. If not, perhaps female officers need encouragement from their supervisors. It is also possible that the promotional process disproportionately screens out female officers. Research shows that the more subjective the process, the less likely women are to pass it. Some safeguards against bias include weighting the process toward "hands-on" tasks (assessment center testing); conducting structured interviews; selecting board members who represent different races and both sexes; and training the board members on interviewing techniques.

AFFIRMATIVE ACTION

The authors acknowledge that there may be some controversial and even legal aspects, in certain states or jurisdictions, to recruitment efforts that target women, as well as ethnically and racially diverse candidates for fire service jobs. Federal law prohibits programs that require meeting specific hiring goals for any particular group except when necessary to remedy discrimination. The discussion of **affirmative action** has been somewhat a divisive issue within the fire service in which firefighters, fire executives, the union, and firefighter associations have held diverse opinions and positions. Affirmative action has also evoked different concerns for the fire service depending upon whether one is looking at it from the legal compliance perspective or the human resources point of view in which it serves as a tool for recruitment. Rather than take any particular sides in the affirmative action debate (e.g., affirmative action as an effective tool for diversifying the workforce, etc.), the authors have articulated the common issues, as highlighted next, and refer any specific issues of affirmative action with regard to the local fire departments to the reader's research at the local level.

affirmative action

Legally mandated programs with the objective to increase the employment or educational opportunities of groups that have been disadvantaged in the past.

In California, Proposition 209—which outlawed governmental discrimination and preferences based on race, sex, color, ethnicity, or national origin—was passed in 1996. In September 2000 justices of the California Supreme Court affirmed that the proposition was legal. In their decision, the justices placed strict limits on employers regarding the types of outreach programs they can legally use to recruit employees; any outreach program that gives minorities and women a competitive advantage is a violation of Proposition 209. Many other states followed California's example by enacting similar legislation. Agencies need to research what strategies are legal and appropriate within their state and jurisdiction.

COURT ACTION

In 2003, the Supreme Court of the United States ruled that the University of Michigan's law school admissions policy, which uses race as a factor, was not in violation of the Constitution's Fourteenth Amendment. The court actually issued two rulings. One, a 5–4 decision, was that race can be considered in the admissions process to the school of law. The other, a 6–3 decision, was that the University of Michigan's undergraduate admissions policy, which used a point system to give minority students a considerable edge, was unconstitutional. The court ruled that the university could not use "rigid systems that seem like quotas, and must adopt race-neutral admissions policies as soon as practical." Only time will tell what impact, if any, this decision will have nationwide and, therefore, what impact it will have upon the recruitment of a diverse fire service workforce.

The practice of tracking race, ethnicity, and gender via questionnaires is considered controversial. Ward Connerly, an African American who championed the end of affirmative action in California, argues that the practice is divisive, especially in California where no single race holds a majority and where people increasingly consider themselves multiracial ("Connerly Starts Push to End Tracking Race," 2001). Opponents of this view contend that collecting such data is crucial to preventing discrimination and allocating state resources. Affirmative action and consent decrees have been only moderately successful in achieving parity in the hiring of women and individuals from ethnically diverse backgrounds. There has been even less success with promotions of these groups to command ranks. In fact, studies have found that affirmative action has produced uneven results by race, gender, and occupation across the nation.

Mini-Case Study: What Would You Do?

You are the new personnel sergeant responsible for a department with 60 career firefighters. The department has one Hispanic man, two African American men, and two women officers. There has never been a large number of minority or women candidates applying for permanent positions in the department, and your department does not reflect the demographics of the city. Your city has an affirmative action program, but to date, no outreach programs have been initiated to recruit women and minorities. Your chief has asked that you provide a proposal on what strategies you would suggest to recruit women and minorities. Develop a list of what you would propose. Make a list of what other departmental processes should take place prior to applicant testing.

An unfortunate problem that can be associated with the promotion of women and nonwhites is that doubts are raised about their qualifications: "Are they qualified for the job or are they products of affirmative action?" Peers may subtly or even explicitly express to each other that the promotion was not the result of competence, and the promoted candidate may feel that his or her success is not based entirely on qualifications. Consequently, employees may experience strained relationships and lowered morale. There is no denying the potential for a strong negative internal reaction in an organization when court orders have mandated promotions. Some white employees feel anger or frustration with consent decrees or affirmative action–based promotions. Clearly, preventive work must be done to avoid these problems.

Failure to promote qualified candidates representative of the diverse populations agencies serve, including women, can result in continued distrust of the fire service by the communities. Underrepresentation within fire departments also aggravates tensions between the fire department and the community. Some scholars and public service experts argue that underrepresentation at all levels within fire service agencies hurts the image of the department in the eyes of the community. At the same time, representation of women and minorities at all levels within fire service agencies sends a positive inclusion message to the community.

Bett Clark—one of only 25 women who head fire departments in the country —just wants to be known as a good fire chief for Bernalillo County. "I'm not as proud of being a female chief," says Clark, 41, from her spacious third-floor office at the Bernalillo County Fire and Rescue headquarters on Second Street Northwest. "I'm a chief. I don't put that qualifier in there." Clark last fall was sworn in as president of the International Association of Fire Chiefs' Southwestern division. For one year, she'll lead an organization of fire executives from New Mexico, Texas, Oklahoma, Arkansas and Louisiana. . . . Her achievements since becoming the county's first female chief include drafting the department's first strategic plan and increasing staff and efficiency. Volunteers have increased from 15 in 2001 to 45, she said. And she's a year-and-a-half into a three-year plan to hire 76 paid firefighters. Included in those growing numbers are women: four new paramedics, a deputy fire marshal, seven new firefighters and another four in the academy. While the diversity is encouraging, women aren't necessarily targets of her recruitment efforts, she said. "I don't hire and promote these people. They get hired and

Mini-Case Study: How Would You handle It?

You are a male lieutenant in charge of the fire rescue unit. The first female officer will be assigned to your specialized unit shortly. The fire rescue team members are voicing negative opinions about a female officer being assigned to the unit. They are complaining that there will be a lowering of standards in the unit because women are not as physically fit as men for this assignment. When you have overheard these conversations, or when they have been addressed to you directly, you have refuted them by pointing to the women in the department who are in outstanding physical shape. This strategy has not been effective, as the squad members continue to complain that women firefighters can't do the job and that the male squad members' personal safety might be in jeopardy. How would you go about alleviating their concerns and helping to promote change in their perspective of women firefighters?

promoted," she said. "I want firefighters. White, black, male, female—good firefighters." But she hopes that being a woman in a position largely associated with men makes her a visible example to others hoping to break into public safety. "The nice thing is that because I'm female, that they can say, 'Look, she's done it. I can do it, too,'" she said (Siemers, 2006 p. A2.).

ROLE IDENTIFICATION AND ROLE MODELING

The lack of **role identification** and of **role models** have been indicated as barriers to the diversification of the fire service. For example, if one had come from a family in which many family members were firefighters or had belonged to a group of peers in which many peers became firefighters, such role identifications are certainly facilitative of individuals joining the fire service. We do not consider such role identifications as being barriers to a diverse workforce. Role identification helps to maintain a constant and usual stream of potential firefighter employees and helps the recruitment process pull from similar past sources. However, we do consider the importance of communication, marketing, and recruitment strategies as the bases for developing a diverse workforce and for helping to create role models of the future for the fire service and for the role identification of individuals from diverse communities.

Additionally, stereotypes of roles at work have had an impact upon the fire service. Clearly, stereotypes about work roles and functions would be inaccurate, whereas proper "job analysis" for the different role functions has been most useful in delineating work roles. For example, within the fire service, some of the information obtained through using certain physical tests and paper-and-pencil tests had proven to be not associated with the work based upon the job analysis.

role identification
The possibilities to identify with roles in which many family members were firefighters or had belonged to a group of peers in which many peers became firefighters; such role identifications are facilitative of individuals joining the fire service.

role modeling
Leadership and positive examples seen among members senior in the fire service to facilitate positive actions and behaviors of all fire service personnel.

A PARADIGM SHIFT

With changes in the hiring, screening, and promotional policies and practices of fire service agencies comes an unprecedented opportunity to build a future in which differences are valued and respected in our communities and our workforces. Progressive fire service executives must strive to secure the most qualified employees to serve the public. To do so, they must not only be committed to the challenges of affirmative hiring (where legal) but also be capable of educating and selling their workforce on the legitimate reasons, both legal and ethical, for such efforts. Fire service agencies, to be competitive in the market for qualified employees, must develop new ways to recruit targeted people or risk losing highly skilled potential employees to other occupations.

paradigm shift
What occurs when an entire cultural group begins to experience a change that involves the acceptance of new conceptual models or ways of thinking and results in major societal transitions (e.g., the shift from agricultural to industrial society).

A transition is taking place as the fire service continues to move from the aggressive, male-dominated (and predominantly white) fire departments and culture of the past. Most fire service observers agree that with the **paradigm shift** to a community/professional model of fire service, women can thrive. Fire service agencies must overcome the common perception that fire service is a male-oriented profession that requires only physical strength (Blair, 2005). Fire departments need to begin looking for people who are community oriented and have good interpersonal skills, and they are finding increasingly that women meet these qualifications. Fire service executives should seek minority employees for the same reasons. Although it contributes to better fire service–community relations, improvement in protected class representation in fire departments and other related professions (e.g., EMS) alone will not resolve misunderstandings. Increased numbers of diverse staff members provide only the potential for improved dialogue, cooperation, and problem solving within both the organization and the community it serves. Community and fire service officials should remember that serving multicultural and multiracial neighborhoods can never be the sole responsibility of workforce members from diverse ethnic and racial groups. In most jurisdictions, their limited numbers make this level of responsibility unrealistic. All staff should be prepared to understand and relate to diverse groups in a professional and sensitive manner.

SUMMARY

Recruiting, hiring, retaining, and promoting a diverse workforce will remain important issues in fire service for a long time. Many of the issues of equity and diversity have not been resolved in the fire service workplace, by the courts, or even in the legislative and executive branches of the U.S. government. The number of people available and qualified for entry-level jobs will continue to decrease; more employers, both public and private, will be vying for the best candidates. This seems to be true regardless of the economic condition of federal, state, or local governments. Hiring and promoting women and minorities for fire service careers are achievable goals when agencies use strategies outlined in this chapter. In this chapter we presented effective strategies and successful programs for implementing changes in recruitment and the promotion of a diverse workforce. Fire service will have to use innovative and sophisticated marketing techniques and advertising campaigns to reach the population of desired potential applicants and must develop fast-track processes for hiring these candidates.

Discussion Questions and Issues

1. **Institutional Racism in the Fire Service.** Fire service agencies typically operate under the pretense that all their members are one color and that the uniform or job makes everyone brothers or sisters. Many members of diverse ethnic and

racial groups, particularly African Americans, do not agree that they are consistently treated with respect and believe that there is institutional racism in fire service. Caucasians clearly dominate the command ranks of fire service agencies. Discuss with other students in your class whether you believe that this disparity is the result of subtle forms of institutional racism or actual conscious efforts on the part of the persons empowered to make decisions. Consider whether tests and promotional processes give unfair advantage to white applicants and whether they discriminate against department employees of other races and ethnicities. Do officers from diverse groups discriminate against members of other different cultures?

2. **Employment of a Diverse Workforce and Fire Service Practices.** How has the employment of a diverse workforce affected fire service practices in your city or county? Is there evidence that significant changes in the ethnic or racial composition of the department alter official fire service policy? Can the same be said of gay and lesbian employment? Does employment of protected classes have any significant effect on the informal fire service subculture and, in turn, firefighter performance? Provide examples to support your conclusions.

Website Resources

Visit these websites for additional information about recruitment and related issues:

Asian Firefighters Association (AFA): http://www.asianfire.org
This website provides information about the Asian Firefighters Association, with the mission to create and maintain equality of opportunity for Asians and other minorities within the fire service throughout the San Francisco community.

International Association of Black Professional Firefighters (IABPFF): http://www.iabpff.org
This website features fact sheets, news articles, and publications for cultivating and maintaining professional competence among firefighters, establishing unity, and keeping alive the interest among retired members for the purpose of both improving the social status of African Americans and increasing professional efficiency.

International Association of Fire Chiefs (IAFC): http://www.iafc.org
An online resource for fire service executive and leadership issues and publications developed by the IAFC.

International Association of Firefighters (IAFF): http://www.iaff.org
A website for information about the range of issues confronting firefighters and their departments.

National Association of Hispanic Firefighters (NAHF): http://www.nahf.org
A website for information about the recruitment, retention, and advancement of the Hispanic firefighter by developing and conducting national unbiased and cultured awareness programs in these areas.

National Career Academy Coalition (NCAC): http://www.ncacinc.org
A website for information concerning career academies and career development of professions.

National Fire Academy: http://www.usfa.fema.gov/nfa
A website for information about training, careers, jobs, publications, and research for the fire service.

U.S. Fire Administration: http://www.usfa.fema.gov
A source of fire service resources, programs, and statistics collected and published by the Federal Emergency Management Administration (FEMA).

Women Chief Fire Officers Association (WCFOA): http://www.womenfireofficers.org
A website for information to provide a proactive network that supports, mentors, and educates current and future women chief officers.

Women in the Fire Service, Inc. (WFS):
http://www.wfsi.org
A website for information about training
conferences, careers, jobs, publications, and

research pertinent to networking women in
today's firefighting world.

- -

References

Arado, M. (2002, March 26). "Conference Focuses on Diversity in Police, Fire Departments." *Chicago Daily Herald,* p. C1.

Bergman, M. (2004). "Census Bureau Projects Tripling of Hispanic and Asian Populations in 50 Years; Non-Hispanic Whites May Drop to Half of Total Population." *U.S. Census News (CB04-44).* Washington, DC.

Bernstein, M. (2003, April 18). "Poll Raps Fire Bureau's Culture." *The Oregonian,* p. D1.

Blair, M. D. (2005). "Fire Service Selection in the New Millennium." 29th Annual IPMAAC Conference on Personnel Assessment. Orlando, Florida.

Brown, C. (2005, October 7). "Diversity on Job Challenged." *Star Tribune (Minneapolis, MN)*, Metro Edition, p. 1B.

Coleman, S. (2003, October 9). "Randolph Urged to Diversify, Town Work Force Plan Has Been Stalled Five Years." *Boston Globe,* p. 1.

"Connerly Starts Push to End Tracking Race." (2001, March 27). *Contra Costa Times,* pp. A1, A12.

Davis, E. D. (2006). "Message from the President." National Association of Hispanic Firefighters (NAHF), http://www.nahf.org/html/index.php

Dettweiler, J. (1993, September). "Women and Minorities in Professional Training Programs: The Ins and Outs of Higher Education," *Cultural Diversity at Work*, 6(1): 3.

Edsall, T. B. (1991, January 15). "Racial Preferences Produce Change, Controversy," *Washington Post.*

Federal Emergency Management Administration. (FEMA). (1999a, September). *Many Faces, One Purpose: A Manager's Handbook on Women in Firefighting.* Washington, DC.

Federal Emergency Management Administration. (FEMA). (1999b, November). *Many Women Strong: A Handbook for Women Firefighters.* Washington, DC.

FEMA (U.S. Fire Administration) and NFPA (National Fire Protection Association). (2002). "Needs Assessment of the U.S. Fire Service." A cooperative study authorized by U.S. Public Law 106-398, Washington, DC.

Fulbright, L. (2004, February 7). "Female, Minority Firefighters Scarce: Diversity Attempts Have Failed to Reach Target Populations." *Seattle Times,* p. B3.

Hewitt, A. S. (2006, January 26). " Fire Department Audit Finds Hazing, Harassment, Retaliation." Copley News Service, California Wire.

Kershaw, S. (2006, January 23). "Answering the Fire Bell in the Company of Women." *New York Times*, p. A1.

Little, J., and Levins, H. (2002, June 2). "Black Firefighter Reflects on Changes in City over 38 Years." *St. Louis Post-Dispatch* (Missouri), p. A10.

Marrujo, R., and Kleiner, B. (1992). "Why Women Fail to Get to the Top." *Equal Opportunities International*, 2(4).

Micari, M. (1993, September). "Recruiters of Minorities and Women Speak Out: Why Companies Lose Candidates," *Cultural Diversity at Work*, 6(1): 1.

Morrison, A. M. (1987). *The New Leaders: Leadership Diversity in America.* San Francisco: Jossey-Bass.

O'Neal, M. (2004, May 11). " Fire Department Makes Gains in Minority Hiring," *Chattanooga Times Free Press* (Tennessee), p. B1.

Runnett, L. (1999). "What Color Is a Band-Aid? A Look at Cultural Diversity in the Fire Service: A Strategy for Improving Diversification." Paper for the Charlottesville, Virginia, Fire Department.

Schaitberger, H. (2006, January 31). "IAFF Makes Progress in Diversity." Ernest "Buddy" Mass Human Relations Conference—"Leading through Diversity." International Association of Firefighters (IAFF), San Antonio, Texas.

Shannon, J. M. (2003, June 4), Testimony of James M. Shannon, president and CEO of the NFPA, before the Committee on Science, United States House of Representatives. Washington, DC.

Siemers, E. (2006, January 16). "Just a Chief." *Albuquerque Tribune* (New Mexico), p. A2.

Snyder, D. (2004, December 2). "Minority Firefighter Recruitment Climbs." *Washington Post*, p. T2.

Watts, P. (1989). "Breaking into the Old-Boy Network." *Executive Female*, 12(32).

Women in Fire Service, Inc. (WFS). (2005). "Status Report—2005," quoted from http://www.wfsi.org/women_and_firefighting/status_report.php

CHAPTER 3

Multicultural Awareness Training in Fire Departments

Key Terms

awareness, p. 92   bias, p. 76

Overview

Every human resources program in fire departments should have a substantial component that deals with culture and its impact on human behavior. This multicultural awareness is important for students in fire science programs, new and veteran firefighters, and nonfirefighters employed by or who volunteer with fire departments across the United States (Figure 3.1). Ongoing training in this area is a process that leads to cultural competency. The authors of this book recognize the ongoing value of continuing education and training. We specifically want all readers to be well versed in the material addressed in this chapter because it benefits all trainers, instructors, and learners in the collaborative efforts of multicultural awareness in the classroom. More now than ever, either new employees are co-teaching and/or even in some cases teaching because of the subject matter expertise they possess. This is one more reason why this chapter talks about training and provides information to the reader that usually would not be included in general textbooks.

In this chapter we discuss why there has been resistance to and apprehension about multicultural awareness training in fire departments. We provide basic information in how to conduct a needs assessment as it relates to multicultural awareness training. The importance of identifying the learning goals and objectives for the training is addressed and how it benefits the student and trainer. We examine effective strategies and methods in training for the student and trainer. In addition, information is included on the selection of multicultural awareness trainers, instructors, and consultants. Last, there is the accountability mechanism in determining how you know whether the training is paying off for your students and your fire department.

Objectives

After completing this chapter, participants should be able to:

* Identify components and behavior in the training of multicultural awareness that causes resistance to the training.

FIGURE 3.1 ◆ Multicultural awareness instructors need to demonstrate the classic signs of enthusiasm and organization, as reflected in this group presentation to a panel of adult learners.

- Describe how to conduct a needs assessment as it relates to multicultural awareness training.
- Explain what are the learning goals and learning objectives of multicultural awareness training.
- List effective strategies and methods of learning for multicultural awareness training.
- Evaluate measurements to determine whether the training is working in the firehouse and in operations.

LEARNING TASKS

Knowing the requirements of a training course is essential for the trainer and the student. Successful training programs identify the learning goals and objectives for all classes being taught. Learning about ethnic, racial, and cultural groups in one's workforce and service area is critical toward understanding the multicultural challenges and to becoming culturally competent in the fire service. Talk with your agency, and find out how the learning goals and objectives are decided in the multicultural awareness training classes. Be able to:

- Find out how the agency overcame resistance to the training for your firefighters and nonfirefighter personnel.
- Identify how the agency determined what teaching strategies work for the different types of adult learners who are receiving the training.

- Determine what type of measurement tools the agency uses to determine whether the multicultural awareness training is working.
- Ascertain what action is taken when the multicultural awareness training is not working to its desired objectives(s).

PERCEPTIONS

The importance of multicultural awareness training for the fire service is highlighted in the following quotation from Shreveport, Louisiana's Assistant Fire Chief Brian A. Crawford, who is also a resident instructor at the National Fire Academy.

> In breaking down some of the barriers to diversity in the fire service; human relations and diversity training for (firefighters) officers are a must. After the leadership is trained, all other members should follow suit, including a segment incorporated in basic training. That training should be maintained throughout a firefighter's career, not just at one setting. The problem didn't occur overnight, and one three-hour training session isn't going to work constructively toward the solution (Crawford, 2004).

Further, *The Fire Chief's Handbook*, 6th edition, recognizes the value of employee relations and cultural awareness training when it states:

> The type of leadership skills needed for incident command is vastly different than the leadership skills necessary to deal with personnel during the rest of the day. All too often mid-level fire officers are selected and promoted based on their ability at emergency incidents. As individuals move up the promotional ladder, they are expected to learn how to manage and lead firefighters through trial and error on the job. The ability to fight fire and to manage an emergency incident through strong knowledge of the incident command system does not teach new fire officers at any level how to lead and manage people, except on the fire ground. In order to be an effective leader in a fire department, the knowledge and skills necessary to manage personnel when NOT at emergency incidents must be learned. If the fire department does not provide this education, it is imperative that fire officers take the initiative to attain this information on their own (Barr and Eversole, 2003, p. 229).

Cultural awareness and employee relations are the most serious issues in the workplace today, yet neither most employers nor their managers are prepared to deal with them. However, fire departments that are progressive and forward thinking are taking the lead of Assistant Fire Chief Crawford by offering cultural diversity classes to their personnel as part of continuing education for their employees, even on a 24-hour basis. One such fire department is Clackamas County Fire District 1 in Clackamas, Oregon. Faced with scheduling challenges, Assistant Fire Chief Marc Crain contracted with a cultural diversity training consultant who produced a customized training tape for its urban and rural 100-plus fire department (Olson, 2004). The video, *Seeking Excellence in Employee Relations,* was specifically formatted on DVD and VHS tape so a facilitator in the firehouse could show it, pause it, and lead lesson plan discussions when his or her crew were not on runs or alarms. The training tape with the facilitator lesson plan proved to be an effective and useful tool.

There have been race and diversity-related riots in the United States in the 1960s, 1970s, 1980s, 1990s, and 2000s. The public, scholars, and independent study commissions have paid considerable attention to causes of the strife, and much of the focus has centered on law enforcement procedures and training. Historically, due to the nature of the work, fire departments have not been put in the adversarial role law enforcement has when it comes to the tensions associated with ethnic and minority group encounters for the allegations of police brutality and racial profiling. Though this is an accurate statement, we need to point out that firefighters have risked, and continue to risk their personal safety as emergency responders when there are property and person crimes associated with protests, demonstrations, riots, and assaults on public safety officers. Further, one of the authors (Olson, 2007), a retired Oregon State Police patrol supervisor, knows from firsthand experience that firefighters will come to the aid of a police officer who needs help in restraining a violent person who is a danger to self or others.

The focal point of our discussion is that fire departments, like law enforcement and other traditional professions, have had workforce challenges especially with the increasing numbers of women, nonwhites, and other ethnic groups joining fire departments prompted by the passage of the Equal Employment Opportunity Act of 1972. Some agencies reacted by implementing or revising race relations and human relations training; a few instituted "sensitivity" classes using outside facilitators (Figure 3.2). Many agencies would and continue to have staff from human resources teach affirmative action and sexual harassment, thinking this would meet the need. As educators and professional trainers on multiculturalism, we offer our definitions and comments on the terms *human relations, race relations,* and *sensitivity training.*

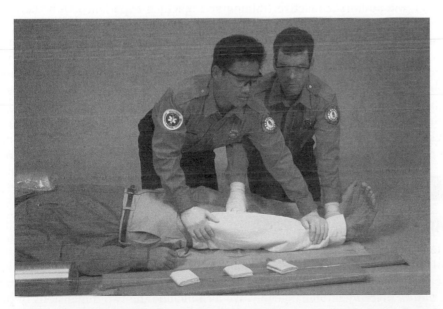

FIGURE 3.2 ◆ Multicultural awareness instructors and students need to recognize the same relevant application of cultural competence, as illustrated in this training of first aid treatment.

Human relations is a rather ambiguous term that applies to the interaction of people. It can mean everything from a study of organizational behavior and productivity of its employees to the positive relationship skills an employee uses to provide services to the community. Human relations as an all encompassing term includes cross-cultural and race relations.

Race relations as a training topic was frequently taught in the 1970s, but it primarily dealt with the interactions of different races and ethnic groups. It spent a lot of time on the history of prejudice and discrimination. Some outside trainers would purposely "bash" white males or encourage nonwhite students to join in the bashing for past acts of discrimination. This stance would create a hostile learning environment for all students. Most race relations classes did not touch on women or gender issues.

Sensitivity training was a class that was frequently taught in the 1970s and 1980s. It focused on ways to increase awareness on the differences between minority groups and explored outward behavioral strategies to bridge misconceptions based on the other person's perceptions. Unfortunately, many of these training classes had a tendency of being too emotionally based in the affective domain and were seen by some to be "touchy-feely" training sessions when taught by outside training consultants. As a profession, firefighters, police officers, 9-1-1 dispatchers, and all other emergency medical services personnel must be even tempered and nonemotional in the performance of their duties. Outside nonemergency medical services trainers failed to recognize this profession-specific trait when using the affective domain (Bloom, 1979) (emotional side) of learning. Another problem with sensitivity training was it encouraged bashing by the trainer and other students in the classroom sessions.

Private organizations, government agencies, textbooks, educators, and trainers use various terms to describe such training: *cultural awareness*, *cultural competency*, *cultural diversity*, *diversity*, *human relations*, *multicultural awareness*, *race relations*, and *sensitivity training*. Some of these terms are used frequently and some less frequently. Some of the training and curriculum programs dealing with this subject are effective and many are ineffective.

From the authors' collective training and teaching experience with the professions of fire service, law enforcement, corrections, and 9-1-1 personnel, we advise all students and instructors to expect resistance from some students to the topic of cultural diversity. A sentiment from some students may be, "Treat everyone fairly and with respect. Don't treat people differently just because of their background." Our response to this sentiment is to quote the immortal words of the late Dr. George Santayana who said, "Those who cannot remember the past are condemned to repeat it" (Santayana, 1905). Remember, cultural diversity training and awareness is a process and, like all processes, it is ongoing.

WHY THE RESISTANCE TO THE TRAINING?

WHAT IS CULTURAL AWARENESS?

We define cultural awareness as the understanding that an individual has about different cultures. The term is often expanded to include age, physical disability, ethnicity, gender, race, religion, and sexual orientation. We use the terms *cultural awareness* and *cultural diversity* interchangeably. We also include the two dimensions of diversity: the primary and secondary dimensions of diversity discussed in Chapter 1. Cultural awareness and cultural diversity go beyond race, ethnicity, and gender. They

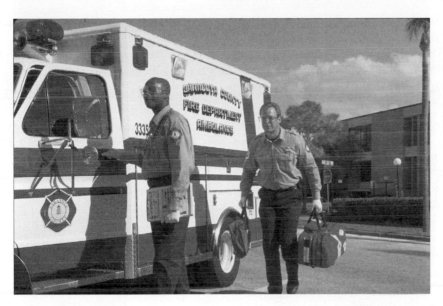

FIGURE 3.3 ◆ Putting their training into use, firefighter EMTs coming off their ambulance at this call for assistance.

include education, employment, social status, and specifically the organizational values of fire departments.

REASONS WHY FIREFIGHTERS ARE APPREHENSIVE ABOUT THE TRAINING

Firefighters from different fire departments were interviewed and listed several reasons for their apprehension about cultural awareness training. The information was collected in a series of separate and collective interviews (Figure 3.3). To encourage candid and honest comments, the respondents were told their name and agency would not be disclosed. The reasons listed follow:

1. The instructor forced his beliefs on us, saying we don't have a choice in what we are being taught.
2. The instructor did not have a clue to what we did in our professions and the real world.
3. The instructor's only reference point was academia and the ivory tower.
4. Firefighters function as a team and most firefighters don't know when they cross the line. I remember one instructor saying if you don't say it to your mother, don't say it to a female firefighter. You've got to be kidding.
5. A number of students in the class get turned off to the instructor's methods of in your face and say there is nothing wrong with our old culture in the firehouse.
6. There is white male bashing from nonwhite instructors who say white males are the problem and have to change their culture in the firehouse.
7. The nonwhite instructor calls white males racists and sexists. Female students have called male students sexists too.
8. Too much time on being "touchy feely." This is the wrong audience to be too "touchy feely." Use straight talk instead of therapeutic counseling communication.
9. Outside training consultants have canned material and it needs to be customized to our agency or profession. Some of the group scenario exercises are goofy and unrealistic.

10. We talk about the same subject, sex and race; nothing more. Let's talk about where we are from and the other categories of diversity.
11. The instructor is abrasive and takes on an "us against them" approach.
12. The instructor is abrasive to people who disagree with him.

The twelve reasons listed for why firefighters are apprehensive about cultural awareness training are no new phenomena. Law enforcement, other emergency service professionals, and college students have their own misgivings and negative experiences to tell as well. Anytime any group is belittled or verbally attacked by an instructor or another student, creates a hostile learning environment and derails the purpose of training or instruction. Instructors teaching cultural diversity have a responsibility to maintain a safe learning environment. Instructors must monitor their own **biases** so they do not come across being one-sided and obsessed about promoting a political or personal agenda. Although academic freedom allows instructors to express their opinions, it does not permit them to be tyrants over their students. Instructors who ignore the ethics of teaching and who are disrespectful to the students in the classroom should be relieved of their teaching responsibilities. It is vital that the instructor creates a positive learning environment to maximize fire service receptivity to the training.

bias

Preference or inclination to make certain choices that may be positive (bias toward excellence) or negative (bias against people), often resulting in unfairness.

SEVEN GUIDELINES TO FOLLOW IN CULTURAL DIVERSITY CLASSES

We have developed seven guidelines that enhance the value of cultural diversity instruction for the learner and the instructor. When used, these guidelines help eliminate the resistance to cultural diversity training. Students and instructors should be aware of the following guidelines and discuss them at the onset of any new class on cultural diversity.

1. Listen fully to each person's opinion, resisting the temptation to interrupt or disagree.
2. Establish from the beginning of the class that students and the instructor must give each other respect for diverse viewpoints; the instructor should be the first to model this respect.
3. Establish that students and the instructor can agree to disagree, but students and the instructor must control the temptation to confront or accuse one another.
4. All of us in the classroom (students and instructor) are a diverse group (remember the primary and secondary dimensions of diversity). Let's talk about each person's individual dimensions for different perceptions, experiences, and reality.
5. Point out whether the students and the instructor have a difficult time acknowledging the diversity of viewpoints in their class in a professional manner; they may tend to become easily inflamed with citizens. Calm and controlled communication in the class is a good practice.
6. Require that, as a skill that can be transferred to day-to-day interactions, all students and the instructor work on using effective communication skills to defuse conflict, rather than to escalate it.
7. Remind ourselves of the following if we do not recognize the importance of cultural diversity and its impact on fire service and other emergency services:
   a. Dealing with citizens whose native tongue is not English requires special awareness and skill for communication to be effective. Even if one is not bilingual, reality calls for the necessity of communication with English as a second language or limited English speakers.
   b. The ability to get along with multicultural coworkers is important.

c. Cultural diversity training, along with many other topics, is a high-priority subject.

d. We will be at a higher level when firefighters and other emergency services take it for granted that cultural diversity deserves a regular place in the curriculum.

Cultural diversity training for emergency medical services (especially for firefighters) can be informative, interesting, and pragmatic. Firefighters can benefit from the training in the firehouse (interpersonal relationships) and on runs and alarms. As the population of new immigrants and refugees in our communities continues to grow, firefighters can use their cultural diversity training skills in assessing critical cultural issues that need to be taken into consideration when dealing with these groups from other countries.

CONDUCTING A NEEDS ASSESSMENT

HOW DO WE DETERMINE THE TRAINING?

To be effective, cultural awareness training must be agency and community specific and must encompass all aspects of multicultural relations, both internal and external to the organization and its firefighters. This is important to know but how do we determine the training? What steps are taken to acquire this information and who makes this assessment? We will explore these questions and determine that this process extends to all training topics.

ASSESSING THE TRAINING NEEDS

In order to develop an effective training program or curriculum, we must first determine the basic purpose of the organization (Figure 3.4). Basic questions would include: "What does the organization do, what services does it offer, and what is its plan

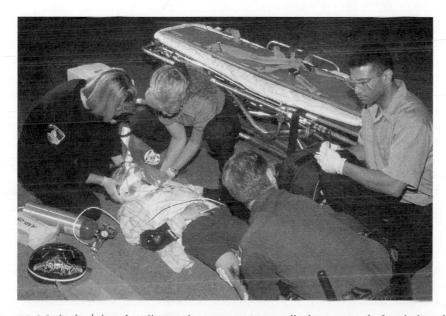

FIGURE 3.4 ◆ An injured cyclist receives emergency medical treatment before being placed on a gurney.

of operations?" This procedure is called assessing your organization, or conducting a needs assessment. A broad definition of a needs assessment is "a systematic set of procedures undertaken for the purpose of setting priorities and making decisions about program or organizational improvement and allocation of resources. The priorities are based on identified needs" (Witkin and Altschuld, 1995, p. 4). A more practical and specific type of needs assessment is called a training needs assessment (TNA) provided by Rossett (1987):

> Training needs assessment (TNA) is an umbrella term for the analysis activities trainers use to examine and understand performance problems or new technologies. You've heard it called problem analysis, pre-training analysis, figuring things out, needs assessment and front end analysis. What you call it doesn't matter. What matters is whether you get the information you need to effectively solve problems in the corporation or agency. That detailed information, from the purpose of the various sources or stakeholders, is the purpose for TNA (p. 14).

Needs assessments usually focus on three levels of analysis: organization, position or task, and individual or person (Goldstein, 1993, McGehee and Thayer, 1961; Moore and Dutton, 1978; Sleezer, 1991). In this undertaking, all functions of the organization, positions or tasks, and individuals or persons must be identified and analyzed. In fire departments, usually a staff member from the training division conducts this needs assessment and is called the assessor. Consultants have been known to be used if the fire executive wants an outside assessor to complete the assessment. The main reasons for having an outside assessor conduct the needs assessment are usually they have the proficiency and experience in such matters, they are an external resource, and they are perceived as a neutral entity with no preconceived opinions.

In examining the needs of the organization, we need to review the mission statement and the vision and value statements. Agency written policies and procedures should be examined as well. Most organizations have a strategic five-year plan and emergency operations plans. All of these documents need to be studied, and any obsolete or outdated ones must be updated. Technology, politics, funding sources, and reorganizations have a dramatic effect on organization needs. A major obstacle to conducting a needs assessment is if there is no mission statement or if the vision and value statements do not echo the mission statement. If this is the case, the immediate response should be to correct the problem. "By starting at the organization, a needs assessment is most likely to lead to well-designed interventions with a very good chance of solving real performance problems. Evaluation after the problem is easy because the criteria for each level were determined before the interventions were designed" (Phillips and Holton, 1995, p. 2).

In regards to positions or tasks, the objective is to look at each position in the organization and determine what that position does or what its tasks are. The best source is the position description or job description. Efficient organizations have accurate position descriptions and update them whenever there is a change in the duties of the position. It is best and recommended to annually review and update changes to the position description at the same time when yearly personnel evaluations take place.

Now when it comes to individuals or persons in the needs assessment, this is a very important part in the evaluation (Figure 3.5). People are not always consistent in performing the tasks for their positions because of a variety of reasons. Circumstances such as poor health, a family crisis, workload stress, a hostile work environ-

FIGURE 3.5 ◆ A team of rescue firefighters pull the roof off of a wrecked car propped on its side and a paramedic treats the victim.

ment, or a combination of other outside issues could adversely impact employee performance. On the flip side, a person may have the aptitude and skills to perform the job admirably or be inept and a total failure in performing the job or certain tasks related to the job. The individual must be able to perform the tasks associated with his or her position description if the organization is going to benefit from the individual's performance.

THE ASSESSOR'S FIVE-POINT CHECKLIST FOR THE NEEDS ASSESSMENT

In having experience in conducting needs assessments, we offer the following checklist of five points to aid assessors in fire departments in performing a training needs assessment:

1. List the *standard* of performance or knowledge for the task or skill.
2. Identify the *existing* performance or knowledge for the task or skill.
3. Obtain *feedback* from the participants on the standard for performance.
4. Ask for *causes* of all problems associated to the standard for performance.
5. Ask for *solutions* for all problems linked to the standard for performance.

The assessor should complete a form on each individual participant from the checklist items. Even students should have the opportunity to respond to items 3 to 5 on a course survey question. The information from the forms should be tabulated on one table and then analyzed by the assessor. The company training officer should sit in each class and collect the information from each training session.

The checklist for a training needs assessment can be used while observing a class that is being taught or extracting information from training records (Figure 3.6). The assessor should request that the instructor of the class complete the form as well. If the assessor is extracting information from training records, he or she should review course evaluations related to the class that is examined. If information is lacking on the course evaluations, the assessor should consider recontacting the students who

Checklist for the Training Needs Assessment		1. Standard to Perform: CARE Approach in 20-minute group scenario exercise on cultural diversity class on secondary dimensions and behavior in the firehouse	
2. Existing Skills	**3. Feedback**	**4. Causes**	**5. Solutions**
10 students performed the task to standard 2 students were hostile to the group exercise and started bashing volunteer firefighters from a rural county who were attending the training	See attached forms 8 students thought the scenario was realistic; 2 students thought the scenario was distasteful; 2 hostile students thought white males were the problem	See attached forms 8 students thought the scenario was realistic and thought the 2 hostile students have a political agenda 2 students thought that the cultural diversity scenario was a waste of time because they do not want to get along with other people who are not from a large fire department 2 students did not respond to the question	See attached forms 8 students thought that the 2 hostile students should have a one-on-one remedial training consultation with the fire department company training officer on being professional around other firefighters from different departments 2 students did not like the group exercises and prefer lecture and videotapes 2 hostile students think that the cultural diversity training should be discontinued
Name of Company Training Officer:			
Name of Instructor:			
Date and Completed By:			

FIGURE 3.6 ◆ An Example of a Training Needs Assessment on a Cultural Diversity Class
Source: Aaron Olson

completed the training and request they fill out a follow-up survey capturing the required information for the training needs assessment.

TYPES OF DATA FOR NEEDS ASSESSMENTS: QUANTITATIVE AND QUALITATIVE

The two basic types of data are quantitative and qualitative. Quantitative data consist of numerical data and are based on real numbers. Statistics, percentages, and score ranking surveys are examples of quantitative data. Qualitative data consist of interviews, focus groups, committees, narratives, and written reports. Surveys and questionnaires with a "please explain response" are examples of qualitative data. There are advantages and disadvantages with both types of data for needs assessments, as pointed out by Rust-Eft and Preskill (2001):

> Quantitative analyses, as compared with qualitative analyses, usually involve
> less personnel time, are almost always performed using computers, and often

result in substantially lower data analysis costs. In some cases and for some stakeholders, "hard numbers" are seen as more credible than the text-base analyses of qualitative data, even if they are not any more valid. On the other hand, the powers of individuals' words from a qualitative analysis often "speak" more loudly to some stakeholders.

Analysis of qualitative data, on the other hand, can be more expensive and time-consuming than quantitative analysis because of the amount of time it takes to read, categorize, and code transcripts that result from interviews or observations and open-ended items from surveys. In addition, the process of analyzing qualitative data usually requires a greater level of training or expertise on the part of the persons analyzing the data to obtain valid and credible results (p. 316).

The bottom line about quantitative and qualitative data is that if you want to save time in collecting information and save money, use quantitative data. If you have time and a large budget for analysis and want to receive comments on questions, use qualitative data.

OTHER DATA COLLECTION METHODS

Extant data consist of written documentation or a manuscript. They are excellent sources of information, which can be easily accessed or retrieved for data analysis. Listed here are examples of extant data:

1. Lesson plans
2. Training manuals
3. Training records
4. Training schedules
5. Position or job descriptions
6. Memorandums
7. Course lesson plans
8. Employee performance appraisals
9. Employee personnel records
10. Incident reports
11. Accident reports

Several other data collection methods can be used by the assessor as well. The following data methods may be time consuming, but the data received are invaluable for needs assessments:

1. Action research
2. DACUM (development of a curriculum)
3. JTA (job task analysis)
4. Nominal group technique
5. Observation
6. Subject matter analysis
7. Work sampling

A FINAL COMMENT ON NEEDS ASSESSMENTS

As you can see from this list and from all the information presented on needs assessments, the assessor has several options to choose from. The most important questions that all assessors must be able to answer for their organization are (1) how much time do they have to conduct a needs assessment and (2) what are their resources. Assessors must also remember that each process has its strengths and weaknesses. Once the information is received about the needs assessment, trainers need to make a decision about what the next step is. Are they going to spend more time on training, or are they going to make modifications to the required standards of performance? They also need to determine whether an employee aptitude problem or an organization policy issue exists.

WHAT ARE YOUR LEARNING OBJECTIVES?

When the training is over, what is it that we want every student to be able to do as a result of it? The precise answer is we want them to be able to demonstrate the knowledge, skills, and abilities to achieve the required standards of performance (Figure 3.7). The training needs assessment asked these questions and ultimately these are the job performance requirements. It is necessary that each individual performance requirement have a learning objective, which in academia is also known as an intended outcome. These learning objectives or outcomes must be developed by the instructional design representative responsible for the training of the fire department. The instructional designer may be the chief, a member of the training staff, or an instructor. It is essential that this person identify what is to be accomplished and how. The steps taken or procedures to accomplish a learning objective are called tasks. It is common to have students perform several tasks to accomplish one learning objective. Learning objectives are also called training objectives and even performance objectives, because ultimately they are the performance standards required to do the job in fire service.

ESTABLISHING CREDIBILITY

Effective training programs are based on solid educational design systems and practices. To comprehend the value of this statement, we briefly look at the two terms

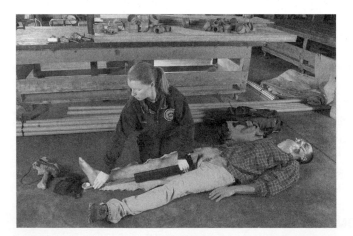

FIGURE 3.7 ◆ A paramedic is applying a Sager splint to the leg of a supine man with a non-rebreather mask over his face.

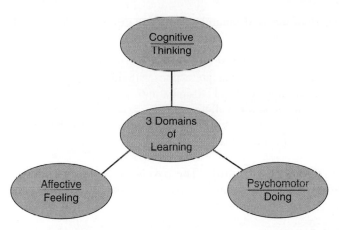

Figure 3.8 ◆ Interrelationship of the Three Domains of Learning *Source*: Aaron Olson

pedagogy and *andragogy*. Pedagogy is best described as the art and science of teaching. It is inclusive of all ages of learners. Andragogy is more specific in that it is the art and science of teaching adult learners. Instructional designers, instructors, and students should be familiar with these two terms because these terms are embedded in the learning process and educational systems. To illustrate our point, the American Council on Education provides an evaluation program called the College Credit Recommendation Service. It evaluates your fire department's training and education program compared to known standards within the academic venue and offers college credit recommendations. The service's recommendations can be submitted to colleges for potential acceptance of college credit. This feature provides additional credibility to your training program, if college credits are awarded.

BLOOM'S THREE DOMAINS OF LEARNING

Educational institutions and training programs all over the world recognize the contribution of Dr. Benjamin Bloom as it relates to learning. This is of particular importance as we look at writing learning performance objectives. Students and instructors who recognize and act upon these learning domains will optimize the learning process. The three domains of learning are cognitive, affective, and psychomotor. Figure 3.8 illustrates their interconnection and relationship.

Cognitive Domain (Learning)   The cognitive domain of learning involves the human mind in the ability to think, reason, and acquire knowledge. This is demonstrated by the following examples:

- Recall and intellectual skills
- Comprehending information
- Organizing ideas
- Analyzing and synthesizing data
- Applying knowledge
- Choosing among alternatives in problem solving
- Evaluation of ideas or actions

Affective Domain (Learning)   The affective domain of learning is best described as a person's emotions and feelings. Some people relate to the "gut" feeling on a topic or issue. Examples of affective domain learning are demonstrated by behaviors indicating the following:

- Attitudes of awareness, interest, attention, concern, and responsibility
- Ability to listen and respond in interactions with others
- Ability to be self-motivated and self-initiating
- Ability to be a team player at all times
- Ability to demonstrate attitudinal characteristics or values that are appropriate to the test situation and the field of study

Psychomotor Domain (Learning) The psychomotor domain of learning involves the physical ability to use the human body to accomplish a physical task or skill. Examples of psychomotor learning are demonstrated by the following:

- Coordination
- Dexterity
- Manipulation
- Strength
- Speed
- Fine motor skills to operate tools, equipment, and machinery

BLOOM'S TAXONOMY FOR LEARNING

The stair steps in Figure 3.9 show the progression for building taxonomy for categorizing levels of conceptual complexity as it applies to learning and the relationship of measurement. Dr. Bloom created the systematic classification and Bloom's Taxonomy has worldwide acceptance in educational and training institutions. The authors of this textbook are the first to introduce the stair-step progression and illustration for Bloom's Taxonomy.

TIPS: WRITING EFFECTIVE LEARNING OBJECTIVES

In Figure 3.10, we have provided action verbs to use when writing learning objectives for a class you are teaching at all six levels. If instructors and students were clear, succinct, and specific in their communication on multicultural awareness issues, as illustrated in this figure, ambiguity and misunderstandings would disappear.

Verbs to Avoid Using We recommend avoid using passive verbs because they are too general, vague, and difficult to measure. There may be circumstances when you would use them, but you still need to have clear and measurable learning/performance

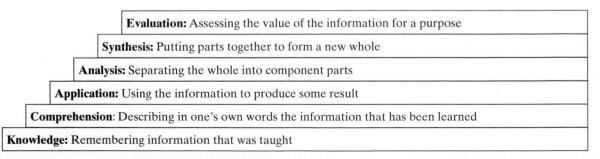

Evaluation: Assessing the value of the information for a purpose

Synthesis: Putting parts together to form a new whole

Analysis: Separating the whole into component parts

Application: Using the information to produce some result

Comprehension: Describing in one's own words the information that has been learned

Knowledge: Remembering information that was taught

FIGURE 3.9 ◆ Taxonomy Stair-Step Illustration *Source:* Aaron Olson

Taxonomy Level	Action Verbs to Use
1. Knowledge	arrange, define, describe, list, memorize, name, organize, recall, recognize, relate, repeat, reproduce, state
2. Comprehension	classify, describe, discuss, explain, express, identify, indicate, locate, report, restate, review, select, sort, translate
3. Application	apply, choose, demonstrate, illustrate, interpret, operate, prepare, sketch, solve, use
4. Analysis	analyze, appraise, categorize, compare, contrast, diagram, differentiate, distinguish, examine, inventory, question, test
5. Synthesis	arrange, assemble, construct, create, design, formulate, organize, plan, prepare, set up, synthesize
6. Evaluation	appraise, assess, choose, compare, defend, estimate, evaluate, judge, rate, select

FIGURE 3.10 ◆ Taxonomy Level and Action Verbs to Use *Source:* Aaron Olson

objectives so the students understand what is expected of them. Because passive verbs generally deal with feelings and internalization, the affective domain seems a likely domain for appealing to this sense of learning. In Figure 3.11 is a list of passive verbs that we have provided to new trainers to avoid when writing effective learning objectives.

An interesting commentary about cultural diversity classes is that the majority of the classes use passive verbs for the learning/performance objectives, which contribute to the apprehension and ambiguity of the classes. Instructors should make good use of the students' time in writing learning objectives that consist of action verbs.

Writing a Lesson Plan   Writing a lesson plan is the final step in putting together the performance objectives for a course to be taught (Figure 3.12). Regardless of whether the course lasts 2 hours or 40 hours, instructors must write a lesson plan, as a road map in outlining what the learning goals and learning/performance objectives are in the class they

Passive Verbs to Avoid	
◆ Accept	◆ Grasp
◆ Acknowledge	◆ Internalize
◆ Appreciate	◆ Realize
◆ Be aware of	◆ Recognize
◆ Believe	◆ Sense
◆ Comprehend	◆ Understand
◆ Enhance	◆ Value

FIGURE 3.11 ◆ Passive Verbs to Avoid in Writing Performance Objectives
Source: Aaron Olson

FIGURE 3.12 ◆ Two paramedics from the fire department attend to an accident victim as two other paramedics carry a stretcher out from their ambulance.

are teaching. Specifically, a lesson plan is a document that sets forth the learning objectives students will attain in a single lesson, the content they will learn, and the methods by which they will achieve those objectives. Students expect their instructor to be organized and prepared to teach the learning/performance objectives the students will learn. The lesson plan, similar to the blueprint a builder uses to construct a house, must be detailed and complete. As the following information will benefit instructors who teach, it will benefit students as well. Students have the right to know what constitutes effective training and teaching. The lesson plan should be no mystery in their learning endeavor.

The Lesson Plan Outline   The lesson plan is similar to an outline a speaker would use if providing an extemporaneous speech. An effectively delivered extemporaneous speech is when the speaker uses excellent communication skills, has thorough knowledge of his or her topic, and only glances at the outline to keep on track for the presentation (Gronbeck, McKerrow, Ehninger, and Monroe, 2000). The lesson plan an instructor uses accomplishes the same result as an extemporaneous speech outline but details more information and processes because it references a class that is being taught as opposed to a speech being presented. Listed here is the skeleton of a basic lesson plan we have used in our teaching and training experiences in teaching to public safety students. The lesson plan must be preapproved by the fire department training section before any class is instructed. The training section also gets a copy of the lesson plan and of all attachments listed in the lesson plan.

Lesson Plan Format

1. **COVER PAGE**
    a. Title (topic) for the lesson
    b. Length of the lesson
    c. Who prepared the lesson plan (instructor and organization)
    d. Teaching methods and strategies to be used

    e. Date the lesson was prepared or updated

    f. Target group or audience

2. STATEMENT OF OBJECTIVES

    a. Instructional goal: a statement written in broad, general terms that describes what you hope to accomplish by teaching this lesson (desired learning outcome).

    b. Learning/performance objectives: a series of statements (one for each learning activity) describing specifically what you expect the student to learn and do after the block of instruction, to what standard, and under what well-defined conditions.

3. CRITERION TEST

    a. Learning/performance objectives and test questions should match. There should be at least one test question for each learning/performance objective.

    b. Place test question answers on a separate sheet of paper following the test questions.

    c. Test questions should be applicable to job behaviors.

    d. Test questions can consist of performance-oriented tests such as participating in a role-play scenario, demonstrations the students must perform, and other hands-on situations.

    e. Tests can be standard written tests (i.e., multiple choice, true and false, fill in the blank, matching, short answer, and essay).

    f. The tests should always incorporate at least four levels of Bloom's Taxonomy for learning. Sometimes the tests would include all six levels.

4. INSTRUCTIONAL ITEMS AND MATERIAL

List all the materials you need to teach the lesson. This list may include items such as

    a. Textbooks

    b. Pamphlets and handbooks

    c. Overhead projector and screen

    d. Overhead transparencies (described and numbered)

    e. PowerPoint, projector, and screen

    f. Models

    g. Equipment for each student for hands-on practice/demonstrations

    h. Audio equipment and cassette/compact discs

    i. Audiovisual equipment and videotapes/digital videodiscs

    j. Television/cable

    k. Computers, monitors, printers, and so on

    l. All handouts

    m. Others

5. OUTLINE OF ACTIVITIES

    a. Can use any format or combination (sentences, paragraphs, or topical outline).

    b. Use a consistent set of symbols and format to show main headings and subheadings.

    c. Be consistent with document indentation.

    d. Use the same format for instructions to yourself.

    e. Use the same format for questions to students (different from directions to yourself).

    f. Introduction:

        1. Greeting and a statement/activity to get their attention.

        2. Introduce yourself and the name of the topic.

        3. Needs statement: shows a problem exists and sets up students' need to learn.

        4. Thesis statement: tells students general goals/content of the course and what is expected of them in the class.

            a. Must be a complete sentence

            b. Must be clear and specific, with a definite purpose

        5. Preview of main points: briefly state the topical areas the lesson will address.

        6. Administrative issues: go over breaks, safety, and so forth.

    g. Body:

        1. List all main points and supporting points (all information you plan to teach).

        2. Include illustrations and examples.

        3. Show the activities you will use to convey your information (questions to ask the students, questions to start discussions, when to use training aids, notes to self).

        4. Describe demonstrations and practical problems.

            a. Purpose

            b. Sequences to cover and procedures

            c. Safety precautions

            d. Directions to the students (directions to yourself)

            e. When to pass out equipment and tools needed

    h. Summary:

        1. Restate the thesis or learning goal.

        2. Give a summary of the lesson (learning objectives taught).

        3. Ask for questions or comments.

        4. Respond to questions.

        5. Finish by thanking them and ending with a quote or important point.

6. REFERENCES

    a. List all the sources you used in developing the information portion of the lesson plan.

    b. Use the proper citation of your sources.

7. ATTACHMENTS

The training section of your fire department needs to have a copy of all of the items listed in your attachment:

    a. Overheads

    b. Handouts

    c. Compact disc

    d. Videotapes

    e. Digital videotapes

    f. PowerPoint

    g. Others

EFFECTIVE STRATEGIES AND METHODS IN TRAINING

AGENCY-AND COMMUNITY-SPECIFIC TRAINING

Cultural diversity trainer Joyce St. George (1991) lists the following nine points that every course design for cultural diversity should address. The information she provides has relevance for fire departments. It benefits the course, the students, and the instructor.

1. Be multidimensional: Students must learn via such courses to assess what they encounter from several perspectives—organizations, community, and legal—to allow the totality of the circumstances encountered to be considered.

2. Be relevant: Training programs must be structured to satisfy the specific interests and needs of the audience and the community. Needs, goals, demographic and cultural information, and issues participants currently face should be specific to the training audience.

3. Is behavior based: The course should focus on how students express their attitudes through their actions and behaviors, which are governable and can be legally and departmentally controlled.

4. Be empathetic: The program must ask students to empathize with the feelings and concerns of minority community members, but it should also address the feelings and concerns of students.

5. Be practical: Programs must offer officers practical ways to assess and confront human dynamics—the tools students will need to handle persons from diverse backgrounds and lifestyles.

6. Allow for controversy: Courses on such topics often stir up controversy among students. The program design must allow students to openly question the materials presented, especially if they are being asked to evaluate their own beliefs and behavior.

7. Be experiential: The use of exercises (e.g., role-playing activities, game playing) in a controlled workshop environment is important. This training medium allows students time to practice and/or refine their communication, conflict management, and confrontation skills. Participants practice using the culture-specific information they have received while following appropriate officer safety (operations) procedures.

8. Provide follow-up support: Organizations should implement follow-up training on an ongoing basis through briefings and the like.

9. Identify potentially hostile employees: Trainers conducting these programs should identify employees who are aggressive or adversarial toward members of diverse groups. Employees who might be a threat to good community relations (also firehouse) should be identified. They should receive additional training and be made aware of the potential for discipline if their behavior is unacceptable.

Joyce St. George's nine points for teaching cultural diversity classes have application at college classes, fire academies, and in-service classes. Her approach to a class that deals primarily with the affective domain of learning is sound and practical. She uses different methods and strategies to appeal to students' ethics, logic, and feelings. Recognizing that all students may not have the same opinion on topics, St. George encourages open discussion even if there is controversy and disagreement. She integrates group activities in which students can use their communications and conflict resolution skills in a controlled environment. Further, as an outside consultant, she stresses the importance of the training being specific to the needs of the audience. Students will have a difficult time relating to training that does not apply to their duties or work environment. The message here is that instructors and trainers must always make cultural diversity training relevant and realistic to the scope of fire service.

STREETWISE TRAINING HINTS FOR COMMAND OFFICERS

Veteran firefighters, authors, and national trainers Rick Kolomay and Robert "Bob" Hoff (2003) understand and teach the importance of proactive training for firefighters. Their book *Firefighter Rescue & Survival* lists 10 common sense training hints for company command officers, which have application for all instructors and trainers who teach any topic. Students will benefit too. The 10 training hints follow (p. 14):

1. Keep your training sessions brief—10 to 30 minutes unless the firefighters demand more.
2. Always attempt to mix academic with "hands-on."
3. Relate the training to its context. Stay realistic!
4. Require that each firefighter participate at some point during the training session.
5. Present the opportunity for challenge, not embarrassment.
6. Focus on behavior rather than personality. It is the appropriate action, not attitude, that will extinguish the fire or succeed in a rescue.

FIGURE 3.13 ◆ A patient signs a refusal of treatment form for a firefighter EMT-B with a police officer witnessing.

7. Train at the company level in small groups of four to eight individuals in order to maintain involvement and interest.
8. Design training for results, not processes. Enable the firefighters with the training and education needed to accomplish their goal safely and effective without getting bogged down in procedures. Firefighting is about one result—getting the fire extinguished (Figure 3.13).
9. As long as the training objective has been achieved, let the group run with it!
10. It's okay to have fun, too.

DEVELOPING THE TRAINING PROGRAM

In this section we outline the following six steps to accomplish before any training program can take place. Your particular fire department may require more steps initially and must be collaborative in design to work.

1. Develop a mission statement incorporating the fire department's vision, values, and philosophy concerning the fire service and cultural diversity awareness.
2. Research what and how other fire service agencies and academies are teaching cultural awareness, including who is providing the instruction.
3. Appoint a cultural awareness training facilitator. With the approval of the fire chief, this person selects the cultural awareness trainers from within the organization. The facilitator might also contract with a cultural diversity training consultant who offers train-the-trainer sessions to existing or new department cultural awareness trainers. The fire department's cultural awareness trainer(s) assist you in these six steps.
4. Design a training needs assessment form and disseminate to all personnel with the department. Analyze the training needs based on the results of the assessment survey forms.

Involve members of the various ethnic and racial communities to assist in the program design. (Appendix B contains examples of needs assessment forms.)

5. Select community-based organization representatives of the various cultural, racial, and ethnic groups to assist in the program design. Include fire service community relations topics. Get as much input as you can from the community.

6. Develop and implement the program.

TRAINING DELIVERY METHODS AND STRATEGIES

Training in multicultural fire service should be a continuous and natural extension of the firefighting profession. The training emphasizes human relations skills, which most firefighters understand as being necessary. Cultural awareness instructors and facilitators should utilize a mixture of teaching approaches and strategies, such as combining lectures (academics) with skills development and self-appraisal techniques (simulations, role-playing, small-group exercises). Critical incident case studies, experiential assignments, and some cross-cultural videotapes or digital video discs are highly useful.

The *Multicultural Fire Service Instructor's Manual* that accompanies this text is particularly useful for instructors and facilitators of cultural awareness programs because it contains detailed information and resources for conducting such courses. A listing of consultants and resources is provided in Appendix C of this textbook. Also, the Website Resources at the end of this chapter offer additional information about training and related issues.

FOUR CRITICAL PHASES IN TEACHING CULTURAL DIVERSITY AWARENESS

FIRST PHASE: THE RELATIONSHIP BETWEEN CULTURAL AWARENESS AND FIRE SERVICE PROFESSIONALISM

After an overview of the topics is presented, the instructor provides the rationale for such training. The objective is to emphasize that knowledge of diversity and good fire service community and interpersonal relations is both professional and practical. It is important to emphasize that the knowledge and skills gained will improve the firefighter's effectiveness in working with diverse communities and coworkers, and reduce complaints and lawsuits. The instructor's goal of the first phase is to motivate the course participants to want to learn, and only an effective, prepared instructor is capable of accomplishing this objective. Many instructors' experiences have shown that achieving this objective is best facilitated by small-group discussion and exercises.

The class should be divided into small groups and requested to list the reasons why cultural diversity (fire service community and interpersonal relations with coworkers) awareness and skills training are important to the participants. When the groups come back together, the spokesperson for each records his or her list on a flip chart at the front of the room. Discussion is facilitated by the instructor on those items that reinforce the training objective. This phase is extremely important; and we have used this approach in teaching cultural awareness training and instruction in college classrooms, public safety academies, and in-service venues. The results have been overwhelmingly effective to get the learner to buy in to the training. A cautionary note: If an instructor does not successfully convey the importance of the training, firefighters may remain skeptical throughout the sessions.

SECOND PHASE: RECOGNITION OF PERSONAL PREJUDICES

awareness

Bringing to one's conscious mind that which is only unconsciously perceived.

One of the most difficult and sensitive aspects of the training course or curriculum deals with participants' **awareness** and acceptance of their own personal prejudices and biases. Once students have this awareness, the instructor can develop participants' understanding of how those prejudices and biases affect behavior—their own as well as that of others. This instruction should take place through a brief lecture, small-group exercise, and open class discussion when the small groups report back to the entire class. Because firefighters or any participants resent and react negatively to confrontational instructional approaches, this is a taboo in teaching and these methods must be avoided. The instructor and/or facilitator must be highly skilled and sensitive to this aspect of training and demonstrate the relevance of performance skills to the participants. That is, the instructor must approach the material from a practical perspective rather than a politically correct one and avoid dogmatic and extreme, militant behaviors.

THIRD PHASE: FIRE SERVICE–COMMUNITY RELATIONS

The block of instruction on fire service–community relations is critical to the success of the training (Figure 3.14). It involves the instructor and community members, who can provide participants with current information on and a historical overview of minority community perceptions about the fire service. Some instructors may have the ability to impart on what community perceptions are, but if they do so, they must not alienate the class. Some courses have effectively used minority community members to discuss culture- or race-specific information, but this approach must be closely monitored and controlled. The same is true of classes on the gay and lesbian community.

FIGURE 3.14 ◆ A multicultural community group having a conversation about public safety concerns.

Community member presenters must be carefully selected. Many community people are unable to handle the defensiveness of students during cultural awareness training classes. Specifically, some do not understand the fire service subculture or the major demands of the profession. The community member selected should work with the fire service instructor in the development of his or her own lesson plan so that there are no inappropriate surprises.

Outsiders must receive guidance so they will become familiar with the precise goals and objectives of the training before they participate. Understanding the goals will ensure that their involvement contributes to, rather than detracts from, the overall intent of the training and curriculum. An honest discussion of issues between firefighters and spokespersons from different cultural communities is useful; however, the intimidation of firefighters is nonproductive. Instructors representing ethnic, racial, or other cultural groups who are hostile toward firefighters will alienate the audience and should therefore not be part of the training program. This forum is not an opportunity for community representatives to vent their frustrations by berating firefighters, the fire department, or society for alleged insensitivities. Representatives of community-based organizations who want to assist will avoid confrontation and can provide valuable interaction and discussion about relevant topics. They must be prepared for firefighters to be direct, critical, and even confrontational if the firefighters feel even slightly attacked. If this forum is used, the fire service training officer is encouraged to have ongoing meetings with the community-based presenters to ensure that the goals and objectives of the course are being met. Modifications of presentations are appropriate if problems are identified in evaluations or during fire department class observations.

FOURTH PHASE: INTERPERSONAL RELATIONS SKILLS TRAINING

The fourth phase of training deals with the development or review of interpersonal relations skills that can help participants reduce tension and conflict in their interaction with community members and coworkers. The goal is to improve students' verbal and nonverbal communication skills. The approach is based on the idea that students learn how their behavior affects feelings and, in turn, the behavior and response of persons with whom they come in contact. Through brief lectures, role-playing exercise (videotaped with discussion), and other individual or small-group exercises, participants can learn how to control their own verbal and nonverbal behavior appropriately. The training can cover such subjects as barriers to effective communications, conflict resolution skills, the power of words (e.g., slurs, abusive, racial), and problem solving. It is our experience as professional trainers and educators that most participants will actually be eager to receive tools for the job—those immediate and practical skills that will help them deal more effectively with different cultural community members and coworkers.

Cultural awareness training courses involve imparting a great deal of information and skills to participants in a short period of time. To ensure learning and maximum retention, the instructor or facilitator, in addition to using a variety of proven teaching methods, must provide the class with several basic principles to remember at the end of each session. These basic principles should be related directly to students' job assignments. The instructor has several media available (e.g., PowerPoint, dry erase boards, overheads, and flip chart) to summarize the key points learned. The best approach to cultural awareness training is to link it to other curricula too, such as ethics, communications, and safety.

EXTERNAL CONSULTANTS AND TRAINERS

The observations of Dr. George Simons (1991) still have application today in that most consultants hired by organizations follow a four-stage pattern:

1. Needs assessment: The consultant, depending on the size and scope of the task, may conduct meetings and interviews or initiate an extensive in-depth assessment with instrumentation designed for the situation or circumstances in which the organization operates. The consultant may also be able to assist the organization leader with the "internal selling" of the project.

2. Project design and pilot: In collaboration with his or her clients, the consultant produces a design or sequence of events and activities aimed at achieving the results that the needs assessment has found critical. The design is customized to meet the needs of the organization and is normally tested in a pilot program with a limited group of people to determine whether the results desired can actually be achieved by the proposed design. The pilot program process may be repeated until the program is fine-tuned and ready for presentation. Packaged programs should be chosen carefully to ensure that the material meets the training objectives and needs of the organization. (We discourage the use of most packaged programs.)

3. Implementation, program delivery, and administration: The tested program is then put into service by either organizational trainers or the consultant or consultant team. Consultants can also train organization trainers and evaluate the results for the agency employing them.

4. Evaluation and follow-up: The consultant can also assist in the design and implementation of a training evaluation program.

CONSULTANT SELECTION

According to Simons (1991), organizations frequently do not know how to select a consultant and wonder whether the firm selected should include minority persons. He believes that in a very mixed target population, it may be useful, at least at the outset or for longer-term programs, to employ a team that is visibly diverse and able to model working together successfully. Simons further relates that the decision whether to use internal or external consultants is usually dictated by two factors: (1) the resources of the organization, and (2) the balance between objectivity toward the issues and the need for familiarity with the organization. To minimize the risks associated with hiring a consultant, Simons provides the following tips:

- Know roughly what objectives the organization intends to address through use of the training consultant.
- Interview thoroughly from a wide selection of candidates or services.
- Expect to pay for these services, including the expenses of the interviewing process.
- Watch how the potential consultant or team interacts with the diverse population of your interviewing team or your organization.
- Make sure that the candidate sees the relationship of diversity to the big picture of the organization and the community in which it operates.

There are many diversity and multicultural consultants throughout the United States. The Internet is an excellent source for locating cultural awareness training consultants and resources. We cannot stress enough the importance of selecting a consultant who understands the fire service profession, firefighters, and the nature of their work. Prior experience working with fire service agencies is important. Ultimately there is no better resource than word-of-mouth referrals from organizations that have had positive experiences with training consultants.

CULTURAL AWARENESS TRAINER OR FACILITATOR SELECTION

The selection of an academy and in-service trainer or facilitator is crucial to the success of cultural awareness training programs. Successful training programs of this nature have involved the chief or leadership team in the selection process. The facilitator must be respected, have credibility with the organization, and be a good role model for the philosophies and values that will be taught. This person must possess effective communication and good interpersonal skills. A trainer or facilitator who is not dedicated to the subject or is a novice can cause more damage and create more dissension than existed before the cultural awareness training.

Academy instructors and in-service trainers must attend technical courses to develop their teaching skills. A few states have developed a certified train-the-trainer course for cultural awareness instructors. The master trainer trains new instructors on the specifics of cultural awareness training. The new instructors must then earn the respect and gain the credibility of their audience.

TIPS FOR THE COURSE PARTICIPANT

As professional trainers and educators, we know what works to enhance the training for the adult learner (Figure 3.15). The course participant or student needs to understand the expectations of the trainer and the class. To create a positive learning experience, the participant needs to do the following:

- Must be able to express himself or herself openly and freely in front of a group and not be afraid to do so.
- Can offer constructive criticism of the information being taught.
- Should keep an open mind and be willing to learn about other cultures, races, religions, and all the dimensions of diversity.

FIGURE 3.15 ◆ Professional business consultants are equipped to offer their expertise to benefit the fire service on multicultural awareness and diversity strategies.

- Should make an effort to learn about the community and neighborhoods in which he or she works.
- Should make a positive effort to build an understanding of the diverse members of the community and workforce.
- Should strive to be open to learning about interpersonal and cross-cultural communication skills that contribute to better relationships.
- Will not take remarks by trainers or community presenters as a personal attack.
- Will share factual knowledge and experiences about his or her dimensions of diversity as needed.
- Will accept that the primary purpose of the training is to improve his or her interactions with coworkers and the public.

HOW DO YOU KNOW WHETHER THE TRAINING IS PAYING OFF?

IMMEDIATE EXAMINATION

Training as an ongoing process requires constant examination by the training staff and those in leadership positions in the fire service. The topic of multicultural awareness is no exception. As we review the results of our training, we need to ask the following questions:

1. Did the students (academy or in-service) perform the learning/training objectives?
2. Did the tasks in the training sessions meet the learning objective measurements?
3. Did the testing measurements (written or hands-on) align themselves with the training objectives?
4. Were all resources utilized effectively for the delivery of the training?
5. Were course and instructor evaluations completed by the students?
6. Were instructors allowed to complete their own evaluation on the training?
7. Did the training staff meet with the instructors to conduct a debriefing on questions 1 through 6?

NATIONAL FIRE ACADEMY

Again the question is, "How do we know whether the training is paying off?" The National Fire Academy uses Kirkpatrick's four levels of evaluation to answer this question on all training as it applies to long-term evaluation of the student and what the student learned and currently performs (Barr and Eversole, 2003). This sequential method of evaluation measures the effectiveness of the training program over a period of time. It is extensive and covers both the training program and the individual's application of what has been learned toward the job. Kirkpatrick's (1998) four levels of evaluation follow:

1. Reactions
   - Immediate evaluation of the participant's reaction to the training program.
2. Learning
   - This evaluation moves beyond participant's reaction to assessment of participant's advancement in skills, knowledge, or attitude. Requires the use of tests conducted before the training and after the training.
3. Transfer
   - This evaluation looks at the participants' use of the knowledge obtained in their everyday activities.

4. Results
- This evaluation requires the feedback from the participants' department or supervisor as to how well they are performing based upon the knowledge learned in the program.

Education and training that help familiarize firefighters with multicultural groups in their community and workforce are a high priority for fire service. Fire service personnel can effectively address the community's needs only if they understand the cultural traditions, moral climate, and values of that community. This understanding is essential to the on-the-job awareness skills and to formulating collaborative problem-solving strategies.

In this chapter we have dealt with cultural awareness training for fire service agencies in a myriad of multicultural settings. This information provides both the fire service executive and the trainer strategies and program ideas on awareness training that can be implemented in agencies of any size and in any community. We have stressed that executives must take a leadership role in the development of community partnerships and the training of employees to become culturally competent. Effective training models concentrate on cultural education, communications skills, interpersonal skills, and conflict resolution techniques. To be successful, courses must be designed to fit participant and community needs.

Discussion Questions and Issues

1. **Personal Action Plan.** Complete a short-term and long-term personal goals summary that makes use of concepts learned during cultural awareness training, and explain how you will apply them at work.

2. **Instructors.** You are a cultural awareness trainer. What skills have you learned that will assist you in teaching others about prejudice reduction on the part of your audience? Make a list of teaching methods that you would utilize to teach about prejudice, bias, and their effects on behavior. How can prejudice be revealed in a nonthreatening way in a training course?

3. **Training Goals and Objectives.** You have been selected as the training facilitator for your fire department. Make a prioritized outline of the steps and processes that you will need to complete to design and implement a cultural

awareness training course. After the course is planned, make a to-do list of things that must be accomplished three weeks before, one week before, and after the program.

4. **Fire Executive.** You intend to implement cultural awareness training at your department. List the steps you will need to take to ensure that an effective training course is prepared, implemented, and accepted. How will you determine what cultural groups to include and discuss in the training sessions? How will you budget for the training and integrate this in your scheduled in-service training program?

5. **Course Evaluation.** You are the cultural awareness trainer or facilitator. Make a list of the various methods you might employ to evaluate the effectiveness of the training class that is implemented at your department or in your academy.

■■■

Website Resources

Visit these websites for additional information about training and related issues:

American Council on Education (ACE): http://www.acenet.edu
Founded in 1918, the American Council on Education (ACE) is the nation's unifying voice for higher education. ACE has approximately 1,800 accredited, degree-granting colleges and universities and higher education–related associations, organizations, and corporations. To access its Credit Recommendation Service, go to the Program and Services link of its website. ACE's area of focus includes access, success, equity, and diversity; institutional effectiveness; lifelong learning; and internationalization.

International Association of Fire Chiefs (IAFC): http://www.iafc.org
It is an online resource for fire service executive and leadership issues and publications developed by the IAFC.

International Association of Fire Fighters (IAFF): http://www.iaff.org
It is a website for information about the range of issues confronting firefighters and their departments.

National Career Academy Coalition (NCAC): http://www.ncacinc.org
It is a website for information concerning career academies and career development of professions.

National Coalition Building Institute (NCBI): http://www.ncbi.org
A nonprofit leadership training organization based in Washington, DC. Since 1984, NCBI has been working to eliminate prejudice and intergroup conflict in communities throughout the world. It is a website for information about training programs, accomplishments, chapters, discussion groups, and campus and new programs. Leadership teams embody all sectors of the community, including elected officials, emergency responders, government workers, educators, students, business executives, labor union leaders, community activists, and religious leaders.

National Fire Academy: http://www.usfa.fema.gov/nfa
It is a website for information about training, careers, jobs, publications, and research for the fire service.

Teaching Tolerance: http://www.tolerance.org
Founded in 1991 by the Southern Poverty Law Center, Teaching Tolerance provides educators with free or low-cost materials that promote respect for differences and appreciation of diversity in the classroom and beyond.

U.S. Fire Administration: http://www.usfa.fema.gov
A source of fire services resources, programs, and statistics collected and published by the Federal Emergency Management Administration (FEMA).

Women Chief Fire Officers Association (WCFOA): http://www.womenfireofficers.org
It is a website for information to provide a proactive network that supports, mentors, and educates current and future women chief officers.

■■■

References

Barr, R. C., and J. M. Eversole. (Eds.). (2003). *The Fire Chief's Handbook,* 6th ed. Tulsa, OK: PenWell.

Bloom, B. (1997). *Taxonomy of Education Objectives, Handbook 1: The Cognitive Domain.* New York: Longman.

Crawford, B. A. (2004, September 1). "Patchwork Force," *Fire Chief.*

Goldstein, I. L. (1993). *Training in Organizations,* 3rd ed. Pacific Grove, CA: Brooks/Cole.

Kirkpatrick, D. (1998). *Evaluating Training Programs: The Four Levels.* San Francisco: Berrett-Koehler.

Kolomay, R., and R. Hoff. (2003). *Firefighter Rescue and Survival.* Tulsa, OK: PenWell.

McGehee, W., and P. W. Thayer. (1961). *Training in Business and Industry.* New York: John Wiley.

R. E. McKerrow, B. E. Gronbeck, D. Ehninger, and A. H. Monroe. (2000). *Principles and Types of Speech Communication,* 14th ed. New York: Longman.

Moore, M. L., and P. Dutton. (1978, July). "Training Needs Analysis." *Academy of Management,* 532–45.

Olson, Aaron T. (2004). *Seeking Excellence in Employee Relations.* Oregon City, OR: Willamette Falls Television Studio.

Olson, Aaron T. (2007). Professor of criminal justice at Portland Community College, organizational consultant and retired Oregon State Police patrol supervisor, Portland, Oregon, personal communication, February 2007.

Phillips, Jack J., and Elwood F. Holton III. (1995). *Conducting Needs Assessment.* Alexandria, VA: American Society for Training and Development.

Rossett, A. (1987). *Training Needs Assessment.* Englewood Cliffs, NJ: Educational Technology Publications.

Russ-Eft, D., and H. Preskill. (2001). *Evaluation in Organizations: A Systematic Approach to Enhancing Learning, Performance, and Change.* New York: Basic Books.

Simons, G. F. (1991). *Diversity: Where Do I Go for Help? A Guide to Understanding, Selecting, and Using Diversity Services and Tools.* Santa Cruz, CA: George Simons International.

Sleezer, C. M. (1991). "Developing and Validating the Performance Analysis for Training Model." *Human Resource Development Quarterly,* 2: 355–72.

St. George, J. (1991, November 30). "'Sensitivity' Training Needs Rethinking." *Law Enforcement News,* 7(347): 8–12.

Witkin, B. R., and J. W. Altschuld. (1995). *Planning and Conducting Needs Assessment.* Thousand Oaks, CA: Sage.

CHAPTER 4

Communication for Firefighters

Key Terms

cross-cultural, p. 120

Overview

This chapter provides information in distinguishing the communication differences from operational alarms and calls to the firehouse setting (Figure 4.1). We discuss the first responder's challenges of language barriers when English is a second language and provide a list of tips when translation services are not available. We discuss commonly held attitudes about nonnative English speakers and explain the difficulty involved in second-language acquisition. The next sections cover verbal and nonverbal communication, proxemics, and facial expressions, eye contact, posture, gestures, high and low context, the communication process, and cross-cultural differences. Next, the chapter provides an overview of specific issues that fire service professionals encounter with regard to communication in a diverse environment. We discuss several common reactions that people have when communicating with people from different backgrounds, including defensiveness, overidentification, denial of biases, and the creation of "we–they" mind-sets. In addition, we discuss responses to citizens' accusations of racial or ethnic discrimination as well as communication post-9/11. We then present information on key issues and skills required for interviewing and gathering data, particularly across cultures. The final section presents male–female communication issues, particularly within the fire service.

Objectives

After completing this chapter, participants should be able to:

- Identify the differences between the communication in operational alarms and calls compared to the firehouse setting.
- Describe the first responder challenges of language barriers when English is a second language.
- List tips for communicating in situations in which English is an individual's second language.
- Describe verbal and nonverbal communication, the elements of voice communication, and the forms of body language.

FIGURE 4.1 ◆ A news reporter interviews a firefighter as two teen boys watch.

- ◆ Describe high and low context, the communication process, and cross-cultural differences.
- ◆ List effective strategies to communicate across cultures and deflect accusations of insensitivity or discrimination.
- ◆ Describe the male–female communication issues, particularly within the fire service.

LEARNING TASKS

Communication is an ongoing process for all professions, organizations, groups, and individuals. Successful organizations use a variety of communication media to converse and relay information. Learning about the myriad of different communication methods as it relates to the fire service in a multicultural society and workforce provides the knowledge, skills, and abilities to be competent in the communication process. Talk with your agency, and find out how it uses communication in its external and internal operations. Be able to:

- ◆ Identify the methods of communication the agency uses to relay information to firefighters and nonfirefighter personnel.
- ◆ Identify the methods of communication the agency uses to relay information to multicultural populations in its fire service district.
- ◆ Determine what existing communication challenges the agency has for the exchanging of information between firefighters, management, and nonfirefighting personnel.
- ◆ Determine what existing communication challenges the agency has for the exchanging of information with the multicultural community it serves.
- ◆ Ascertain what ongoing strategies the agency has in place or is developing to improve communications with all employees and the multicultural community it serves.

PERCEPTIONS

The challenges of communication across cultures for firefighters and all first responders are multifaceted. First responders talk and listen to speakers from their own and other languages. They also deal with the changing culture within their agencies and the sensitivities required in a multicultural workforce. As experienced multicultural educators and trainers, we know it is important not to assess a person's speech patterns based solely on what we are hearing. An untrained listener can easily misunderstand when a speaker is not fluent in English. The listener should use several indicators for interpreting correctly what the nonfluent English speaker is saying. Some examples of indicators are facial expressions, tone of voice, gestures, body language, and, if needed, pointing at pictures.

INTRODUCTION

Communication is an essential function for all firefighters to perform. The National Fire Protection Association, the International Association of Fire Chiefs, and several other professional fire service organizations recognize the importance of effective communications. Extensive training is dedicated to using communications equipment, learning radio codes, communicating verbally and in written form, and using other media of communication to perform the duties required of a firefighter and for all firefighters to work together as a team. The National Fire Academy offers classes on fire service communications as part of its Management Science Curriculum. The National Fire Protection Association Standard 1001 delineates the communication requisites for knowledge and skills for firefighters at Levels I and II. The information imparted by both organizations is excellent and useful; however, it does not address the complete aspects of communication and its multicultural impact to the fire service as this chapter will. This straightforward comment aligns itself to the response to the question "What advice do you have for rookie firefighters?" as given by John Salka, battalion chief, Fire Department of New York, New York.

> Listen up! You don't learn a single thing by talking; when you listen to your officer, other firefighters, and citizens, you will learn. Make a conscious effort to limit your talking and maximize your listening. Since you are a new member of the fire department, most of the other firefighters don't care what you think anyway, so pay attention to what they are doing and saying, and learn. Don't get me wrong: Everything you see and hear in the fire station will not be correct, but you must be listening to figure out what advice is good and is bad (Salka, 2005, p. 32).

COMMUNICATION FROM ALARMS AND CALLS TO THE FIREHOUSE

Vehicle rescue and extrication, fire suppression, search and rescue, and emergency medical care calls are critical job tasks performed by firefighters. No other public safety professionals in the United States work 24 hours on and respond to these types of calls during their shift. No other public safety professionals return to the firehouse

FIGURE 4.2 ◆ Firefighter emergency medical technicians demonstrate their communication skills as they inventory their rescue equipment on an ambulance.

between calls and make their workstation their home away from home during their shift. Police officers working a major crime scene investigation may work past their eight- or ten-hour shift but do not routinely work a 24-hour shift, as firefighters do. One of the authors (Olson, 2007), a retired Oregon State Police patrol supervisor who worked on calls for assistance with firefighters and other emergency responders, and a veteran fire service trainer, shares his personal and professional experiences about communication:

- The type of communication a firefighter uses on calls is directly related to the firefighter's skills and professionalism, and the company's ability to function as a team (Figure 4.2).
- The type of communication a firefighter uses between calls and back at the firehouse is more relaxed. To deny firefighters or any first responder the time to rest and relax with team members is unrealistic.
- Communication tends to be relaxed between calls and back at the firehouse. The communication between firefighters and all first responders is personal and inclusive.

LANGUAGE BARRIERS AND FIRST RESPONDERS

Nationwide, changing demographics have resulted in the need for the fire service to deal increasingly with a multicultural population who are speakers of other languages and do not have the equivalent skills with English. For citizens on vehicle rescue and extrication, fire suppression, search and rescue, and all emergency medical care calls, there are no absolute assurances that first responders will understand them. Firefighters who are not bilingual are justifiably frustrated by language barriers and find it difficult to do their jobs the way they have been trained to do them. Many first responders modify their English so that they will be better understood by the citizen or noncitizen. Some first responders are noticeably impatient when they deal with nonfluent or non-English-speaking individuals. The reality is that there are over 150 languages, and just because a firefighter is bilingual doesn't guarantee his or her multiple language skills will be used on the next call.

Sensitivity to the difficulties of those who do not speak English is in order, but that is only a partial solution to the problem. In attempting to cope with the problem of non-English-speaking citizens and noncitizens, many public safety agencies not only have increased the number of bilingual employees in their workforce but also have begun to utilize translation services. Some services provide over 150 languages 24 hours a day. Most emergency 9-1-1 centers have access to these translation services in which a translator will be patched into the 9-1-1 emergency lines within seconds to assist the emergency telecommunications call taker to process the emergency call. Having access to translation services and referrals is a first step in addressing the challenge of communication with those who speak no or limited English, but this is not a long-term solution.

Having bilingual firefighters constitutes a more direct method of addressing the problem. Many public safety agencies across the country offer language classes with tuition reimbursement and incentive pay for second-language proficiency. Training personnel are encouraged to look for classes specially designed for public safety and first responders. If these classes do not exist in your area, a few selected firefighters and language educators should form a partnership so that firefighters can guide language teachers in the development of fire service–specific second-language curriculum. David B. Dees is one such person who developed a curriculum and teaches Spanish at fire departments, sheriff's departments, and community colleges throughout California. He has created specialized Spanish instruction programs for police, firefighters, and other emergency responders. In his book with an audio compact disc, *Quick Spanish for Emergency Responders,* Dees provides essential words and phrases for firefighters, paramedics, and EMTs in real-life emergency situations (Dees, 2006). When translation is not available, firefighters have no choice but to rely on English and their other skills. In doing so, the tips listed in Figure 4.3 on modifying one's language will be helpful.

ATTITUDES TOWARD NON-ENGLISH OR LIMITED ENGLISH SPEAKERS

Most first responders are aware that a citizen or noncitizen with few or no English skills is not necessarily an illegal immigrant. However, sometimes first responders have seen English difficulties as a sign that a person is an illegal immigrant and the consequences have been unfortunate. Donya Fernandez (2000), a language rights attorney, cites the case of a Spanish speaker with extremely limited English skills whom police turned over to the former Immigration and Naturalization Service (INS). The INS held him for 48 hours, despite the fact that the man was a U.S. citizen who had been born in the United States but had spent most of his life in Mexico.

Any person's constant use of a second language and the accompanying frustration for first responders can be overwhelming. In general, whether the society at large (or the fire service as a microcosm of society) is concerned about a particular group's use of its native language seems to be directly related to the population size of that group. For example, when large groups of Cubans or Puerto Ricans speak Spanish, there is often a higher level of anxiety among the dominant white population than when a few Vietnamese speak their native language. Virtually every immigrant group is said to resist learning English, yet the pattern of language acquisition among the generations of immigrants follows a predictable course. Members of the second and third generations of an immigrant family almost always become fluent in English,

1. Speak slowly and articulate your words clearly.

2. Face the person and speak directly to him or her, even when using a translator.

3. Avoid constant eye contact if the person is not making constant eye contact with you.

4. Avoid jargon, slang, idioms, or reduced forms (e.g., *gotta, wanna, wouldja, shoulda*).

5. Avoid complex verb tenses (e.g., "If you will please indicate to me by describing which part of your body the injury is located or where the pain is distinguishable upon the cessation of my question").

6. Rephrase key questions or key points.

7. Avoid closed-ended questions with "yes" or "no" answers, but use open-ended questions that by the person's narrative response shows he or she understands your question.

8. Use short, simplistic sentences; and pause between your sentences.

9. Use only one idea or question per sentence.

10. Use visual cues such as gestures, demonstrations, and short written phrases, if needed.

11. Use active rather than passive verbs (e.g., "I need your help," [active] rather than "Your help is needed" [passive]).

12. Have written materials available in bilingual format.

13. Allow the person time to respond to your question, because English as a second language speakers need time to translate in their minds and formulate their responses in English.

14. Monitor the person's comprehension by asking the person to summarize in his or her own words what you said or asked.

15. Do not speak louder; it will frustrate the person and yourself.

16. Concentrate more on using active listening skills if the English is broken or absent.

17. Exercise patience, because you could be in a similar situation in another country in which the government official doesn't speak English.

18. Encourage and provide positive feedback on the person's ability to communicate.

FIGURE 4.3 ◆ Tips for Communicating When English Is a Second Language

Source: Shusta, Levine, Wong, Olson, & Harris (2008)

whereas many of the first-generation immigrants (the grandparents and the parents) struggle, sometimes partly learning English and sometimes not learning it at all. Many immigrants, however, are extremely motivated to learn English and become productive members of society. Newcomers are fully aware that, without English, they will never be able to integrate into society. This is why there is explosive growth in English as a second language (ESL) classes throughout the United States. These classes often have long waiting lists and have not been able to keep up with the demand in many urban communities with large numbers of immigrants.

Some people, including established immigrants, have a tendency to overgeneralize their observations about new immigrants and refugees. It is true that some people do not want to learn English, and even some middle-class U.S.-born Americans do not make efforts to improve their language or communication abilities. How often does one hear that high school graduates who are native English speakers have not learned to write or speak well? Here laziness or lack of high-quality education or both may have contributed to this aspect of illiteracy. In trying to be fair and objective, all groups have a percentage of lazy people, but sometimes people tend to stereotype others. Although not all first-generation immigrants or refugees learn English, there is a great deal of mythology and false assumptions around the "masses" of immigrants and refugees who hold on to their native language.

The native language for an immigrant family is the language of communication for that family. It is not uncommon to hear comments such as, "They will never learn English if they insist on speaking their native language at home." Imagine having been away from your family all day and coming home and interacting in a foreign language. Is it reasonable to expect that one could express affection, resolve conflicts, show anger, and simply relax in another language? Language is an integral part of a person's identity. During the initial months and even years of communicating in a second language, a person does not truly feel like himself or herself. Initially, one often has a feeling of playacting or taking on another role when communicating in a second language.

From a physiological point of view, speaking a foreign language can be fatiguing and exhausting. As a child, when speaking one's own native language, one uses a set of muscles to articulate the sounds of a given language. Changing to another language, particularly as an adult, requires the use of an entirely new set of muscles. This causes mental strain and facial tension, which can result in a person's *shutting down* or *coming against a learning wall*. These terms mean the inability to communicate in English or any other new language. It was no wonder that in the multicultural workforce, clusters of people from different ethnic groups can be seen having lunch together, taking breaks together, and so on. Simply put, it is more relaxing to be able to speak one's own language than to struggle with a new one all day.

Sometimes first responders say, "I know they speak English because they speak it among themselves. The minute I'm on the scene, it's 'No speak English.' Why do they have to play dumb? What do they think I am, stupid?" It would be naïve to say this situation does not occur. There will always be some people who try to deceive others and use or not use English to their own advantage. However, there may be other reasons that people "feign" not knowing English. Several factors affect an immigrant's ability to use English at any given moment. A few of these, in particular, are of special significance to first responders. Generally speaking, an immigrant's ability to express himself or herself in English is best when that person is comfortable with the first responder. So, the more intimidating the contact is with police officers, the higher the likelihood that anxiety will affect the speaker's ability in English. Language breakdown is one of the first signs that a person is ill at ease and stressed to the point of not being able to cooperate and communicate. It is always in the first responder's best interest to increase the comfort level of the citizen or noncitizen, whether a victim, a suspect, or simply a person requiring help. Language breakdown in a person who is otherwise fairly conversationally competent in English can also occur as a result of illness, intoxication, fatigue, and trauma.

Finally, first responders must realize that their attitudes about immigrants and nonnative English speakers, whether positive or negative, may very well affect their interactions with them. This is especially true when a first responder is under pressure and negative attitudes are more likely to surface in communication.

TAKING POSITIVE STEPS

As part of his public safety outreach to new immigrants and refugees, one of the authors, Aaron T. Olson, then a patrol supervisor with the Oregon State Police, established an

FIGURE 4.4 ◆ Sergeant Aaron T. Olson from the Oregon State Police posing with Bosnian and African refugee youth and Immigrant Refugee Community Organization (IRCO) staff at a weekly public safety training meeting in Portland, Oregon. *Source:* Aaron Olson

ongoing partnership in May 2002. The organization called IRCO (Immigrant Refugee Community Organization) hosts classes from Portland, Oregon, in which new immigrants and refugees are oriented on interaction with American police, laws, fire service, and 9-1-1 emergency services (Figure 4.4). Interpreters translate during the presentation and there is time for questions and answers. What is learned from the thousands of immigrants and refugees who have attended the workshops is shared with police officers, firefighters, 9-1-1 personnel, college students, and the general public in college classes and training workshops (Olson, 2007). Olson continues to volunteer his time at IRCO as a training consultant and facilitates the delivery of bicycle helmets and child seats to needy immigrant and refugee families.

VERBAL AND NONVERBAL COMMUNICATION

Some instructors and trainers who teach multicultural training workshops or classes ignore the important features and fundamentals of verbal and nonverbal communication. Too often they focus only on the communication differences between various cultures instead of first talking about the themes, concepts, and principles that serve as a common thread. Unfortunately when these fundamental features are omitted, it makes it difficult to compare and contrast variances for cross-cultural communication. In recognizing this problem, we purposely address what all cultures have in common for verbal and nonverbal communication. Once this foundation is laid, then we can add to our framework of knowledge any features or characteristics of cross-cultural communication for different cultures. (See Chapter 5 for specific cultures.)

Communication comes in all forms and methods of delivery but is divided primarily into two categories—verbal and nonverbal. Verbal is using your voice to communicate and nonverbal is using your body to communicate (McKerrow, Gronbeck, Ehninger, and Monroe, 2000). The verbal method includes words, intelligibility, and vocal variety. The nonverbal methods include facial expressions and gestures, touch, and proxemics. *Vocalics* is the study of nonverbal cues of the voice. Some of these cues are tone, pitch, and volume, but for clarity to the reader, they will be addressed under voice communication for the verbal method. It should be noted many scholars in this discipline usually use a strict meaning of the term *verbal*, indicating "of or concerned with words," and do not use *verbal communication* interchangeably for oral or spoken communication.

One of the most widely recognized studies on communication was conducted by Dr. Albert Mehrabian, a psychologist at the University of California at Los Angeles (UCLA). In the study he examined the relative importance of verbal and nonverbal messages when people communicate their feelings (emotions) and attitudes (negative or positive). As a result of his study, he identified three elements of communication that account differently for the message: 7 percent for words, 38 percent for tone of voice, and 55 percent for body language, known as the "7%–38%–55% Rule" (Mehrabian, 1971). In order for the communication to be effective and provide value, the three elements of the message need to support each other in what they mean. The meaning provides congruency in that each component agrees with the message. When any of the components do not match each other, there is the likelihood the message will lose its meaning to the receiver of the communication. In other words, the listener is paying attention to what he or she sees from the speaker (55 percent), how the speaker sounds (38 percent), and the actual words the speaker is using (7 percent). Figure 4.5 shows Dr. Mehrabian's three elements of communication for feelings and attitudes.

Dr. Rick Brinkman and Dr. Rick Kirschner, who are naturopathic physicians, professional speakers, and corporate trainers, further add:

> This is what we affectionately refer to as the "55, the 38, and the 7," or if you prefer, "Numbers of Meaning." We believe these Numbers of Meaning are important for understanding communication in general. . . . Now, in many ways, these percentages should come as no surprise. After all, common expressions such as "Seeing is believing," and "Action speaks louder than words" point to the stronger influence of the 55 percent, the visual component of nonverbal communication. . . . The 38 percent of communication, the way someone sounds when they talk to you, usually reflects their emotional state and sends

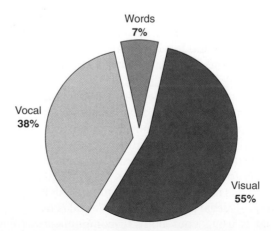

FIGURE 4.5 ◆ Three Elements of Communication *Source:* Herbert Wong and Aaron Olson

an ego message about you, and plays a significant role in how you make sense of mixed messages. . . . And while the actual words we use may constitute only 7 percent of the meaning of a particular communication, we all know that just one small word can serve as a trigger, or "buzzword" that sets entire chains of reaction in motion! . . . The greatest value of knowing about the 55, the 38, and the 7 is in helping you to remember the order of priority by which people make sense of each other, and how it is possible that mixed messages produce misunderstanding (Brinkman and Kirschner, 2002, pp. 192–93).

CAUTION OF MISINTERPRETATION OF MEHRABIAN'S RULE

Some people misinterpret the 7%–38%–55% Rule, thinking that it applies to all communication situations in which nonverbal cues are more important than the meaning of the words. This overgeneralization, from the very specific conditions in his experiment studies, is the basic mistake around Mehrabian's rule because emotions and attitudes are not always conveyed by nonverbal cues. Mehrabian (2005) provides clarity to this rule on his webpage, which reads:

> Total Liking = 7% Verbal Liking + 38% Vocal Liking + 55% Facial Liking: Please note that this and other equations regarding relative importance of verbal and nonverbal messages were derived from experiments dealing with communications of feelings and attitudes (i.e., like–dislike). Unless a communicator is talking about their feelings or attitudes, these equations are not applicable. Also see references 286 and 305 in *Silent Messages*—these are the original sources of my findings (*http://www.kaaj.com/psych/smorder.html*).

VOICE COMMUNICATION

INTELLIGIBILITY

Effective voice communication consists of two important elements: intelligibility and vocal variety (Figure 4.6). Intelligibility is when the listener can understand what the speaker is saying. The understanding of the speaker is dependent upon his or her volume, rate, enunciation, and pronunciation of words. Dialect can influence intelligibility as well. A brief explanation of each term and its application follows:

1. *Volume* is the degree of loudness or intensity of a sound. Firefighters on a run or alarm adjust their volume for speech based on the ambient noise or their distance at the scene. The volume for speech at the firehouse will probably be less because of less distance between coworkers.
2. *Rate* is the speed at which a person speaks words per minute. When one is providing vital or important information, it is wise to slow down one's rate of words per minute to make sure the listener comprehends.
3. *Enunciation* is the crispness and precision with which a person articulates words. Good enunciation of words is important when communicating technical and precise information.
4. *Pronunciation* is the acceptable standard of saying the word. To be understood by the listener, it is important to say the proper consonant sound for the word.
5. *Dialect* is a factor because it is a regional variety of language distinguished by features of vocabulary, grammar, and pronunciation unique to the listener. A person who is bilingual, in which English is a second language, may have a native country accent that interferes

FIGURE 4.6 ◆ A company of firefighters climb a ladder at their station in New York City.

with his or her speech. The listener has to make a more concentrated effort to understand a speaker with a different dialect from the Northwest, South, East Coast, Midwest, or from another country.

VOCAL VARIETY

Vocal variety is produced by changes in a person's rate of speech, pitch, stress on words, texture of the voice, and tone of voice. We have discussed how the rate of speech impacts the intelligibility of the speaker's message, but it does play into vocal variety as well. It is not uncommon to have people speak faster when they are nervous, upset, excited, or angry. Likewise, people have a tendency to speak slower when they are serious and want to make a point. A brief explanation of each term and its application follows:

1. *Pitch* is the frequency of sound waves and its loudness. Some people talk in the soprano, alto, tenor, baritone, or bass range. Usually, higher pitches communicate excitement and lower pitches reflect control or solemnity.
2. *Stress* is the method in which sounds, syllables, and words are highlighted or accented. Vocal stress is achieved through emphasis and the use of pauses. It can be accomplished by increasing volume, changes in pitch, or rate. Stress is associated with the emotions of excitement, anger, sadness, and joy. A lack of vocal stress can give the impression that the speaker is just monotone and bored, but could cause misunderstandings of the meaning.
3. *Texture* is the quality of the voice, which could be nasal, hoarse, breathy, scratchy, gravelly, or crackling. Sometimes this texture quality can impact intelligibility or even be a distracter in communication. Listeners must not allow their biases to detour themselves from the message.

4. *Tone of voice* consists of emotional characterizers that show the temperament of the speaker. The speaker could easily be happy, sad, angry, sarcastic, serious, playful, and a plethora of other emotions. It is important that the listener be attentive to these emotions and recognize that the emotional state of the speaker could change as the context of the conversation changes.

BODY LANGUAGE

Body language is nonverbal communication and is nonlinguistic in a technical sense. Through physical behavior, what the voice communicates is either reaffirmed or conversely invalidated by a person's physical behavior (Figure 4.7). Realizing this, we look at components of physical behavior that are associated to body language and serve as an enhancement or detractor to communication. These components are proxemics, facial expressions, eye contact, posture, and gestures.

PROXEMICS

We define proxemics as the spatial distance that separates two or more people from each other in various social, work, family, and interpersonal situations. Further, proxemics has relevance as it relates to environmental and cultural influences. Anthropologist Edward T. Hall did extensive research on proxemics and introduced its meaning and usage in his book *The Hidden Message*, which serves as a helpful measurement for distances in all human interaction. We have taught proxemics in all of our classes and workshops for fire service personnel at the academy and

FIGURE 4.7 ◆ Grade school children listen to a firefighter at the Brooklyn Fire Department.

in-service levels. The four zones of proxemics are intimate, personal, social, and public (Hall, 1966):

1. Intimate ranges from 0 to 18 inches. Some examples would include touching, embracing, holding hands, or whispering.
2. Personal distance ranges from 18 inches to 4 feet. Some examples would include friends or close coworkers.
3. Social distance ranges from 4 to 12 feet. This would include impersonal business interactions or interactions with a stranger or acquaintances.
4. Public distance measures from 12 feet to more. Usually this is the distance between a public speaker or in a performance at a play or auditorium.

We agree with Hall that different cultures vary their distances for the different zones of proxemics. Although personality and context influence interpersonal distance, some cultures and individuals encounter strangers in their intimate and personal distances, whereas others feel that these uninvited people, under these conditions, are invading their space and become uncomfortable. When someone violates this space, a person may feel threatened and back away. In general, Latin Americans and Middle Easterners are more comfortable at closer distances than are northern Europeans, Asians, or the majority of second- or third-generation Americans. The importance of feeling comfortable rests upon the social situation, gender, and individual liking. Having the awareness and recognition of these cultural differences improves cross-cultural interactions and aids in eliminating discomfort people may experience if the interpersonal distance is too much (staying away) or too small (intrusive).

FACIAL EXPRESSIONS

Facial expressions are a form of body language and result from one or more motions or positions of the muscles of the human face. The expressions can be voluntary or involuntary. Facial expressions are closely associated with all human emotions (Figure 4.8). A person must remember not all facial expressions mean the same thing across cultural backgrounds and contexts. A smile or giggle from Southeast Asian cultures can cover

1. Anger	13. Intense
2. Bored	14. Joy
3. Concentration	15. Laughter
4. Contempt	16. Melancholy
5. Crying	17. Pout
6. Depression	18. Puzzled
7. Disgust	19. Sarcastic
8. Eager	20. Shock
9. Excited	21. Smile
10. Fear	22. Sneer
11. Glad	23. Suspicious
12. Happy	24. Etc.

FIGURE 4.8 ◆ Facial Expression Examples
Source: Aaron Olson

up pain, humiliation, and embarrassment. Some women (e.g., Japanese, Vietnamese) cover up their mouths when they smile or giggle. Upon hearing something sad, a Vietnamese or Laotian person may smile. The smile does not mean that the person is trying to be sarcastic or rude; it is simply a culturally conditioned response. The same observation has been noted with new immigrants and refugees who have a tendency to smile more often when they are unable to speak English as a second language. Further, in an individual context, a person crying may be crying because of sadness, but those tears may be tears of joy and laughter too.

First responders need to able to discern facial expressions in all situations so they can assess situations correctly (Figure 4.8). This applies to working with coworkers in the firehouse or on emergency calls. The degree to which people show emotions on their face depends on the context of the situation and on their cultural background. To say a certain group is more expressive in a facial sense is stereotyping and putting people in a "box." However, it has been noted that first-generation immigrants, refugees, and most cultures have a tendency to mirror their country, region, or socioeconomic level when in the United States. Generally, people from the East Coast (New York); first-generation Latin Americans, Mediterranean, Arabs, Israelis, Italians, Eastern Europeans immigrants; and socioeconomically depressed blacks and whites have a propensity to show emotions facially. On the other side, Asian cultural groups, socioeconomically affluent whites, and other groups tend to be less facially expressive. We call the less expressive face a "poker" face because it is difficult to read the person's emotions or mind-set. Persons in management positions are often criticized of having a poker face.

EYE CONTACT

Eye contact is when two people look at each other and their eyes meet at the same time. As a form of nonverbal communication, eye contact can have strong emotional impact, or people attempt to avoid or limit their eye contact altogether. In U.S. culture, consistent eye contact establishes a sense of trust between two people, and friendly eye contact shows the other person you are interested in him or her and what that person has to say. The eye contact intensifies into a gaze when Americans like each other and they appear to cooperate. Conversely, when people tend to make less eye contact or avoid eye contact, it is implied they dislike each other, they disagree, or someone is being deceitful. In stoic, reserved, and formal cultures, a lack of eye contact usually displays respect.

In many other countries, eye contact is avoided with authority figures. In parts of India, for example, a father would discipline his child by saying, "Don't you look me in the eye when I am speaking to you"; whereas an American parent would say, "Look me in the eye when I'm speaking to you." To maintain direct eye contact with a teacher or parent in some cultures would be disrespectful. First-generation Hispanic and Asian Americans are taught to show limited eye contact with authority figures out of respect, which could cause a problem with first responders who are unaware of these cultural differences. Further, out of cultural customs, most Native Americans show limited eye contact, and when a person stares at a Native American, that eye stare means an invitation to a physical altercation.

POSTURE

Human posture is part of nonverbal communication and the position of the body has significant cultural impacts. In most cultures, it is expected for a person seated to stand up when greeting a friend, guest, family member, or person of authority. Also, it is considered insulting if a person maintains a slouched position in a chair or seat when another person of authority is speaking, teaching, or present. Sitting and baring the sole of one shoe in an upward position would be deemed offensive to Southeast Asian and Arab cultures. To show one's foot in many cultures outside of the United States is insulting because the foot is considered the dirtiest part of the body.

GESTURES

A gesture is a component of nonverbal communication that is demonstrated with a part of the body. A gesture can be used in combination with verbal communication or by itself. The infinite array of body gestures provides people the opportunity to express a myriad of feelings and thoughts. These feelings and thoughts can range from contempt and hostility to approval and affection. The majority of people use gestures and other forms of body language in addition to the words they speak. Some cultural groups and languages use gestures more than others do, and the degree of gesturing that is deemed culturally acceptable varies from one culture or region to the next. What seems to be consistent with the use of gestures is that they are natural, descriptive, spontaneous, and match the message.

A few American gestures are considered offensive in other cultures. For example, the "OK" gesture with the fingers and one hand is obscene in Latin America; the good luck gesture is offensive in parts of Vietnam; and the "come here" gesture (beckoning people to come with the palm up) is very insulting in most of Asia and Latin America (Levine and Adelman, 1992). The point to remember is that a single gesture can have a very diverse significance in different cultural contexts, ranging from complimentary to highly offensive.

HIGH- AND LOW-CONTEXT COMMUNICATION

Up until this point in the chapter, we have discussed how verbal and nonverbal communication relate across cultures and their relevance to the fire service. All firefighters and civilians in the fire service understand the meaning of "linear" or "direct" communication, which translates to getting to the point, being straightforward, and being matter of fact (Figure 4.9). These phrases are no surprise to all in public safety regardless of whether they are fire, police, or medical personnel. The demands and duties of the profession require a context level of direct communication that relies upon these professionals to get to the point and get the job done. Operationally, dispatchers and call takers in 9-1-1 emergency centers use this same context of communication, which is "low context"; however, some citizens or noncitizens who need public safety services are nondirect and "high context" in their communication, and it may take longer to communicate. First responders and 9-1-1 emergency center personnel who have adapted their communication style to high context in the "high-context and low-context" encounter are able to keep the communication fluid and get the job done. Those who have not adapted their communication style in high-context and low-context exchange generally express frustration, and the call for service may take longer or be of a lesser quality. In his book, *Beyond Culture*, Edward T. Hall discusses the significance as well as the typology of high-context cultures being more common in eastern cultures than in western cultures (low context) and in countries with low racial diversity (Hall, 1976).

FIGURE 4.9 ◆ A firefighter directs traffic with a flare at the scene of a motor vehicle accident.

An understanding of the continuum of high-context and low-context communication will contribute to first responders' understanding of direct and indirect communication (or explicit and implicit communication). People or cultural groups who tend toward the high-context end of this communication continuum may exhibit the characteristics in Figure 4.10. People or cultural groups who tend toward the

1. The group's needs are more important than the individual's.

2. The culture relies on the common background to explain situations rather than using words.

3. The culture has a strong sense of tradition and history, and has changed little over time, such as the Maori of New Zealand and the Native Americans.

4. The static culture keeps the high context through future generations.

5. Professional lives and personal lives intertwine.

6. More inclined to ask questions rather than attempt to work out the solution independently.

7. Tendency toward avoiding saying "no" and avoiding conflict.

8. Difficulty in responding to "closed-ended questions," such as "yes" or "no" responses.

9. More comfortable with "open-ended questions," such as narrative responses.

10. Focus on "wide context" of interaction—past events, tone, nonverbal communication, relationship, and status of speakers; good readers of implicit communication.

11. Circular in communication, indirect, subtle, gets to the point slowly, and uses metaphors.

12. Concerned about saving face (theirs and others).

13. Has to adapt or be accommodated when shifting to a low-context culture.

14. Unwritten laws, policies, rules, and procedures.

15. Prefers interdependence and maintains lasting relationships.

FIGURE 4.10 ◆ Higher-Context Communication Tendencies
Source: Shusta, Levine, Wong, Olson, & Harris (2008)

1. The group has a greater diversity of backgrounds, and the individual's needs often conflict with those of the group.

2. Uses more words to explain a position or policy.

3. The value of tradition and historical lessons is often ignored with negative consequences.

4. The culture changes drastically from one generation to the next, such as the United States.

5. Professional lives and personal lives are kept separate.

6. Has a tendency to work the solution out collaboratively if unable to do so independently.

7. At ease with direct communication and responding directly to conflict.

8. "Yes" equals "yes" and "no" equals "no."

9. Comfortable with responding to "closed-ended" questions, such as "yes" and "no" responses.

10. Focus more on words and what is verbalized than what is not communicated implicitly— less focus on tone, nonverbal, relationship, status, past events, implicit communication.

11. Is linear, to the point, and direct; "beating around the bush" is seen as a weakness.

12. Saving face (theirs and others) not as important as saying the truth, even though corporate criminals go against this attribute.

13. Has to adapt or change mind-set when shifting to a high-context culture.

14. Uses written procedures, rules, condified laws, and contracts.

15. Demands independence and goes through many relationships.

FIGURE **4.11** ◆ Lower-Context Communication Tendencies
Source: Shusta, Levine, Wong, Olson, & Harris (2008)

lower-context end of this communication continuum may exhibit the characteristics in Figure 4.11.

As fire service personnel know, rapport building with all citizens is essential in building trust, which is preliminary to citizens opening up. This is especially true when communicating with people who have higher-context communications styles, but the same rapport building is equally true for lower-context communication styles. Typically, these styles are more characteristic of cultural groups coming from Asian and Latin American cultures, and there are gender differences even in lower-context cultures as well (i.e., women tend to have a higher-context style than men). Chapter 5 provides several examples of higher-context communication in the fire service context and makes several recommendations for interaction with citizens and coworkers and for fire service personnel in a public safety approach to specific cultures.

COMMUNICATION PROCESS

Communication is an interactive process between two people that is conversational and participatory. When there are more than two people, communication changes into a group process. Even with formal command systems in place, poor communication is the reason why leadership and productivity fail in organizations. Communication is a skill and like all skills can be learned, improved upon, and mastered in all

FIGURE 4.12 ◆ A fire and rescue paramedic reports to headquarters while speaking into a CB radio receiver in the back of an ambulance.

multicultural interactions (Figure 4.12). Fire service personnel will benefit from using these skills at all times when on the scene of an emergency call; helping citizens; and working with their coworkers and other public safety personnel, during training, and when at the firehouse. There are five elements in the communication process; two involve people and three involve tasks; each component will be discussed fully. Regardless of whether the communication is under operational (formal) conditions or personal (informal) conditions, the communication process is useful and an effective design for both. The five elements of the communication process are speaker, message, listener, paraphrasing, and feedback.

SPEAKER

The speaker is the person who is speaking, and this role can and does frequently change in the communication process of two or more people. *Transmitter* and *sender* are other words that are also familiar in the communication discipline. The speaker is the source of the message and brings his or her perspective, identity, and experiences to the communication transaction (German, Gronbeck, Ehninger, and Monroe, 2006). It is recommended that the speaker consider the ambient noise level of the environment and make any volume adjustments with his or her voice before speaking. This is particularly true if the speaker is unable to find an alternate location when on the scene of an emergency call. If in the firehouse, the speaker and the listener should move to a different room if ambient noise is a problem. Moving to an alternate room applies to other conditions as well if heating, privacy, and the like don't make it conducive to have a conversation.

MESSAGE

The message is composed of the speaker's ideas or information. The features of the message consist of content, structure, and the speaker's verbal and nonverbal communication style. The content is the topic and details that the speaker discusses with the listener. If the topic is substantive and has value, the speaker should indicate this in the message. The structure is the organization of the message and how it is presented. We recommend referring to notes if there is technical information or several points or subpoints to address. We also recommend allowing the listener to paraphrase and provide feedback after each point.

1. Determine the communication conditions, whether formal (operational) or informal (personal).

2. The communication conditions dictate the preparation necessary and the use of verbal and nonverbal communication styles.

3. Avoid using sexist, racial, obscene, and offensive language.

4. Maintain desirable personal and oral hygiene.

5. Respect the listener's personal distance and time.

6. Treat the listener the way you desire to be treated.

7. Avoid distractive mannerisms, such as picking your nose, playing with an object while talking, talking with a full mouth of food, and the like.

8. Avoid using overused words repeatedly (*awesome, fantastic, interesting, check*, etc.).

9. Avoid verbal pauses and nonwords (*okay, right, ya'know, uh, ah, um, err*, etc.).

10. Monitor the listener's body language for visual cues, and clarify question.

11. Actively listen when the listener is paraphrasing and providing feedback.

12. It is okay to disagree, but do not make personal attacks or become hostile.

FIGURE 4.13 ◆ Suggestions to Speakers During the Communication Process
Source: Aaron Olson

Usually, most people do not give any thought to their verbal and nonverbal communication styles, particularly in personal or informal conversations. The degree of self-consciousness increases when the speaker has to provide information in an operational or formal meeting. The speaker should want his or her message to be received in a credible manner, and this is where the content and structure selected are now ready to be delivered by the style of communication chosen. As in any task to be performed, we recommend using voice communication and body language that are commensurate to the speaker's repertoire and suitable for the listener in the meeting. Figure 4.13 provides suggestions to speakers during the communication process, which are obvious but often ignored.

LISTENER

The listener is the person who is receiving the message from the speaker. The listener is a participant in the communication process, and he or she temporarily changes to speaker when asking questions, engaging in a discussion, or paraphrasing what was heard and providing feedback. Most importantly, the listener can possibly assume the role of speaker if another message surfaces in the informal and even formal meeting. The aspects of effective listening are to actively listen, paraphrase, and provide feedback. This is reinforced every day as first responders listen to coworkers on vehicle rescue and extrication, fire suppression, search and rescue, and all emergency medical care calls. In the communication process of this chapter, paraphrasing and providing feedback are addressed separately because of the integral role the speaker has in these two elements to complete the communication process with the listener.

Active listening requires the listener to be 100 percent focused on the message the speaker is giving. This means to eliminate distracters such as reading a magazine or checking e-mail when another person is talking. In active listening, a listener should

1. Attempt to be open and objective as the speaker delivers the message.
2. Show respect to the speaker and treat him or her the way you want to be treated.
3. Respect the speaker's personal space and distance.
4. Actively listen to the speaker by concentrating on the message by paying attention to his or her voice communication and body language.
5. Do not perform other tasks (i.e., reading: newspaper, e-mail, report; listening: to voice mail, headphones, radio; watching: television, DVD, looking out the window; etc.).
6. Do not finish the speaker's sentences if he or she pauses between words.
7. Do not interrupt the speaker.
8. Acknowledge nonverbally (i.e., nodding your head or other acceptable form).
9. Acknowledge verbally (i.e., saying "yes," "uh-huh," etc.).
10. Paraphrase the message, ask questions, and wait for any clarification.
11. Provide objective and constructive feedback.
12. Respond to questions on feedback.

FIGURE 4.14 ◆ Suggestions to Listeners During the Communication Process
Source: Aaron Olson

use all senses to take in and evaluate the meaning of the speaker's information (Farrell and Weaver, 2000). Effective negotiators, facilitators, and counselors practice active listening and acknowledge the speaker's message by giving verbal and nonverbal cues in the communication process. The listener should allow the speaker to finish his or her sentence and not interrupt the speaker. If the listener has a question, the listener should ask it when the speaker has finished his or her sentence. Figure 4.14 provides suggestions to listeners during the communication process.

PARAPHRASING

Paraphrasing is when the speaker has completed the message and now the listener summarizes or synthesizes what he or she just heard from the speaker. The listener uses his or her own words to show comprehension of the message. Repeating back by verbatim is not paraphrasing. As the listener uses his or her own words to paraphrase what the message was, the speaker will be able to assess whether he or she got the point across to the listener. If the point was not clear, the speaker will discern this when listening to the listener's paraphrasing, and the speaker will have to restate or rephrase the point.

FEEDBACK

Feedback is the response from the listener about the ideas and information the speaker presented (Figure 4.15). Listeners get their chance to comment and elaborate what they liked or disliked about the speaker's message. Listeners should be encouraged to be forthright and objective in their responses. To make this work, the attentive speaker must practice active listening and not interrupt the speaking listener. When the speaking listener finishes, the listening speaker may need to paraphrase what he or she heard said and then provide a response in the form of feedback. This process has a life of its own, continuing until there is no longer a need to have feedback.

FIGURE 4.15 ◆ A smiling uniformed African American female firefighter provides her feedback in the communication process.

CROSS-CULTURAL COMMUNICATION ATTEMPTS AND REACTIONS

cross-cultural

Involving or mediating between two cultures.

Sometimes it can be less challenging to interact with people from one's own background than with those from a different group. Communication can be strained and unnatural when there is no apparent common ground. The "people are people everywhere" argument and "just treat everyone with respect" advice both fall short when one learns that there can be basic differences in the areas of behavior and communication across cultures. Some first responders feel that an understanding of **cross-cultural** communication is unnecessary if respect is shown to every person. Yet, if first responders have had limited contact with people from diverse backgrounds, they may inadvertently communicate their lack of familiarity. In the next few sections we exemplify typical ways people attempt to accommodate or react to cultural or racial differences and how they may cover up their discomfort in communication across cultures.

USING LANGUAGE OR LANGUAGE STYLE TO BECOME JUST LIKE ONE OF "THEM"

Nothing is more disingenuous or phony than when a person attempts to imitate another person's accent, dialect, or speech style. Some coworkers may take this as humor, and others may take this as ridicule. Similarly, firefighters attempting to establish rapport with citizens or noncitizens should not pretend to have too much familiarity with another language, culture, or group, or use words selectively to demonstrate how "cool" or "politically correct" they are. Some examples would include using "*señor*" with Spanish-speaking people, calling an African American "my man," saying to a woman "sister," or referring to a Native American as "chief." The people of one cultural background may find themselves in situations in which an entire crowd or family is using a particular dialect or slang. If the first responder lapses into the manner of speaking of the group, he or she will likely appear to be mocking that style. Ultimately, the first responder should be sincere and natural. "Faking" another style of communication can have extremely damaging results.

WALKING ON EGGSHELLS

When in the presence of different cultural backgrounds and when in office or classroom settings, some people find that they have a tendency to work hard not to offend. Consequently, they are unable to be themselves and may do what they would normally not do under those circumstances. Some observers call these attempts being "politically correct" or "walking on eggshells." In a cultural diversity training session for firefighters, several male and female firefighters commented that the communication in the firehouse among coworkers is usually open, relaxed, and direct. When asked the same question about responding to emergency calls, the response was still open and direct, but there was a greater emphasis on being focused and professional because of being on a call. When asked under what conditions they felt uncomfortable, the same group of firefighters responded when a person is confrontational to them, challenging their values, beliefs, and actions in an office or classroom sitting. Many of the same group commented on a sexual harassment class or cultural diversity class in which a "civilian" from human resources was giving the class. The mix of "civilians" described were either white, black, male, or female. The underlining message was "you are put on the defensive and you are guarded in your words."

In another interview, a male firefighter conveyed he was very cautious in his words to a subordinate female firefighter whom he described as a loner in the firehouse and very militant in her political views. When asked if they were still able to function as a company, the lieutenant said yes, but communication was strained in the firehouse. The lieutenant related he is using leadership strategies to improve the situation and is optimistic the situation will improve. It would be naïve for the authors of this textbook to say that there will never be moments when communication is not strained or uncomfortable. It is not within the scope of this discussion to analyze this in cross-cultural encounters. A person must attempt to recognize his or her tendencies to reach the goal of communicating in a sincere and authentic manner with people of all backgrounds.

"SOME OF MY BEST FRIENDS ARE . . ."

In an attempt to show how tolerant and experienced they are with members of different ethnic, racial, or other groups, many people often feel the need to demonstrate their tolerance strongly by making such comments as "I'm not prejudiced" or "I have friends who are . . . ," or "I know people who are [for example] gay, disabled, Arab/Jewish/Asian/African American." Usually the intention may be to break the ice and establish rapport, but these types of statements often sound patronizing. To a member of a culturally or racially different group, these types of comments come across as extremely naïve. In fact, many people would consider such comments as being counterproductive and the speaker as being prejudiced. People who are of the group being mentioned usually question a need to make a reference to others of the same background when there is no context for doing so. It comes across being too "politically correct" and disingenuous. It is similar to someone walking up to you and saying, "I met a firefighter twenty years ago," when there is no context for doing so. The person saying it means well, but it also shows the person may know little about firefighters if it was 20 years ago.

"YOU PEOPLE," OR THE "WE–THEY" DISTINCTION

Some may say, "I'd like to get to know you people better," or "You people have made some amazing contributions." The usage of "you people" may be another signal of prejudice or divisiveness in one's mind. When someone decides that a particular group is unlike his or her group (i.e., not part of "my people"), that person makes a simplistic division of all people into two groups: "we" and "they." Often accompanying this division is the attribution of positive traits to "us" and negative traits to "them." Members of the "other group" (the out-group) are described in negative and stereotypical terms (e.g., "They are lazy," "They are criminals," or "They are aggressive") rather than neutral terms that describe cultural or ethnic generalities (e.g., "They have a tradition of valuing education" or "They have a communication style that is more formal than that of most Americans"). The phenomenon of stereotyping makes it very difficult for people to communicate with each other effectively because they do not perceive others accurately. By attributing negative qualities to another group, a person creates myths about the superiority of his or her own group. Cultural and racial put-downs are often attempts to make people feel better about themselves.

RESPONDING TO ACCUSATIONS OF RACIAL AND OTHER CULTURAL FORMS OF DISCRIMINATION

Due to the mission and purpose of the fire service, it is not in the same business as law enforcement for enforcing traffic and criminal laws. Firefighters do not make traffic stops, write traffic tickets, or arrest people for crimes. Fire departments do not racially profile motorists, but accusations of racial, ethnic and sexual discrimination and harassment have been reported, investigated, sustained, not sustained or unfounded. The scope of this section is not to probe into the investigative components of these types of allegations but how to respond verbally to these allegations when they are directed at the firefighter, the lieutenant, the captain, and leading up to the fire executive by a citizen.

Firefighters have contact with citizens on all types of calls; this includes but is not limited to fire suppression, emergency medical care, vehicle rescue and extrication, and assisting on traffic control. Due to this volume of calls, firefighters will deal with irate citizens if the police or ambulance personnel don't first. The complaints may range from the motorist's vehicle battery terminals being cut to being delayed because of only one lane being open on the highway. When these or other types of complaints are based on false racial or other cultural reasons, it is important that the first responder remain professional and not escalate a potential conflict or create a confrontation even when the citizen says, "You did this to me because I am [black, Asian, Hispanic, gay, Russian, etc.]." Firefighters should not only try to communicate their professionalism, both verbally and nonverbally, but also try to strengthen their self-control. The best way to deal with these types of remarks from citizens is to work on your own reactions and stress level. People sometimes react to first responders as symbols, using the first responders to vent their frustration.

Let's assume that the first responder's actions are justified at the scene, and there is no racial or cultural motivation. The firefighter needs to respond to the allegation because if he or she remains silent, the citizen will take the silence as a sign of guilt and could become more agitated. Dr. George Thompson—founder and president of the Verbal Judo Institute, Inc.—believes that in these situations, "verbal deflectors" (which are part of his Verbal Judo training course) can be used to deflect the words uttered by the citizen. Dr. Thompson recommends responses such as "I appreciate that, but . . . [I cut your battery terminal because of the threat of combustion]" or "I hear what you are saying, but . . . [traffic is being stopped because of the downed electrical wires]." The characteristics of verbal deflectors are that they (1) are readily available to the lips, (2) are nonjudgmental, and (3) can be said quickly (Thompson, 2004).

COMMUNICATION CONSIDERATIONS—POST SEPTEMBER 11, 2001

Since 9/11, and particularly in the first few months following the terrorist acts (and continuing), there has been what Lobna Ismail, an Arab American cross-cultural specialist, characterizes as collateral damage to the entire Arab American community. Arab Americans may hesitate to call 9-1-1 (police) when they are victims of crimes out of fear of being treated as suspects rather than victims of a crime (Shusta, Levine, Wong, Olson, and Harris, 2008). For this reason, it is especially important to consider the importance of building trust, rapport, and relationship with Arab Americans and people from the Middle East. As for communication with Arab Americans (and this applies to communication with other cultural groups as well), first responders should strive to be culturally competent and responsive at all times. The end result will be effective communication and professional fire service.

INTERVIEWING AND DATA-GATHERING SKILLS

Interviewing and data-gathering skills form the basic techniques for communication and intervention work with multicultural populations (Figure 4.16). For the fire service first responder, the key issues in any interviewing and data-gathering situations are as follows:

- Establishing interpersonal relationships with the parties involved to gain trust and rapport for continual work.
- Bringing structure and control to the immediate situation.
- Gaining information about the problems and situations that require the presence of the fire service first responder.
- Giving information about the workings of the fire service guidelines, resources, and assistance available.

FIGURE 4.16 ◆ 9-1-1 emergency telephone operators and dispatchers at work before rows of computer terminals facing a video screen in a dimly lit room.

1. Be knowledgeable about who is likely to have information. Ask questions to identify the head of family or respected community leaders.

2. Consider that some cultural groups have more of a need than others for rapport and trust building before they are willing to share information. Do not consider the time it takes to establish rapport a waste of time. For some, this may be a necessary step.

3. Provide background and context for your questions, information, and requests. Cultural minorities differ in their need for "contextual information" (i.e., background information) before getting down to the issues or business at hand. Remain patient with those who want to go into more detail than you think is necessary.

4. Expect answers to be formulated and expressed in culturally different ways. Some people tend to be low context and others are high context. There are individual differences in ways of presenting information, as well as cultural differences.

5. It is important to speak in simple terms, bu do not make the mistake of talking down to people. Remember, a person's comprehension skills are usually better than his or her speaking skills.

6. "Yes" does not always mean "yes"; many people use the word "yes" as an acknowledgment or similar to a nod of the head.

7. Remember that maintaining a good rapport is just as important as coming to the point and getting work done quickly. Slow down your pace, if needed.

8. Silence is a form of communication; do not interrupt it. Give people time to express themselves by respecting their silence.

FIGURE 4.17 ◆ Interviewing and Data Gathering in a Multicultural Context
Source: Shusta, Levine, Wong, Olson, & Harris (2008)

- Providing action and assistance, as needed
- Bolstering and supporting the different parties' abilities and skills to solve current and future fire safety problems on their own

Listed in Figure 4.17 are helpful guidelines for providing and receiving better information in a multicultural context.

In the area of data gathering and interviewing, the first responder at the scene of an emergency medical call cannot assume that his or her key motivators and values are the same as those of the other parties involved. Recognizing such differences in motivation and values will result in greater effectiveness. The values of saving face and preserving one's own honor as well as the honor of one's family are extremely strong motivators and values for many people from Asian, Latin American, and Mediterranean cultures. For example, an incident occurred in Oakland, California, in which an Asian rape victim had her niece translate. The values of honor and face-saving prevented the aunt from telling her niece and the police all the details of the crime of which she was the victim. It was later, through a second translator who was a nonfamily member, that the full details were disclosed. Precious time had been lost, but the victim explained that she could not have revealed the true story in front of her niece because she would have shamed her family. The reality is fire service personnel are on the scene with the police on all types of emergency medical calls and usually are on the scene first. As the first responders on vehicle crashes and crimes requiring medical attention, firefighters are put in a position in which they need the same awareness skills as the police.

MALE–FEMALE COMMUNICATION IN THE FIRE SERVICE

With the changing workforce, including increasing numbers of women in traditionally male professions, many new challenges in the area of male–female communication are presenting themselves. Within the fire service, in particular, a strong camaraderie characterizes the relationships mainly among the male members of fire departments, although, in some cases, women are of this camaraderie too. Women allowed into what has been termed the "brotherhood" have generally had to become "one of the guys" to gain acceptance into a historically male-dominated profession. This observation has been noted in law enforcement, in corrections, and even in the military.

Camaraderie results when a group is united because of a common goal or purpose (Figure 4.18). The glue cementing camaraderie is the easy communication among its members. The extracurricular interest of the members of the group, the topics selected for conversation, and the jokes that people tell all contribute to the cohesion or tightness of fire service members. In some fire departments, women find that they or other women are the object of jokes about sexual topics or that there are simply numerous references to sex. Historically, because the majority of fire service agencies have consisted mostly of men, they have not had to consider the inclusion of women on an equal basis and have not had to examine their own communication with each other.

Many women who have entered into the fire service feel that they must tolerate certain behavior to be accepted. Recently in a training session we conducted with new fire fighters, several of the male and female students commented they do joke and kid around with the opposite gender because they consider themselves "all" firefighters. One woman firefighter, in particular, was vocal in commenting she does not want to be treated any differently in speech or in actions from her male counterparts just because she is a woman. As a group though, all acknowledged that sexual innuendoes as well as offensive and sexist language can contribute to a hostile working environment. Hence the questions often asked by men and women: "Where do I draw the line? When does a comment become harassment?" In terms of the commonly accepted definition of sexual harassment, when

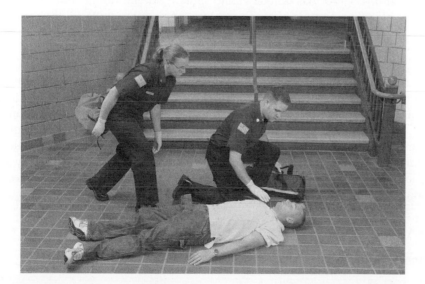

FIGURE 4.18 ◆ Emergency Medical Technicians arrive at the scene of an accident with a victim lying on his back.

1. Use terms that are inclusive rather than exclusive.
   - ◆ Examples: *firefighter* instead of *fireman* and *commendations* instead of *attaboy*
2. Avoid using terms or words that many women feel diminish their professional status.
   - ◆ Examples: *chick* and *babe*
3. Avoid using terms or words that devalue groups of women or stereotype them.
   - ◆ Examples: *dyke* and *bitch*
4. Avoid sexist jokes, even if you think they are not offensive. These are similar to racist jokes.
5. Avoid using terms that negatively spotlight or set women apart from men.
   - ◆ Example: "For a women firefighter, she did a good job." This implies that she is an exception rather than the rule. This can apply to other cultural groups as well.

FIGURE 4.19 ◆ Inclusive Workplace Communication *Source:* Herbert Wong and Aaron Olson

the perpetrator is made aware that his or her comments are uninvited and unwelcome, then that person must be reasonable enough to stop making them. It is not within the scope of this chapter to detail sexual harassment and all its legal implications; however, it should be noted that everyone has his or her own limits. What is harassment to one may be perceived differently by another. When communicating across genders, each party must be sensitive to what the other party considers acceptable or insulting (see Figure 4.19). It is also the responsibility of the individual who has been offended, whether male or female, to make it clear that certain types of remarks are offensive.

SUMMARY

In communication for firefighters, the first responder's own filters and perceptions influence the assessment of each situation and the reactions the firefighter chooses to exhibit. Each firefighter has unique "blind spots" and emotional "buttons" that may negatively affect communication. To explore one's own skills in this regard, the reader is urged to undertake a self-evaluation by filling out the communications inventory in Appendix D.

Firefighters must keep in mind that rapport building is related to trust for many persons of different backgrounds. The more trust firefighters earn with members of multicultural communities, the more helpful these group members will be when fire service agencies need cooperation and assistance. To improve communication across cultures, it is essential that people in the fire service understand the aspects of verbal and nonverbal communication, high-context and low-context communication, and the communication process as it relates to multicultural groups. Further awareness will benefit the special challenges facing men and women in the profession for building upon their multicultural communication skills.

Discussion Questions and Issues

1. **Different Country.** Your 70-year-old grandmother is traveling along with you as you are driving a rental car on a country road in Mexico when you experience a flat right rear tire. You park your car off on the shoulder of the highway and

discover the spare tire is flat also. A Mexican family with their father drives up behind your car and offers help, but the family cannot speak English, and you and your grandmother are unable to speak Spanish. Discuss how you will problem solve your situation.

2. **Firefighters at Crash Scene.** You and your engine company are at the scene of a two-vehicle minor-injury crash. The city police officer on the scene has to interview several witnesses and asks whether you can help him in overseeing the exchange of vehicle registration, driver's license, and insurance card information between both male adult drivers. One driver speaks only Russian and the other male adult driver speaks only English. Discuss how you will problem solve your situation.

3. **Proxemics Comfort Zone.** Identify and describe the four zones of proxemics addressed by Edward T. Hall. Discuss which zones you usually feel comfortable in when interacting with other people. Also, under what social circumstances would you change that distance?

4. **High-Context and Low-Context Communication.** Identify whether you are high context or low context in communication. Provide examples of why you assess yourself as one or the other. Discuss what adaptations a person must make to communicate with a person who is of a different context.

5. **Cultural Observations.** Make a list of your observations for each of the cultural groups with which you have had a substantial amount of contact. After you make your list, try to find someone from that culture with whom you can discuss your observations.

   a. Display of emotions and expressions of feelings
   b. Communication style: loud, soft, direct, indirect
   c. Expressions of appreciation: conventions of courtesy (i.e., forms of politeness)
   d. Need (or lack thereof) for privacy
   e. Gestures, facial expressions, and body language
   f. Eye contact
   g. Touching
   h. Interpersonal space (conversational distance)
   i. Taboo topics in conversation
   j. Response to authority

6. **Accusations of Racial and Other Cultural Forms of Discrimination.** Discuss how you as lieutenant from your engine company would respond to a male adult from a cultural group who makes the accusation "You are a racist because you cut my battery terminals." The context is that his 2006 Tacoma 4 × 4 struck a utility pole head-on and that the gas tank from his vehicle was ruptured and was leaking gas on the city street.

Website Resources

Visit these websites for additional information about communication and related issues:

Association of Public Safety Communications Officials, International, Inc. (APCO International): http://www.apcointl.com
The Association of Public Safety Communications Officials (APCO) International, Inc. is the world's oldest and largest not-for-profit professional organization dedicated to the enhancement of public safety communications. With more than 16,000 members around the world, APCO International exists to serve the people who manage, operate, maintain, and supply the communications systems used by police, fire, and emergency medical dispatch agencies throughout the world. APCO has APCO Institute, which is a not-for-profit educational institute that serves the unique needs of the public safety communications industry.

Association for Women in Communications (AWC): http://www.awcdc.net
The Association for Women in Communications (AWC) Washington, DC, chapter is a professional organization that champions the advancement of women across all

communications disciplines by recognizing excellence, promoting leadership, and positioning its members at the forefront of the evolving communications era. Founded in 1909, AWC has chapters throughout the United States.

Immigrant Refugee Community Organization (IRCO): http://www.irco.org
The mission of the Immigrant and Refugee Community Organization (IRCO) is to assist refugees, immigrants, and multiethnic communities to develop self-sufficiency and cultural awareness while affirming and preserving each culture within an ever-changing environment. IRCO was formed in 1984 from the merger of the Indo-Chinese Cultural and Service Center (founded in 1976) and the SE Asian Refugee Federation, founded in 1980. Originally called the International Refugee Center of Oregon, it became IRCO in 2000. IRCO works with government agencies, business, and nonprofit organizations to accomplish its mission of service. International Language Bank (ILB) translates/interprets in over 50 languages, and IRCO staff reflects the clients it serves, representing over 40 ethnicities and speaking at least 39 languages.

International Association of Fire Chiefs (IAFC): http://www.iafc.org
It is an online resource for fire service executive and leadership issues and publications developed by the IAFC.

International Association of Fire Fighters (IAFF): http://www.iaff.org
It is a website for information about the range of issues confronting firefighters and their departments.

National Communication Association (NCA): http://www.natcom.org
The National Communication Association (NCA) is a nonprofit organization of approximately 7,700 educators, practitioners, and students who work and reside in every state and more than 20 foreign countries. The purpose of the association is to promote study, criticism, research, teaching, and application of the artistic, humanistic, and scientific principles of communication. Founded in 1914, NCA is the oldest and largest national organization to promote communication scholarship and education. Its national headquarters are based out of Washington, DC.

National Fire Academy: http://www.usfa.fema.gov/nfa
It is a website for information about training, careers, jobs, publications, and research for the fire service.

Spanish for Police and Fire Personnel: http://www.spanishforpoliceandfire.com
Spanish for Police and Fire Personnel is a website that serves administrators, instructors, and students who need "workplace Spanish" for law enforcement and emergency responders. It is owned and operated by David B. Dees, of Dees Multilingual Services. Workshops, instructor manuals, and audiocassette tapes/CDs are available to purchase for law enforcement and emergency responders.

Teaching Tolerance: http://www.tolerance.org
Founded in 1991 by the Southern Poverty Law Center, Teaching Tolerance provides educators with free or low cost materials that promote respect for differences and appreciation of diversity in the classroom and beyond.

U.S. Fire Administration: http://www.usfa.fema.gov
A source of fire services resources, programs, and statistics collected and published by the Federal Emergency Management Administration (FEMA).

Verbal Judo Institute: http://www.verbaljudo.com
This website centers on Verbal Judo, a tactical judo, a form of communication used as a tool to generate voluntary compliance from citizens under difficult situations. It lists various types of Verbal Judo courses available as well as books and articles on the subject.

References

Brinkman, R., and R. Kirschner. (2002) *Dealing with People You Can't Stand: How to Bring Out the In People at Their Worst*. New York: McGraw-Hill.

Dees, D. D. (2006). *Quick Spanish for Emergency Responders*. New York: McGraw-Hill.
Farrell, J. D., and R. G. Weaver. (2000). *A Practical Guide to Facilitation: A Self*

Study Guide. Amherst, MA: Human Resource Development Press.

German, K. M., B. E. Gronbeck, D. Ehninger, and A. H. Monroe. (2006). *Principles of Public Speaking,* 16th ed. New York: Pearson, Allyn and Bacon.

Hall, E T. (1966). *The Hidden Dimension.* Garden City, NY: Doubleday.

Hall, E. T. (1976). *Beyond Culture.* Garden City, NY: Anchor Press.

Levine, D., and M. Adelman. (1992). *Beyond Language: Cross-Cultural Communication.* Englewood Cliffs, NJ: Prentice Hall.

Mehrabian, A. (1971). *Silent Messages.* Belmont, CA: Wadsworth.

Mehrabian, A. (2005). *"Silent Messages"—A Wealth of Information About Nonverbal Communication (Body Language).* Retrieved May 28, 2006, from http://www.kaaj.com/psych/smorder.html

McKerrow, R., B. Gronbeck, D. Ehninger, and A. Monroe. (2000). *Principles and Types of Speech Communication,* 14th ed. New York: Longman.

Olson, A. T. (2007). Professor of criminal justice at Portland Community College, organizational consultant, and retired Oregon State Police patrol supervisor, Portland, Oregon, personal communication, March 2007.

Salka, J. (2005, October). "Roundtable: Opinions from Around the County," *Fire Engineering Magazine,* 158(10): 32.

Shusta, R. M., D. R. Levine, H. Z. Wong, A. T. Olson, and P. R. Harris. (2008). *Multicultural Law Enforcement: Strategies for Peacekeeping in a Diverse Society,* 4th ed. Upper Saddle River, NJ: Pearson Prentice Hall.

Thompson, G. (2004). *Verbal Judo: The Gentle Art of Persuasion.* Yucca Valley, CA: Verbal Judo Institute, Inc.

A Public Safety Approach to Specific Cultures

Key Terms

assimilation, p. 138
consent decree, p. 149
dominant culture, p. 145

émigré, p. 140
macro culture/majority or
dominant group, p. 179

micro culture or
minority, p. 1797

Overview

Chapter 5 presents information on African American, Asian/Pacific, Latino/Hispanic, Middle Eastern, and Native American cultural backgrounds (Figure 5.1) with regard to the needs of fire service officers and first responder representatives. We have selected these groups, as opposed to others, for one or more of the following reasons: (1) the group is a relatively large ethnic or racial group in the United States; (2) the traditional culture of the group differs widely from that of mainstream American culture; and/or (3) typically or historically there have been problems between the particular group and fire service departments and officers. Although European Americans would not be considered a minority group in the United States, we have included a section on recent immigrant and refugee groups from European countries because these groups have similar community issues and concerns with the fire service and other public safety agencies as do many of the minority communities in the United States.

In these culture-specific sections, general information is presented in the areas of historical background, demographics, and diversity for each cultural group. Following the introductory information, we present specific details relevant to fire service in the following areas: group identification terms, offensive labels, stereotypes, family structure, and communication styles (both verbal and nonverbal). Each section ends with key concerns for fire officers related to the particular cultural group, a summary of recommendations for fire service officials, and resources for additional information about the cultural group.

FIGURE 5.1 ◆ Multicultural crowds throng an urban fair.

Objectives

After completing this chapter, participants should be able to:

- Summarize the historical background, demographics, and diversity within the African American, Asian/Pacific, Latino/Hispanic, Middle Eastern, and Native American communities in the United States.
- Discuss the implications of communication styles, group identification terms, myths and stereotypes, and family structure of African American, Asian/Pacific, Latino/Hispanic, Middle Eastern, and Native American communities for the fire service.
- Describe the impact of the extended family and community, gender roles, generational differences, and adolescent and youth issues for fire service contact with group members.
- Highlight some of the challenges in providing fire service to recent immigrants and refugees from European countries as similar to those for the African American, Asian/Pacific, Latino/Hispanic, Middle Eastern, and Native American communities in the United States.
- Discuss key fire service concerns and skills, resources, and practices for addressing some of these concerns.

LEARNING TASKS

Knowing the historical background and current demographic information for the ethnic, racial, and cultural groups in your service area is critical toward understanding the multicultural fire service challenges of your area. Talk with your agency and find out how it is involved in the fire service to these diverse communities. Be able to:

- Find out the historical background, demographics, and diversity of the multicultural groups that are most important within your area.
- Describe the group identification terms, myths, and stereotypes for the multicultural groups that you might encounter within your area.
- Identify the communication style and family structure for the multicultural groups that you might encounter within your area.
- Use the seven-part typology model to determine the motivation bases for the multicultural groups that you might encounter within your area.
- Determine the existing challenges that the agency has in dealing with the extended family and gender roles in multicultural communities.
- Describe immigrant and refugee groups that are within your area.
- Find out the differences in generational, youth, and adolescent issues within your area.
- Ascertain the diversity skills, resources, and practices for addressing community fire service in your county and city fire departments.

PERCEPTIONS

The following quotes draw attention to the public safety and ethnic/cultural community issues involved for firefighters and first responders within diverse communities:

The 1992 Los Angeles Civil Disturbance Fires that began on April 29, 1992 were, without question, the most devastating and challenging in the history of Los Angeles and the nation [Figure 5.2]. Given the specific conditions at the fire and rescue scenes, how our firefighters managed to limit the fires to the structures known to date is beyond comprehension. I have been informed of

FIGURE 5.2 ◆ Firefighter watches a building burn during the 1992 Los Angeles riots following acquittal of white cops in videotaped beating of black motorist Rodney King.

FIGURE 5.3 ◆ Chinese resident watched a debris-cluttered street in Chinatown amid smoldering, crumbling ruins in the aftermath of the San Francisco earthquake of 1906 with suffocating fires blackening the skies overhead.

numerous testimonies of dramatic and heroic actions committed by or witnessed by our firefighters and paramedics. It is a miracle that only thirty firefighters and paramedics were injured, given the tremendous potential (Statement from Donald O. Manning, Chief Engineer and General Manager, Los Angeles City Fire Department, about the riots following the Rodney King verdict in Los Angeles).

In some respects, the Chinese were even more vulnerable than other San Franciscans [Figure 5.3]. Because of "heathen Chinese" stereotypes and the prevailing racism of the time, they would get little help and less sympathy from hard-pressed city authorities. Traumatic memories of white persecution ran deep, and most Chinese were afraid of seeking food, medical attention or shelter from city aid stations. Soldiers evacuated Chinatown as they did other parts of the city, and the dynamiting began. Lieutenant Freeman, so heroic in other respects, shared the common prejudices of the time. Some Chinese remained behind and the naval officer noted that "at least 20 Chinese, opium fiends and drunks, were blown up by dynamite." Another report laconically mentioned that several mangled Chinese bodies could be seen in the ruins, and that "in at least one building 5 or 6 bodies were thrown 50 feet into the air and back into the flames" ("The Great 1906 San Francisco Earthquake and Fire," 2006).

Rosario Franco and many in his family have fought wildfires across the West for years [Figure 5.4]. His brother and cousin are both firefighters. His father is a contractor for fire crews. Across the country, a growing number of Hispanics are taking on the hot, dangerous and dirty work because the demand is high in season and it usually pays better than farm work. Many—nobody knows how many—are undocumented, a problem Franco claims

FIGURE 5.4 ◆ Hispanic and other firefighters line up a hillside to stage a backfire.

does not concern him. "I think our crews are legal," he said at his home in this Willamette Valley town. "My job is to do my job and that's what I do." Nevertheless, it is clear Hispanics dominate many wildland fire crews ("Documentation, Language Issues Arise for Hispanic Firefighters," 2006, p. B5).

INTRODUCTION

As we look at African American, Asian/Pacific, Latino/Hispanic, Middle Eastern, Native American, and European American individuals, families, and communities, we have developed a seven-part typology that will be useful in understanding and in summarizing some of the differences between individuals within these groups (see Figure 5.5).

From the training and consulting work of one of the coauthors (Wong) with fire service, emergency medical service, and law enforcement, our typology suggests that as fire service and public safety organizations prepare and train their personnel to work with African American, Asian/Pacific, Latino/Hispanic, Middle Eastern, Native American, and European American communities, a focus upon the key differences within each of the typological groups would be most effective. Provided in Figure 5.6 are the *motivating factors* for each of the seven-part typology. These seven motivating factors provide a convenient framework for discussing how the motivational components within each of the groupings may affect the way people respond in a fire service situation.

We will now provide some examples of the motivating perspectives of individuals from the African American, Asian/Pacific, Latino/Hispanic, Middle Eastern, Native American, and European American communities using this seven-part typology with respect to possible interactions with fire service personnel and agencies.

Type I	Recently arrived immigrant or refugee (less than five years in the United States, with major life experiences *outside* of the United States)
Type II	Immigrant or refugee (five or more years in the United States, with major life experiences *outside* of the United States)
Type III	Second generation (offspring of immigrant or refugee *within* the United States)
Type IV	Immigrant (major life experiences *within* the United States)
Type V	Third or later generations *within* the United States
Type VI	Visiting national (anticipates return to home country, to include students, visitors, and tourists with major life experiences *outside* of the United States)
Type VII	Global national (global workplace and residency with major life experiences *outside* of the United States)

FIGURE 5.5 ◆ Typology of African American, Asian/Pacific, Latino/Hispanic, Middle Eastern, Native American, and European American Communities. *Source:* Herbert Wong

Examples of individuals who might be of the *Type I* recently arrived immigrant or refugee group (Figure 5.7) would include the following individuals recently from:

- Haiti, the Congo, the Caribbean (African Americans or Black Americans)
- China, Vietnam, and Korea (Asian/Pacific Americans)
- Brazil, Mexico, and Costa Rica (Latino/Hispanic Americans)
- Egypt, Saudi Arabia, and Israel (Middle Eastern Americans)
- Aleutian Islands (Native Americans)
- Bosnia, Serbia, and Ukraine (European Americans)

Motivating Prespective	*Typology Group*	
Surviving	Type I:	Recently arrived immigrant or refugee (less than five years in the United States, with major life experiences *outside* of the United States)
Preserving	Type II:	Immigrant or refugee (five or more years in the United States, with major life experiences *outside* of the United States)
Adjusting	Type III:	Second generation (offspring of immigrant or refugee *within* the United States)
Changing	Type IV:	Immigrant (major life experiences *within* the United States)
Choosing	Type V:	Third or later generations *within* the United States
Maintaining	Type VI:	Visiting national (anticipates return to a home country, to include students, visitors, and tourists with major life experiences *outside* of the United States)
Expanding	Type VII:	Global national (global workplace and residency with major life experiences *outside* of the United States)

FIGURE 5.6 ◆ Key Motivating Perspectives in Understanding Specific Cultural Groups. *Source:* Herbert Wong

FIGURE 5.7 ◆ Red Cross worker interviewing a Vietnamese refugee following the fall of Saigon. South Vietnamese refugees sought refuge from the invading force from the North in April 1975. American involvement in the Vietnam War came to an end when troops from communist North Vietnam invaded Saigon, the capital of the Republic of Vietnam, resulting in the largest influx of refugees ever to the United States.

The key motivating perspective for individuals in the Type I category is survival. Many members from this category may remember that "people in uniform" such as fire service and public safety officers in their country of origin were corrupt, aligned with a repressive government and the military, and subjected to bribes by those who were more affluent. All activities tend to be guided by this perspective to survive, to get through. This perspective also makes sense in terms of the traumatic ordeals faced by refugees in their journeys to the United States (e.g., political refugees from Afghanistan, immigrants from war-torn areas of the Middle East). Encounters with public safety personnel by these people usually involve saying and doing anything to discontinue the contact because of possible fears of personal harm (e.g., interactions might entail communications of not speaking English, not having any problems, "Yes, I will cooperate!"). As such, fire service officers may find individuals in the Type I category avoiding contact and doing everything to discontinue contact.

For African American, Asian/Pacific, Latino/Hispanic, Middle Eastern, Native American, and European American immigrants or refugees who have been in the United States five or more years (with major life experiences *outside* of the United States—*Type II*), understanding their behaviors should focus on their motivation toward preserving their former homeland or national culture. Because the majority of their life experiences occurred outside of the United States, members are trying to preserve much of the values and traditions of their homeland cultures as it was alive and operating. Much intergenerational conflict between grandparents or parents and youths occurs within this group. Members are inclined to keep to their ethnic communities (e.g., black enclave communities, Little Havanas (Figure 5.8), Little Saigons, Manilatowns, Little Moscows) and have as little to do with fire service or public safety as possible.

Second-generation individuals who are offspring of immigrants or refugees *within* the United States (*Type III*) tend to be those whom we picture as "assimilated" Americans. Those of the second generation work very hard at blending into the American cultural milieu, adjusting, and changing to become a part of mainstream America. Oftentimes, the expectations of second-generation parents are high; parents

FIGURE 5.8 ◆ A Cuban American man talks to a woman while sitting on a motorcycle parked outside a record store in Little Havana in Miami, Florida.

sacrifice so that their offspring will "make it" in their lifetimes. Members may interact primarily with whites and not with members from within their ethnic/cultural groups and take on many of the values and norms of the mainstream society. This group may be considered "marginal" by some in that, try as these individuals may to become like the mainstream ("become white"), they will not be able to change many of the physical features associated with their ethnic groupings (e.g., darker skin, facial features, hair color, etc.) and may be unaware of many of their cultural behaviors that tag them as different from the mainstream.

Uncomplimentary terms heard by one of the coauthors (Wong) in his training workshops for fire service and other public safety personnel to describe Type III members assimilated into the white culture include the following: "Oreos" (for African Americans who are black on the outside but white on the inside); "bananas" (for Asian/Pacific Americans who are yellow on the outside and white on the inside); "coconuts" (for Latino/Hispanic Americans who are brown on the outside and white on the inside); and "apples" (for Native Americans who are red on the outside and white on the inside).

Many from the Type III group try to minimize their contact with public safety personnel and agencies primarily because of the immigration and other experiences relayed to them by their parents' generation (e.g., Japanese Americans' experiences in the internment camps in the 1940s). Individuals of the second generation born in the United States may also have had parents who entered the United States illegally or may have been undocumented aliens. Fears of disclosure of such illegal entries have prevented many from the second-generation groups from cooperating with fire service and with other human and social service agencies and programs. With the emphasis on homeland security in the United States, many from this group have taken extra efforts not to be misidentified with respect to their ethnicity (e.g., South Asians with darker skin tones being mistaken for Middle Easterners) or stereotyped because of their religion (e.g., Middle Eastern Americans who are moderate Muslims, not fundamentalists).

The immigrant (*Type IV*), whose major life experience is within the United States, focuses much of his or her energies on changes (through **assimilation** or acculturation) that have to be made in order to succeed. Although these individuals have tended to continue to value the cultural and ethnic elements of their former homeland, most know that changes are necessary to facilitate assimilation. Members of this group reflect the socioeconomic standings of the respective waves within which they entered into the United States. For example, Middle Eastern Americans began to arrive in the late nineteenth century, with the largest influx between 1900 and 1924, to work in the auto industry. Today, the Detroit–Dearborn area has the largest Arab community in the United States, and many individuals from the Middle Eastern community continue to work in blue-collar positions within the auto industry.

As another example, many Asian/Pacific American individuals entered the United States under the "fifth preference" as professional skilled workers in the 1970s and 1980s. Type IV Asian/Pacific immigrants who are designated as entering the United States under the fifth preference are those who checked the fifth category on the Immigration and Nationalization Service form. This category indicates that the reason for immigration into the United States was because the person had a professional skill in short supply in the United States, and it would be in the best interest of the United States to allow that person to enter (Figure 5.9). This group consists of educated, professional individuals (e.g., the largest numbers of foreign-trained medical doctors and psychiatrists in the United States are from India, the Philippines, and Korea [President's Commission on Mental Health, 1978]). For fire service officers, it

assimilation

The processes by which ethnic groups that have emigrated to another society begin to lose their separate identify and culture, becoming absorbed into the larger community.

FIGURE 5.9 ◆ Asian Indian engineers and others conduct a teleconference meeting with project managers in India at the company's offices in California.

is critical to understand the differences among the immigrant groups (i.e., not to confuse the professional immigrant with immigrants in other socioeconomic groupings).

The third or later generations (*Type V*) category includes individuals who are more able to choose which aspects of the old culture to keep and which of the new culture to accept. The focus is on choosing activities, values, norms, and lifestyles that blend the best of their former homeland culture with American cultures. The importance of being bicultural is a unique aspect of this group. Many may no longer have as much skill with their native language and may rely on English as their primary or only language (thus an individual can be bicultural and not bilingual). Contact by members of this group with fire service personnel may not be any different from contact by any other Americans. This group would include nonimmigrants such as Native Americans who may be choosing between life on the reservation or within the outer communities, the cities, and the suburbs.

For the last two categories, visiting nationals (*Type VI*) and global nationals (*Type VII*), we make a key distinction between those who plan to return to their own country following their stay in the United States (Type VI) and those whose work is truly global, in that individuals may have several residences in different parts of the world (Type VII). Those in the Type VI category (on a work or study assignment in the United States that may last three to five years) maintain their homeland cultural orientation and experiences knowing that when the work assignment is over, they will go back to their home countries again. Within these two groups might be students from different countries and/or global nationals who are in the United States on professional assignments (Figure 5.10).

Fire service and public safety personnel who can understand and utilize the motivating perspectives of this seven-part typology will be better able to communicate and work with African American, Asian/Pacific, Latino/Hispanic, Middle Eastern, Native American, and European American communities in fire service.

FIGURE 5.10 ◆ Employees who are part of the international workforce walk through the terminal at O'Hare International Airport in Chicago, Illinois.

IMMIGRANTS AND REFUGEES

Because a large proportion of the individuals within the diverse communities with whom fire service officers may encounter are born outside the United States, it is important to understand some of the key differences that relate to immigration status. One such difference is that of citizens who are considered refugees and those who are considered immigrants at the time that they enter the United States. Some of the between-group hostilities (within the Latino/Hispanic community, Asian/Pacific community, and among other ethnic minority communities) have been a result of not understanding how refugee status differs from immigrant status in the United States.

émigré

An individual forced, usually by political circumstances, to move from his or native country and who deliberately resides as a foreigner in the host country.

Refugees are sponsored into the United States under the authority of the U.S. government. Although many ethnic groups have come in under the sponsorship of the federal government with refugee or **émigré** status, the largest numbers have come from Southeast Asia as a result of the past upheaval brought on by the Vietnam War. Refugees, because of their sponsorship by the government, are expected to utilize public support services fully (welfare, English as a second language [ESL] programs, educational tuition, job training programs, and case management). It is part of being a "good refugee" to participate fully, and often case managers are assigned to refugee families to ensure that family members are fully utilizing all of the services available. Such participation in public programs may also create dependency and learned helplessness, which can result from having others help or interfere with what many could have done for themselves.

Immigrants enter into the United States under the direct sponsorship of the individual's families. The federal government establishes that immigrants are allowed to enter the United States only if their families can completely support or establish work for the individual. In fact, one criterion for being able to attain permanent residence status (a "green card") is that the immigrant will not become a burden to the government, which means that participation in any publicly funded program may jeopardize that individual's chances for attaining permanent residence status. (Thus, immigrants try very hard to avoid getting involved in public or community programs and services.) In contrast to the refugee, being a "good immigrant" means avoiding any participation in public service programs.

AFRICAN AMERICANS

This section provides specific cultural and historical information on African Americans that both directly and indirectly affects the relationship between fire service officials and citizens. It presents information about demographics and diversity among African Americans, as well as issues related to cultural and racial identity (Figure 5.11). Following the background information is a section on group identification terms and a discussion of myths and stereotypes. Aspects of the family are discussed, including the extended family, the roles of men and women, and single-mother families. A section on cultural influences on communication deals with Ebonics (also known as African American Venacular English) and nonverbal and verbal communication. The summary of the section reviews recommendations for improved communication and relationships between fire service officials and African Americans.

FIGURE 5.11 ◆ A group of African American young men at a Harlem street fair, New York City, 125th Street.

AFRICAN AMERICAN CULTURE

The effects that slavery and discrimination have had on the black experience in America are not to be downplayed and have had a significant impact upon African American culture. In addition, African American culture is in part influenced by African culture and is significantly different from white culture. Many fire service leaders have come to recognize that when there is an influx of immigrants from a particular part of the world, their officers are better equipped to establish trust, good communication, and mutual cooperation if they have some basic understanding of the group's cultural background. However, past history has shown that when it comes to African Americans, cultural differences are seldom considered, even though they can cause serious communication problems between citizens and fire service and public safety officers. Failing to recognize the distinctiveness of black culture, language, and communication patterns can lead to misunderstandings, conflict, and even confrontation. In addition, understanding the history of African Americans (which is related to the culture) is especially important for fire service officials as they work toward improving relations and changing individual and community perceptions.

HISTORICAL INFORMATION

The majority of African Americans or Blacks (the terms will be used interchangeably) in the United States trace their roots to West Africa. They were torn from their cultures of origin between the seventeenth and nineteenth centuries when they were brought here as slaves. Blacks represent the only migrants to come to the Americas, North and South, against their will. Blacks from Africa were literally kidnap victims of Europeans as well as captives purchased by Yankee traders. This has made African Americans as a group very different from immigrants, who chose to come to the United States to better their lives, and different from refugees, who fled their homelands to escape religious or political persecution.

The very word *slave* carries the connotation of an "inferior" being; slaves were counted as three-fifths of a person during census taking. Slave owners inwardly understood that treating people as animals to be owned, worked, and sold was immoral, but they wanted to think of themselves as good religious, moral people. Hence they

Figure 5.12 ◆ Frederick Douglass, an African American, helped a child out of a barrel on a wagon in a drawing of the Underground Railroad.

had to convince themselves that their slaves were not really human, but a lower form of life. They focused on racial differences (skin color, hair texture, etc.) as "proof" that black people were not really people, after all. Racism began, then, as an airtight alibi for a horrifying injustice. The notion of the slave (and by extension, any African American) as less than human has created great psychological and social problems for succeeding generations of both black and white citizens. Slavery led to a system of inferior housing, schools, health care, and jobs for black people, which persists to this day. The institution of slavery formally ended in 1863, but the racist ideas born of slavery have persisted. These ideas continue even now to leave deep scars on many African Americans. Today, particularly in the lower socioeconomic classes, many blacks continue to suffer from the psychological heritage of slavery (Figure 5.12), as well as from active, current discrimination that still prevents them from equal opportunity in many realms of life.

Until recently, the "public" history that many Americans have learned presented a distorted, incomplete picture of black family life (emphasizing breakdown) during the slave era, which had crippling effects on families for generations to come. This version of history never examined the moral strength of the slaves or the community solidarity and family loyalty that arose after emancipation. There is no doubt that these strengths have positively affected the rebuilding of the African American community.

The more accurate version of history counters the impression that all slave families were so helpless that they were always torn apart and could never reestablish themselves, or that their social relationships were chaotic and amoral. The resolve of large numbers of blacks to rebuild their families and communities as soon as they were freed represents an impressive determination in a people who survived one of the most brutal forms of servitude that history has seen. African American survival and, consequently, African American contributions to American society, deserve a high level of respect and testify to a people's great strength. It is not within the scope of this section to discuss African American contributions to society, but suffice it to say that frequently the perceptions of some people in fire service and public safety are often conditioned by their exposure to problems inherent within the black underclass.

DEMOGRAPHICS: DIVERSITY AMONG AFRICAN AMERICANS

Currently, Blacks comprise approximately 35 million people, or about 13 percent of the population of the United States (U.S. Bureau of the Census, 2003). Until the last few decades, the vast majority lived in the South. Between 1940 and 1970, over 1.5 million blacks migrated initially to the North and then to the West Coast, largely seeking better job opportunities. Historically, over 50 percent of the black population has lived in urban areas. Since 1980, that number has decreased slightly, to where 12.5 percent live in nonmetropolitan areas and 36 percent live in suburban areas, up from 21.2 percent in 1980 (Frey, 1998). The rural black population has decreased and the shifting flow has contributed to the increase in numbers in suburban areas (some rural blacks may migrate directly into the inner city while others may move to more suburban areas; the flow is not actually known). In any case, urban core areas experience cycles of repopulation, mainly by blacks, Hispanics, and various new immigrant groups. The most vivid example is the city of Detroit, where approximately 85 percent of the population is black, while most whites and immigrant groups have settled in its outlying areas. Similarly, other cities, such as Washington, DC, St. Louis, Chicago, and Cleveland, are populated mainly by blacks and new immigrants, creating layers of tension where diverse groups with conflicting values and customs suddenly find themselves crowded into the same urban neighborhoods.

Between 1970 and 1980, 3.5 million blacks and Hispanics moved to cities that had been abandoned by more than 3 million white residents (Lohman, 1977). These population shifts have created "two Americas." One America is the world of the suburbs, where schools, recreational facilities, and community resources are far better. The other is the inner city, where many African Americans (as well as Latinos and other diverse groups) have much poorer access to educational and job opportunities and where living conditions are often closer to those of cities in the ravaged developing world than to those of America's comfortable suburban environment.

Although many African Americans belong to the lower socioeconomic classes, Blacks are represented in all of the classes, from the extremely poor to the extremely affluent, and have moved increasingly into the middle classes (Figure 5.13). As with all ethnic groups, there are significant class-related differences among blacks, affecting values and behavior. It is no more true to say that a poor black has a great deal in common with an affluent black than to say that poor and rich whites are the same. (Later in this chapter we discuss stereotyping of all blacks based on the black underclass.) However, color, more so than class, often determines how the larger society reacts to and treats blacks. From the consulting and training observations of one of the coauthors (Wong) in fire service and law enforcement, the racial (as opposed to cultural) experience of many African Americans in the United States is similar, regardless of an individual's level of prosperity or education.

The cultural diversity that exists among African Americans is related to a variety of factors. Over the last 400 years, black families have come from many different countries (e.g., Jamaica, Trinidad, Belize, Haiti, Puerto Rico). By far, however, the largest group's forefathers came directly to the United States from Africa. In addition, cultural differences exist among African Americans related to the region of the country from which they came. As with whites, there are "southern" and "northern" characteristics as well as urban and rural characteristics. Religious backgrounds vary, but the majority of American-born blacks are Protestant, and many are specifically Baptist. The first black-run, black-controlled denomination in the country was the African Methodist Episcopal Church (which was created because churches in the

FIGURE 5.13 ◆ Minorities were excluded from white middle-class suburbia. However, in Richmond, California, a group of African American community leaders worked to develop a planned community, named Parchester Village, for middle-class African Americans.

North and South either banned blacks or required them to sit apart from whites). In addition, a percentage of blacks belong to the Black Muslim religion, including the "Nation of Islam" and "American Muslim Mission." (The term *Black Muslim* is often used but is rejected by some members of the religion.) There are also sizable and fast-growing black populations among members of the Seventh-Day Adventists, Jehovah's Witnesses, Pentecostals (especially among Spanish-speaking blacks), and—especially among blacks of Caribbean origin—Santeria, Candomble, Voudun, and similar sects blending Catholic and West African (mainly Yoruba) beliefs and rituals. Rastafarianism has spread far beyond its native Jamaica to become an influential religious movement among immigrants from many other English-speaking Caribbean nations.

ISSUES OF IDENTITY

In the 1960s and 1970s the civil rights and Black Pride movements marked a new direction in black identity. The civil rights movement removed many barriers to educational and employment opportunities and to active political involvement. Some adults marched in the civil rights movement knowing that they themselves might never benefit directly from civil rights advances: they hoped that their efforts in the struggle would improve the lives of their children. The middle-class youths who attended community churches and black colleges became the leaders in the movement for equal rights (McAdoo, 1992).

Many blacks (both American-born and Caribbean-born), inspired by a growing sense of community identification and increased pride in racial identity, have determined to learn more about Africa (Figure 5.14). Despite the great differences in culture

FIGURE 5.14 ◆ African American second-grade teacher preparing for Black History celebration in her classroom.

between African Americans and Africans, blacks throughout this hemisphere are discovering that they can take pride in the richness of their African heritage, including its high ethical values and community cohesiveness. The combination of the Black Pride movement of the 1960s and 1970s and the more current focus on cultural roots has freed many African Americans from the "slave mentality" that continued to haunt the African American culture long after emancipation. A new pride in race and heritage has, for some, replaced the sense of inferiority fostered by white racial supremacist attitudes.

GROUP IDENTIFICATION TERMS

Several ethnic groups, including African Americans, in a positive evolution of their identity and pride, have initiated name changes for their group. This is not confusion (although it can be confusing), but rather represents, on the part of group members, growth and a desire to name themselves rather than be named by the **dominant culture**. Until approximately the early 1990s, the most widely accepted term was *Black*, this term having replaced *Negro* (which, in turn, replaced *colored people*). *Negro* has been out of use for at least two decades, although some older blacks still use the term (as do some younger African Americans among themselves). To many, the term *Negro* symbolizes what the African American became under slavery. The replacement of *Negro* with *Black* came to symbolize racial pride. (The exception to the use of Negro and colored is in titles such as "United Negro College" and "National Association for the Advancement of Colored People [NAACP].") *African American*, a term preferred by many, focuses on positive historical and cultural roots rather than on race or skin color.

In the 1990s, the usage of the term *African American* grew in popularity. (It is the equivalent of, for example, Italian American or Polish American.) Many feel that the word *black* is no more appropriate in describing skin color than is *white*. (Yet, some Americans who are black do not identify with African American, as it does not fully represent their background, which may be Caribbean or Haitian; and they may not identify with the African part at all.) Since the 1980s and early 1990s,

dominant culture

Refers to the value system that characterizes a particular group of people that dominates the value systems of other groups or cultures. See also *macro culture/majority or dominant group.*

the term *people of color* has sometimes been used, but this catchall phrase has limited practical use for fire service officers, as it is used to describe anyone who is not white and can include Asian/Pacific Americans, Latino/Hispanic Americans, as well as Native Americans. Indeed, there is much controversy and history associated with broad, collective terms that attempt too quickly to categorize people.

The use of the derogatory "nigger," "boy," and "coon" is never acceptable at any time, especially in fire service and public safety, no matter how provoked a fire service officer may be. (Although you may hear black youths using "nigger" to refer to each other, the word is absolutely taboo for outsiders—especially outsiders wearing badges.) Fire service officers hearing these types of labels used in their own departments, even when they are not being used to address somebody directly, should remind their peers of the lack of professionalism and prejudice that those terms convey. Officers who fall into the habit of using these words in what they think are harmless situations may find that they are unable to control themselves during more volatile situations with citizens, as seen in the following example:

> On Feb. 2, 2004, Scheuneman was returning from a run when he got angry at an African-American motorist driving a vehicle with Wisconsin plates. Speaking to a supervisor, Scheuneman used the n-word three times, telling his boss there was "nothing worse than a cheesehead n——." Scheuneman was apparently unaware that the microphone on his radio was open and that his inflammatory remarks were being broadcast over fire radio. . . . Two weeks later, Scheuneman was yanked out of Austin and reassigned to racially mixed Bridgeport at the request of the local alderman, Emma Mitts. He was also slapped with a 90-day suspension (Spielman, 2006, p. 40).

MYTHS AND STEREOTYPES

Many of the impressions that people form about African Americans within society (and, consequently, within the fire service form about African Americans) come from their exposure to messages and images in the media about African Americans. Even the fact of a crime rate that is disproportionately high among young black males does not justify sweeping statements about all African Americans. Certainly, 36 million African Americans cannot be judged by a statistic about the criminal element. Unfortunately, for the vast majority of the African American population, some whites do base their image of all blacks largely on the actions, including criminal behavior, of the black members of America's "underclass." African Americans have to contend with many myths and stereotypes. A review of the recent research revealed the following five workplace-related stereotypes about African Americans:

1. **Viewing African Americans as being in jobs as a result of affirmative action.** This stereotype of African Americans would be coupled with the assumption that the person in the position were less qualified and primarily hired as a result of affirmative action and/or equal opportunity efforts (Aquino, Stewart, and Reed, 2005).

2. **Viewing African Americans as less competent and less serious about work.** When compared to other groups in the workplace, African Americans were stereotyped to be less serious, competent, and polite within their workplace interactions (Gilbert, Carr-Ruffino, Ivancevich, and Lownes-Jackson, 2003).

3. **Viewing that there are more African American men in jail than in college.** Two implications about this stereotype are posed: (a) African American men are criminals, and (b) African American men are not interested in higher education. As Baraki (2004) notes,

If you choose the age demographic that falls between 18 to 24 years of age then there are more Black men in college of this age range than in jail. This is nothing to be excited about, but it does provide a clearer perspective and lets us know that Black men are on the right track. By highlighting this disparity without clarification, Black men who are in college and not involved in the criminal justice system receive no credit.

4.   **Viewing African American young people as not career oriented.**   African American youths are often stereotyped as lacking ambition and career orientation. As noted by Baraki (2004), they are "criticized for their style of dress, choice of music, way of expressing themselves and if you really think about it, for their audacity in existing at all."

5.   **Viewing African American women as welfare recipients.**   Yoder and Berendsen (2001), in their study of stereotypes of African American women firefighters, found two prevailing and demeaning stereotypes of African American women in fire service professions. They were stereotyped as (a) welfare recipients and (b) "beasts of burden" (when it came to doing some of the heavier work involved in fire service).

It is important for fire service officers to be aware of the different stereotypes of African Americans. The key to effectiveness with any ethnic or racial group is not that we completely eliminate myths and stereotypes about these groups, but that we are aware of these stereotypes and can monitor our thinking and behaviors when the stereotypes are not true of those persons with whom we are interacting.

THE AFRICAN AMERICAN FAMILY

African American families generally enjoy very strong ties among extended family members, especially among women. Female relatives will often substitute for each other in filling family roles; for example, a grandmother or aunt may raise a child if the mother is unable to do so. Sometimes several different family groups may share one house. When there is a problem (i.e., an incident that has brought an officer to the house), extended family members are likely to be present and to want to help. A fire service officer may observe a number of uncles, aunts, brothers, sisters, and friends who are loosely attached to the black household. Enlisting the aid of any of these household members (no matter the relationship) can be beneficial.

The Role of the Man and the Woman   A widespread myth holds that African American culture is a matriarchy, with one or more women heading the typical household. Historically, it is true that women did play a crucial role in the family, because of repeated attempts to break down "black manhood" (Bennett, 1989). However, in a true matriarchy (female-ruled society), women control property and economic activities. This is not generally true of black America, but outsiders often assume that African American women are always the heads of the household. In the 1960s and 1970s, for instance, the media stereotyped the black mother as loud and domineering, clearly the boss. In addition, African American women have, in a sense, been considered less of a threat to the status quo than has the African American male and, consequently, have been able to be more assertive in public. Historically, in contacts with fire service and public safety personnel, an African American man may feel that, unlike a woman, he risks mistreatment if he "talks back," as might be seen in the following situation:

A federal lawsuit filed Thursday compares the alleged actions of two North Lake firefighters to "the Ku Klux Klan at the height of its activities" because fire and Town of Merton officials created an environment that allowed an

African-American fisherman and his family to be accosted at a pond last spring. Fire Chief Terrence J. Stapleton and firefighter Mark J. Weber are facing hate-crime charges in Waukesha County Circuit Court that accuse them of taking part in the April 19 incident that shook the Town of Merton. Weber is accused of pointing a handgun at Mark Bratton of Milwaukee and cocking the trigger while shouting racial slurs at him for refusing to leave the Monches Mill Pond. Stapleton is accused of helping Weber, chasing Bratton with a leashed German shepherd. Stapleton later told a reporter that he used the animal "because colored people don't like dogs." . . . The suit names Stapleton, Weber, the town and the Fire Department as defendants. It says the department and the town "created the environment in which action just like those described in this complaint could occur." "The North Lake Fire Department firehouse was like a clubhouse for firefighters," the lawsuit says. "It was a place to gather, drink alcoholic beverages, play games and otherwise seek advice and help from fellow firefighters." (Enriquez, 2006).

African American fathers, regardless of income, usually view themselves as heads of the household (Hines and Franklin, 1982), and thus any major decisions regarding the family should include the father's participation. It is insulting to the father when, with both parents present, fire service officers automatically focus on the mother. An assertive mother does not mean that the father is passive or indifferent, and a father's silence does not necessarily indicate agreement.

The Single Mother   Feeding the stereotype of the female-headed black household is the fact that in many urban African American homes, no father is present. In 1965, about 25 percent of African American families nationally were headed by women. Between 1970 and 1997, births to single African American women increased by 93 percent, resulting in 68.4 percent of black births being to single women in 2001 (National Center for Health Statistics, 2003). Public safety officers observed that in some urban core areas (particularly in housing projects) in 2000, it was not uncommon to find nearly 90 percent of African American women living alone with children.

The single mother, particularly in the inner city, does not always receive the respect that she is due; outsiders may be critical of the way she lives—or the way they think she lives. She is often stereotyped by fire service officers who doubt their own effectiveness in the urban black community. For instance, an unmarried welfare mother who has just had a fight with her boyfriend should receive the same professional courtesy that a married suburban mother is likely to receive from a fire service officer in the same situation. In practice, this is not always the case, and being an unmarried mother on welfare and other characteristics associated with the underclass may be accompanied by a devaluing of the individual by fire service and other public safety officers. A complaint frequently heard in the African American community by women is that white men, in particular, treat the African American woman poorly. This oftentimes means that they feel comfortable using profanity in the woman's presence and treat her with little respect.

"EBONICS" OR AFRICAN AMERICAN VERNACULAR ENGLISH

"The use of [what has been called] black language does not represent any pathology in blacks. . . . The beginning of racial understanding is the acceptance that difference is just what it is: different, not inferior. And equality does not mean sameness" (Weber, 1991). There are many varieties of English that are not "substandard," "deficient," or "impoverished" versions of the language. Instead, they often have a complete and consistent set of grammatical rules and may represent the rich cultures of the groups that use them. For example, Asian Indians speak a variety of English somewhat unfamiliar to Americans, whereas the British speak numerous dialects of British English, each of them containing distinctive grammatical structures and vocabulary not used by Americans. Similarly, some African Americans speak a variation of English that historically has been labeled as substandard because of a lack of understanding of its origins.

Many African Americans use (or have used at least some of the time) what has been called "black English," African American Vernacular English, or Ebonics. Although some African Americans speak this variety of English, many enjoy the flexibility and expressiveness of speaking Ebonics among peers and switching to "standard English" when the situation calls for it (e.g., at work, in interviews, and with white friends).

NONVERBAL COMMUNICATION: STYLE AND STANCE

Black social scientists have been studying aspects of black nonverbal communication, which have been often misunderstood by people in positions of authority. Psychologist Richard Majors at the University of Wisconsin termed a certain stance and posturing as the "cool pose," demonstrated by many young black men from the inner city. "While the cool pose is often misread by teachers, principals and fire service officers as an attitude of defiance, psychologists who have studied it say it is a way for black youths to maintain a sense of integrity and suppress rage at being blocked from usual routes to esteem and success" (Goleman, 1992).

Majors explains that while the cool pose is not found among the majority of black men, it is commonly seen among inner city youth as a "tactic for . . . survival to cope with such rejections as storekeepers who refuse to buzz them into a locked shop." The goal of the pose is to give the appearance of being in control. However, a storekeeper, a passerby, or a fire service officer may perceive this stance as threatening, so a negative dynamic enters the interaction (e.g., the officer seeing the cool pose feels threatened and then becomes more authoritarian in response). This form of nonverbal communication may include certain movements and postures designed to emphasize the youth's masculinity. The pose involves a certain way of walking, standing, talking, and remaining aloof facially. The pose, writes Majors, is a way of saying, "[I'm] strong and proud, despite [my] status in American society" (Goleman, 1992).

EFFORTS TOWARD A POSITIVE RELATIONSHIP BETWEEN FIRE SERVICE AND THE AFRICAN AMERICAN COMMUNITY

Many fire service officers believe that they are putting their lives in danger when going into certain communities, particularly those in urban areas where poverty and crime go hand in hand (Figure 5.15). As a result, some segments of the African American community feel fire service and other public safety agencies are not protecting them.

Communities and fire departments differ widely in the ways that they build positive relationships with each other. On the one hand, **consent decrees** have forced a change of inappropriate (prejudicial or unjust) behavior or actions. A consent decree is an out-of-court settlement whereby the accused party (i.e., accused of inappropriate behavior or actions) agrees to modify or change behavior rather than plead guilty or go through a hearing on charges brought to court (CSIS, 2003). It is beyond the scope of this chapter to delve into the details and the controversy surrounding consent decrees; however, many inappropriate fire service actions within African American communities have resulted in the implementation of consent decrees. Some fire service

consent decree

An out-of-court settlement whereby the accused party agrees to modify or change behavior rather than plead guilty or go through a hearing on charges brought to court. (Source: CSIS Project Glossary, http://csisweb. aers.psu.edu/glossary/c.htm)

FIGURE 5.15 ◆ Guardian Angels stand in line during a civil rights rally in New York to protect the community.

leaders believe that a consent decree is not a symbol of a positive relationship; at the same time, many community leaders have supported them.

> One leader mentioned that the consent decree "gave people an opportunity to feel as though they had a voice." Other community leaders felt that some white and black officers were working better together and that they were responding more quickly to calls for service from the African-American community. "It seems they are trying to ask more questions and there is more respect for the community," she said (Davis et al., 2002).

Although controversy exists about consent decrees, and some officers complain that they add too much paperwork and bureaucracy to their already demanding jobs, they serve a purpose. They are an attempt, on the part of the federal government, local and state fire service departments, and communities, to try to improve community–fire service interaction in a quick and dramatic manner. Some outcomes can be influenced informally and others by consent decrees. From the perspective of one of the coauthors (Olson), as a training officer with fire service and public safety, the key to improved relationships deeply rests upon all of the following:

- Leadership
- Vision
- Respect
- Goals
- Strategies
- Mutual benefits for fire service and community
- Effective communication and practices of both fire departments and the community

Added to all of the above, community members, leaders, and fire officers need to rely upon private means—one-on-one and face-to-face interaction, as well as goodwill—to improve relations.

SUMMARY OF RECOMMENDATIONS FOR FIRE SERVICE IN THE AFRICAN AMERICAN COMMUNITY

1. The experience of slavery and racism, as well as cultural differences, has shaped African American culture. Patterns of culture and communication among black Americans differ from those of white Americans. In face-to-face communication, fire service officers should not ignore or downplay these differences.

2. For many African Americans, particularly those in the lower socioeconomic rungs of society, the history of slavery and later discrimination continue to leave psychological scars. Fire service officials may represent a "system" that has oppressed African Americans and other minorities. To serve properly in many African American communities across the nation necessarily means that fire service officers will need to go out of their way to establish trust and win cooperation.

3. The changing terms that African Americans have used to refer to themselves reflect stages of racial and cultural growth, as well as empowerment. Respect the terms that African Americans prefer to use. *Negro* and *colored* are no longer used and have been replaced by *Black* for the purposes of many fire service communications. However, in speech, many people prefer the term *African American*. Fire service officers can learn to become comfortable asking a citizen which term he or she prefers if there is a need to refer to ethnicity in the conversation. Fire service and public safety officers are advised to stop each other when they hear offensive terms being used. Not only does this contribute to the first step of making a department free of overt prejudices, it also helps the individual officer to practice control when he or she is faced with volatile citizens. An officer in the habit of using offensive terms may not be able to restrain himself or herself in public situations. Therefore, officers monitoring each other will ultimately be of benefit to the department.

4. African Americans react negatively to stereotypes that they hear about themselves. Many of the stereotypes about African Americans stem from ignorance as well as an impression people receive from negative aspects of the Black underclass. Fire service officers, in particular, must be sensitive to how their own perceptions of African Americans are formed. The disproportionately high crime rate among African American males does not justify sweeping statements about the majority of African Americans who are law-abiding citizens.

5. Because many African American households are headed by women, particularly in the inner city, coupled with the social myth of the woman as "the head of the household," fire service officers may dismiss the importance of the father. Despite common myths and stereotypes regarding the woman, fire service officers should always approach the father to get his version of the story and to consider his opinions in decision making. If the father is ignored or not given appropriate attention, he is certain to be offended, making any tasks of the fire service officer more difficult.

6. The use of African American varieties of English does not represent any pathology or deficiency and is not a combination of random errors, but rather reflects patterns of grammar from some West African languages. Many people have strong biases against this variety of spoken English. Do not convey a lack of acceptance through disapproving facial expressions, a negative tone of voice, or a tendency to interrupt or finish the sentences for the other person. When it comes to black English or an accent, do not fake it in order to be accepted by the group. People will immediately pick up on your lack of sincerity; this, in and of itself, can create hostility.

7. People in positions of authority have often misunderstood aspects of black nonverbal communication, including what has been termed the cool pose. Fire service officers may interpret certain ways of standing, walking, and dressing as defiant. This can create defensive reactions on the part of the fire service officer. In many cases, the fire service officer need not take this behavior personally or feel threatened.

8. A dynamic exists between some fire service officers and African Americans, particularly in poor urban areas, whereby both the officer and the citizen are on the "alert" for the slightest sign of disrespect. The fear that both the citizen and the fire service officer experience can interfere with what may actually be a nonthreatening situation. The fire service officer can be the one to break the cycle of fear by softening his or her verbal and nonverbal approach.

9. In areas populated by African Americans and other racial and immigrant groups all over the United States, there is a need for increased and more effective fire service protection. Bridging the gap that has separated the fire service from African Americans involves radical changes in attitudes toward public safety–community relations. Together with changes in management, even the individual officer can help by making greater efforts to have positive contact with African Americans. The task of establishing rapport with African Americans at all levels of society is challenging because of what the fire service or public safety officer represents in terms of past discrimination. Turning this image around involves a commitment on the part of the fire service officer to break with deeply embedded stereotypes and to have respect and professionalism as a goal in every encounter.

ASIAN/PACIFIC AMERICANS

This section provides specific ethnic and cultural information on Asian Americans and Pacific Islanders. The label *Asian Americans/Pacific Islanders* encompasses over 40 different ethnic and cultural groups. For ease of use, we will use the shortened version, *Asian/Pacific Americans*, to refer to members of these ethnic groupings. We first define this very diverse group and then present an historical overview, focusing on the relationship between fire service and citizens. We present demographics and elements of diversity among Asian/Pacific Americans as well as issues related to ethnic and cultural identity. Following the background information are sections on labels and terms, myths and stereotypes, aspects of the family, and verbal and nonverbal communication. The closing section presents recommendations for improved communication and relationships between fire service personnel and the Asian/Pacific American communities (Figure 5.16).

For the past four decades, the Asian/Pacific American population has experienced the largest proportional increases of any ethnic minority population in the United States (over 100 percent growth for the decades from 1960 to 1990 and 76 percent growth for the decade from 1990 to 2000). The population growth can be attributed to (1) higher emigration from the Pacific Rim countries, (2) greater longevity, (3) higher birthrates, (4) immigrants admitted for special skills and expertise for work in high-technology industries in the United States, and (5) existing ethnic groups added to this population category (including the Asian/Pacific Americans who consider themselves "multiracial"). Growth in major urban areas has been particularly striking, as is seen in New York City, Los Angeles, San Jose (CA), San Francisco, Honolulu, San Diego, Chicago, Houston, Seattle, Fremont (CA), Fairfax (VA), and Quincy (MA). Growth in the population as a whole is most dramatically reflected in terms of increased numbers of Asian/Pacific Americans in politics, community leadership, business, education, and public service areas. Fire service contact with Asian/Pacific people has increased because of their greater presence in communities. Asian/Pacific

Figure 5.16 ◆ Firefighters battle a 7-alarm fire on Essex Street in Boston's Chinatown.

Americans have one of the highest citizenship rates among all foreign-born groups; in 2000, 47 percent of the immigrants from Asian/Pacific countries were naturalized citizens (Barnes & Bennett, 2002).

ASIAN/PACIFIC AMERICANS DEFINED

The term *Asian/Pacific Americans* is actually a contraction of two terms, *Asian Americans* and *Pacific Islander peoples*. Although used throughout this section, Asian/Pacific Americans is, in fact, a convenient summary label for a very heterogeneous group. There certainly is not universal acceptance of this labeling convention, but for practical purposes it has been adopted frequently, with occasional variations (e.g., Asian and Pacific Americans, Asian Americans/Pacific Islanders, Asians and Pacific Islanders). It represents the self-designation preferred by many Asian and Pacific people in the United States, particularly in preference to the more dated (and, to some, offensive) term *Orientals*. The U.S. government and other governmental jurisdictions usually use Asian Americans/Pacific Islanders to refer to members within any of the 40 or more groups comprising this category.

At least 40 distinct ethnic and cultural groups might meaningfully be listed under this designation: (1) Bangladeshi, (2) Belauan (formerly Palauan), (3) Bhutanese, (4) Bruncian, (5) Cambodian, (6) Chamorro (Guamanian), (7) Chinese, (8) Fijian, (9) Hawaiian (or Native Hawaiian), (10) Hmong, (11) Indian (Asian, South Asian, or East Indian), (12) Indonesian, (13) Japanese, (14) Kiribati, (15) Korean, (16) Laotian, (17) Malaysian, (18) Maldivian, (19) Marshallese (of the Marshall Islands, to include Majuro, Ebeye, and Kwajalein), (20) Micronesian (to include Kosrae, Ponape, Truk, and Yap), (21) Mongolian, (22) Myanmarese (formerly Burmese), (23) Nauruan, (24) Nepalese, (25) Ni-Vanuatu, (26) Okinawan, (27) Pakistani, (28) Pilipino (preferred spelling of Filipino), (29) Saipan Carolinian (or Carolinian, from the Commonwealth of the Northern Marianas), (30) Samoan, (31) Singaporian, (32) Solomon Islander,

FIGURE 5.17 ◆ Asian multilanguage signs in Chinatown, New York City, show the diversity and multiethnic nature of the Asian/Pacific American community.

(33) Sri Lankan (formerly Ceylonese), (34) Tahitian, (35) Taiwanese, (36) Tibetan, (37) Tongan, (38) Thai, (39) Tuvaluan, and (40) Vietnamese. Although there are marked differences between the 40 groups listed, individuals within any of the 40 groups may also differ in a vast number of ways (Figure 5.17).

From the viewpoint of fire service, it is important to recognize some of the differences that may cut across all Asian/Pacific ethnic groups. For example:

1. Area of residence in the United States
2. Comfort with and competence in English
3. Generational status in the United States (first, second, third generation, and so forth)
4. Degree of acculturation and assimilation
5. Education (number of years outside and inside the United States)
6. Native and other languages spoken and/or written
7. Age (what is documented on paper and what may be the individual's real age)
8. Degree of identification with the home country and/or region of self or parents' origin
9. Family composition and extent of family dispersion in the United States and globally
10. Extent of identification with local, national, and global Asian/Pacific sociopolitical issues
11. Participation and degree to which the individual is embedded within the ethnic community network
12. Religious beliefs and cultural value orientation
13. Economic status and financial standing
14. Sensitivity to ethnic and cultural experiences and perceptions as an Asian/Pacific person in the United States
15. Identification with issues, concerns, and problems shared by other ethnic-racial groups (e.g., poor community services and discrimination issues voiced by African Americans, Latino/Hispanic Americans, Middle Eastern and Arab Americans, and Native Americans)

It should be noted that the definition itself of the Asian/Pacific group points to and embodies an ever-emerging ethnic mosaic of diverse constituencies. Groups are added and removed based on self-definition and needs for self-choice. In our definition, we have not added immigrants or refugees from the more recently formed Central Asian nations of Kazakhstan, Kyrgyzstan, Tajikistan, Turkmenistan, and Uzbekistan (all were republics of the Soviet Union before that country dissolved at the end of 1991) because of the self-choice issue. As more immigrants from Central Asia settle in the United States, these groups may (or may not) be added to the definition of Asian/Pacific Americans. Clearly, the pooling of separate Asian and Pacific Islander groups under the label of Asian/Pacific Americans emerged, in part, out of the necessity to have a larger collective whole when a greater numerical count may make a difference (especially in political and community issue areas). Merging these 40 ethnic groups into a collective entity allowed for sufficiently large numbers for meaningful representation in many communities and other arenas.

Although one may focus upon "nationality" or "nation of origin" as the basis of identifying Asian/Pacific American groupings (e.g., "Korean" for those whose ancestry is from Korea, "Vietnamese" for those whose ancestry is from Vietnam), there may be other demographic variables of greater importance than that of nationality. For example, whether one is from the Central Asian nation of Uzbekistan, the Southeast Asian nation of Vietnam, or the Pacific Island nation of Micronesia, the demographic variable of ethnicity, if one were Chinese, may be a more important identifier than that of national origin for some persons or groups. As another example, one's "religion" may be a more important "group" identifier than either ethnicity and/or nationality, as might be seen with Muslim (i.e., people who practice Islam). Given that the four largest countries with Muslim populations are in South Asia—that is, Indonesia, Pakistan, Bangladesh, and India (Glasse and Smith, 2003)—some Asian/Pacific American people who practice Islam might see "religion" as a more important identifier than ethnicity and nationality.

HISTORICAL INFORMATION

The first Asians to arrive in the United States in sizable numbers were the Chinese in the 1840s to work on the plantations in Hawaii. Then, in the 1850s, they immigrated to work in the gold mines in California and later on the transcontinental railroad. Of course, the native populations of the Pacific Island areas (e.g., Samoans, Hawaiians, Guamanians, Fijians) were there before the establishment of the 13 colonies of the United States. The Chinese were followed in the late 1800s and early 1900s by the Japanese and the Pilipinos (and in smaller numbers by the Koreans and the South Asian Indians). Large numbers of Asian Indians (nearly 500,000) entered the United States as a result of congressional action in 1946 for "persons of races indigenous to India" to have the right of naturalization. Most immigrants in these earlier years were men, and most worked as laborers and at other domestic and menial jobs. Until the change of the immigration laws in 1965, the number of Asian and Pacific Islander peoples immigrating into the United States was severely restricted (families often had to wait over a decade or more before members of a family could be reunited). With the change in the immigration laws, large numbers of immigrants from the Pacific Rim came to the United States from Hong Kong, Taiwan, China, Japan, Korea, South Asia (e.g., India, Ceylon, Bangladesh), the Philippines, and Southeast Asia (e.g., Vietnam, Thailand, Singapore, Cambodia, Malaysia). After the Vietnam War, large numbers of Southeast Asian refugees were admitted in the late 1970s and early 1980s (Takaki, 1989). The need for engineering and scientific expertise and skills by high-tech and Internet companies resulted in many Asian/Pacific immigrants (under special work visas) immigrating to the United States in the late 1990s and early 2000s.

Asian/Pacific Americans have found the passage and enforcement of "anti-Asian" federal, state, and local laws to be more hostile and discriminatory than some of the racially motivated community incidents that they have experienced. Early experiences of Asians and Pacific Islanders were characterized by the majority population's wanting to keep them out of the United States and putting tremendous barriers in the way of those who were already here. It was the role of immigration, public safety, and community service agencies and officers to be the vehicle to carry out these laws against Asian/Pacific American immigrants.

Anti-Asian Federal, State, and Local Laws Almost all of our federal immigration laws were written such that their enforcement made Asian newcomers feel neither welcomed nor wanted (Figure 5.18). Following the large influx of Chinese in the 1850s to work in the gold mines and on the railroad, many Americans were resentful of the Chinese for their willingness to work long hours for low wages. With mounting public pressure, the Chinese Exclusion Act of 1882 banned the immigration of Chinese laborers for 10 years, and subsequent amendments extended this ban indefinitely. Because of this ban and because the Chinese population in the United States was primarily male, the Chinese population in the United States dropped from 105,465 in 1880 to 61,639 in 1920 (Takaki, 1989). Because the Chinese Exclusion Act applied only to Chinese, Japanese immigration started around 1870 to Hawaii, with larger numbers to the mainland in the 1890s to work as laborers and in domestic jobs on the farms on the West Coast. Similar to the Chinese, public pressure to restrict Japanese immigration ensued. In the case of the Japanese, the Japanese government did not want a "loss of face" or of international prestige through having its people "banned" from immigrating to the United States. Rather, the "Gentleman's Agreement"

FIGURE 5.18 ◆ Former California Secretary of State March Fong Eu worked to ensure that all voters of Asian/Pacific American descent and other ethnic/cultural groups were able to register and to vote.

was negotiated with President Theodore Roosevelt in 1907, which resulted in the Japanese government voluntarily restricting the immigration of Japanese laborers to the United States. Family members of Japanese already in the United States, however, were allowed to enter. Under the Gentleman's Agreement, large numbers of "picture brides" began entering into the United States, resulting in a large increase in Japanese American populations, from 25,000 in 1900 to 127,000 in 1940 (Daniels, 1988). Subsequent laws banned or prevented immigration from the Asiatic countries: the Immigration Act of 1917 banned immigration from all countries in the Pacific Rim except for the Philippines (a U.S. territory). The Immigration Act of 1924 restricted migration from all countries to 2 percent of the countries' national origin population living in the United States in 1890. This "two percent" restriction was not changed until 1965. Moreover, it was not until 1952 that most Asian immigrants were eligible to become naturalized citizens of the United States and, therefore, have the right to vote (African Americans and American Indians were able to become citizens long before Asian/Pacific Americans were given the same rights).

While Pilipinos (preferred spelling for Filipinos) have been immigrating to the United States since the early 1900s (primarily for "practical training" as selected, sponsored, and funded by the United States government), large numbers of Pilipino laborers began entering in the 1920s because of the need for unskilled laborers (and due in part to the unavailability of Chinese and Japanese immigrants, who were restricted entry by law). Similar to previous Asian groups, Pilipino immigration was soon to be limited to a quota of 50 immigrants per year with the passage of the Tydings-McDuffie Act of 1934. Moreover, congressional resolutions in 1935 reflected clear anti-Pilipino sentiment by providing free, one-way passage for Pilipinos to return to the Philippines with the agreement that they not return to the United States.

Anti-Asian immigration laws were finally repealed, starting with the removal of the Chinese Exclusion Act in 1943. Other laws were repealed to allow immigration of Asians and Pacific Islanders, but the process was slow. It was not until 1965 that amendments to the McCarran-Walter Act opened the way for Asian immigrants to enter in larger numbers (a fixed quota of 20,000 per country, as opposed to 2 percent of the country's national origin population living in the United States in 1890). The 1965 amendments also established the "fifth preference" category, which allowed highly skilled workers needed by the United States to enter this country. Because of the preference for highly skilled workers, a second major wave of immigrants from Hong Kong, Taiwan, India, Korea, the Philippines, Japan, Singapore, and other Asiatic countries entered in the mid-1960s. The earlier wave of South Asian immigrants (from India, Pakistan, and Sri Lanka) with expertise to help with America's space race against the Soviets resulted in large numbers of professional and highly educated South Asians immigrating into the United States under the fifth preference after 1965. For example, as part of this second wave of immigration, 83 percent of the Asian Indians who immigrated under the category of professional and technical workers between 1966 and 1977 were scientists with Ph.D.s (about 20,000) and engineers (about 40,000) (Prashad, 2001). With the upheaval in Southeast Asia and the Vietnam War, the third major wave of close to 1 million refugees and immigrants arrived in the United States from these affected Southeast Asian countries starting in the mid-1970s and lasting until the early 1980s (Special Services for Groups, 1983). Most recently, the need for expertise and scientific skills as well as the opportunities present in high-technology, engineering, computer, software, and Internet industries have led to an additional influx of immigrants from India, Pakistan, Singapore, Korea, China (including Hong Kong), Taiwan, and other Asiatic countries in the mid-1990s and early 2000s.

Although many immigrant groups (e.g., Italians, Jews, Poles) have been the target of discrimination, bigotry, and prejudice, Asian/Pacific Americans, like African Americans, have experienced extensive legal discrimination, hindering their ability to participate fully as Americans. This discrimination has gravely affected their well-being and quality of life. Some states had laws that prohibited intermarriage between Asians and whites. State and local laws imposed restrictive conditions and taxes specifically on Asian businesses and individuals. State courts were equally biased; for example, in the case of *People v. Hall,* heard in the California Supreme Court in 1854, Hall, a white defendant, had been convicted of murdering a Chinese man on the basis of testimony provided by one white and three Chinese witnesses. The California Supreme Court threw out Hall's conviction on the basis that state law prohibited blacks, mulattos, or Indians from testifying in favor of or against whites in court. The court's decision read:

> Indian as commonly used refers only to the North American Indian, yet in the days of Columbus all shores washed by Chinese waters were called the Indies. In the second place the word "white" necessarily excludes all other races than Caucasian; and in the third place, even if this were not so, I would decide against the testimony of Chinese on the grounds of public policy. (*People v. George W. Hall,* 1854)

This section about anti-Asian/Pacific American laws and sentiments cannot close without noting that Japanese Americans are the only immigrant group in the history of the American people who have been routed out of their homes and interned without due process. President Roosevelt's Executive Order 9066 resulted in the evacuation and incarceration of 100,000 Japanese Americans in 1942. For Asian/Pacific Americans, the internment of Japanese Americans represents how quickly anti-Asian sentiments can result in incarceration and punishment by law, even if no one was convicted of espionage or a crime.

DEMOGRAPHICS: DIVERSITY AMONG ASIAN/PACIFIC AMERICANS

Asian/Pacific Americans currently number about 12.8 million and represent approximately 4.5 percent of the U.S. population. As stated, the Asian/Pacific American population more than doubled with each census from 1970 to 1990 (1.5 million in 1970, 3.5 million in 1980, and 7.3 million in 1990) and increased by 76 percent from 1990 to 2000 (12.8 million in 2000). While greater longevity and higher birthrates contribute to this population increase, the major contributor to the growth of the Asian/Pacific American population is emigration from the Pacific Rim countries. Since the 1970s, Asian/Pacific American immigration has made up over 40 percent of all immigration to the United States (Barnes & Bennett, 2002). Chinese are the largest group, with 23.0 percent of the total Asian/Pacific American population. Pilipinos are close behind, with 19.9 percent of this population; in the next decade, Pilipinos will be the largest Asian/Pacific American group in the United States. Asian Indians are the fastest growing among the Asian and Pacific American population with 16.0 percent. Vietnamese and Koreans each comprise approximately 10.3 percent of the Asian/Pacific American population. The Japanese American population was the third largest in 1990 and dropped to sixth place in 2000 with 9.7 percent in part because of the lower emigration from Japan. All other Asian/Pacific American groups account for 10.8 percent of this population.

From one of the coauthor's (Wong) field experience in training fire department personnel, the key Asian/Pacific American groups to understand would be the six largest groups: Chinese, Pilipino, Asian Indian, Vietnamese, Korean, and Japanese (considering, in addition, local community trends and unique qualities of the community's populations). Knowledge of the growing trends among this Asian/Pacific American population would also be important for officer recruitment and other human resource considerations. Current Asian/Pacific Americans involved in professional fire service and public safety careers are largely Japanese, Chinese, and Korean Americans. To plan for the changing Asian/Pacific American population base, it is critical to recruit and develop fire service officers from the Pilipino, Vietnamese, and Asian Indian communities.

More than half of all Asian/Pacific Americans are foreign-born (7.2 million) and comprise 26 percent of the foreign-born population in the United States (Bennett, 2002). Five countries contributed the largest numbers of foreign-born Asian/Pacific Americans:(1) China, (2) Philippines, (3) India, (4) Vietnam, and (5) Korea. Foreign-born Asian/Pacific Americans have the second highest percentage of becoming naturalized citizens at 47 percent (only those born in Europe had a higher rate at 52 percent) (Bennett, 2002).

LABELS AND TERMS

As we noted earlier, the term *Asian/Pacific Americans* is a convenient summarizing label used to refer to a heterogeneous group of people. Which particular terms are used is determined by the principle of self-designation and self-preference. Asian and Pacific Islander people are sensitive about the issue because up until the 1960 census the population was relegated to the "Other" category. With the ethnic pride movement and ethnic minority studies movement in the late 1960s, people of Asian and Pacific Islands descent began to designate self-preferred terms for group reference. The terms were chosen over the previous term *Oriental*, which many Asian/Pacific Americans consider to be offensive because it symbolizes the past references, injustices, and stereotypes of Asian and Pacific people. It was also a term designated by those in the West (i.e., the Occident, the Western Hemisphere) for Asian people and reminds many Asian/Pacific Americans of the colonial mentality of foreign policies and its effects on the Pacific Rim countries.

In federal and other governmental designations (e.g., the Small Business Administration), the label used is *Asian American/Pacific Islanders*. Although very few Asian/Pacific Americans refer to themselves as such, the governmental designation is used in laws and regulations and in most reports and publications. For individuals within any of the groups, often the more specific names for the groups are preferred (e.g., Chinese, Japanese, Vietnamese, Pakistani, Hawaiian). Some individuals may prefer that the term *American* be part of their designation (e.g., Korean American, Pilipino American). For fire service officers, the best term to refer to an individual is the term that person prefers to be called. It is perfectly acceptable to ask an individual what ethnic or cultural group(s) he or she identifies with and what he or she prefers to be called.

The use of slurs such as "Jap," "Chink," "Gook," "Chinaman," "Flip," "Babas," and other derogatory ethnic slang terms is never acceptable in fire service and public safety, no matter how provoked an officer may be. The use of other stereotypic terms, including "Chinese fire drill," "DWO (Driving While Oriental)," "Fu Man Chu mustache," "Kamikaze kid," "yellow cur," "yellow peril," "Bruce Lee Kung Fu type," "slant-eyed,"

"Turban man," "Vietnamese bar girl," and "dragon lady," does not convey the kinds of professionalism and respect for community diversity important to fire service and needs to be avoided in public safety work. Fire service officers hearing these words used in their own departments (or with peers or citizens) should provide immediate helpful feedback about such terms to those who use them. Officers who may out of habit routinely use these terms may find themselves (or their superiors) in the embarrassing situation (on the six o'clock news) of explaining to offended citizens and apologizing on behalf of their departments or agencies.

MYTHS AND STEREOTYPES

Knowledge of and sensitivity to Asian/Pacific Americans' concerns, diversity, historical background, and life experiences will facilitate the fire service missions of firefighter officers. It is important to have an understanding about some of the myths, environmental messages, and stereotypes of Asian/Pacific Americans that contribute to the prejudice, discrimination, and bias they encounter. The key to effectiveness with any ethnic or racial group is not the complete elimination of myths and stereotypes about these groups, but rather, awareness of these stereotypes and management of our behaviors when the stereotypes are not true of the person with whom we are dealing.

Some of the stereotypes that have affected Asian/Pacific Americans in fire service include the following:

1. **Viewing Asian/Pacific Americans as "all alike."** That is, because there are many similarities in names, physical features, and behaviors, many fire service officers may make comments about their inability to tell people apart, or they may deal with them in stereotypic group fashion (e.g., they are all "inscrutable," live in overcrowded households).

2. **Viewing Asian/Pacific Americans as successful, "model minorities" or, worse yet, as a "super minority."** Some hold the stereotype that Asian/Pacific Americans are "all" successful, and this stereotype is further reinforced by the media (Lee, 1996). Such stereotypes have resulted in intergroup hostilities and hate crimes directed toward Asian/Pacific Americans and have served to mask true differences and diversity among the various Asian and Pacific Islander groups. Clearly no groups of people are "all successful" or "all criminals." Nonetheless, the "success" and the "model minority" stereotypes have affected Asian/Pacific Americans negatively. For example, because of their implied academic success, fire service organizations may not spend the time to recruit Asian/Pacific American individuals for fire service careers (assuming that they are more interested in other areas such as education and business pursuits). This stereotype also hides the existence of real discrimination for those who are successful, as seen in glass ceilings in promotional and developmental opportunities, for example.

3. **Viewing some Asian/Pacific Americans as possible "foreign" terrorists because of their religious affiliation and cultural dress.** Many Asian/Pacific Americans emigrate from countries with very large populations that practice Islam (e.g., Indonesia, Pakistan, Bangladesh, India, and China), and the majority of these Muslims are from the more moderate wing of Sunni Islam (and not from the more fundamentalist branch). However, it is highly possible for the unfamiliar to group Asian/Pacific Americans who are Muslims into a collective group associated with "fundamentalism" and "terrorism." Moreover, for many immigrants who may stay close to their cultural traditions and cultural dress (e.g., Sikhs who wear turbans and have beards, Figure 5.19), it is easy to misidentify cultural dress and nuances and come to stereotypic conclusions about who might be a "foreign" terrorist. Two examples: (1) One of the first suspects detained for questioning on September 11, 2001 (as seen on the national television news) was the misidentification of a Sikh who was wearing a turban as being an Afghanistan Taliban (who also wear a head covering that is called a "turban"); and (2) in the Midwest South Asians

Figure 5.19 ◆ Sikhs eat a meal at the Sikh Gurdwara Temple in Queens. Leaders in the Sikh community have participated in a training program presented by the U.S. Department of Justice and Community Relations Service in hopes of improving community relations between public safety officials and diverse cultures and religions.

(e.g., Bangladeshis) with darker skin tones were often misidentified as Arabs and detained for questioning on homeland security issues.

4.   Misunderstanding Asian/Pacific cultural differences and practices and viewing differences stereotypically as a threat to other Americans. The more than 40 Asian/Pacific American groups encompass great differences in life experiences, languages, backgrounds, and cultures. It is easy to make mistakes and draw incorrect conclusions because of such cultural differences. Certainly, when one lacks information about any group, it is natural to draw conclusions based upon our own filtering system, stereotypes, and assumptions. Most of the time, these incorrect assumptions and stereotypes are corrected by favorable contact and actual interpersonal relationships with Asian/Pacific American people. From a fire service perspective, the thrust of the community and fire department relationship, as well as cultural awareness training, is to provide the opportunities to modify stereotypes and to provide opportunities to learn about ethnic communities.

THE ASIAN/PACIFIC AMERICAN FAMILY

Obviously, with over 40 different cultural groups under the one label of Asian/Pacific Americans, we find great differences in how families operate within the various subgroups. We would like to share some common characteristics to describe Asian/Pacific American families that might be of value in fire service and community public safety. Asian/Pacific American families generally exhibit very strong ties among extended family members. It is not unusual for three to four generations of the same family to live under one roof. Moreover, the extended family can even have an ongoing relationship network that spans great geographic distances. For example, family members (all of whom consider themselves as one family) can be engaged in extensive communications and activities with members in the same family in the United States, Canada, Hong Kong, and Vietnam, all simultaneously. It is not uncommon for a fire service officer to come into contact with members of the extended Asian/Pacific American family in the course of servicing these communities. One key to the success of fire service officers in working with an extended family network is the knowledge of how best to communicate with an Asian/Pacific American family and which family members to speak to for information, help, and referral, as seen in the following fire service investigation:

> "Somebody did this to me," said Tieu, watching firefighters with his wife and 6-year-old daughter. "Last night, when I went home, I checked everything." The 12,000-square-foot store at 501 E. 12th St. near Fifth Avenue was largely destroyed, with the roof caved in and the interior gutted. Tieu, who was insured, estimated the cost of the damage would be close to $2 million. . . . Nearly 70 firefighters spent about 90 minutes battling the inferno, and kept the flames from burning an adjacent building. The federal Bureau of Alcohol, Tobacco, Firearms and Explosives joined the Oakland Fire and Police departments and the Alameda County arson task force in investigating the fire. In 1997, the Sun Hop Fat 1 Supermarket became one of four Asian markets in the East Lake district damaged by arson in less than a year. Shortly before the 1997 blaze, Tieu and Truong recalled, they received a letter in Vietnamese requesting money. Nothing similar occurred this time, they said. Truong said, however, that her security company called about midnight telling her a front door alarm had been detected; she asked the company to check it out and call her if necessary. . . . Nearby stood one of the market's nearly 20 employees, Tyson Bu, who has worked as a Sun Hop Fat butcher for the last 15 years. Bu, who is married with two children, 6 and 14 years of age, looked stunned (Hoge, 2005, p. B1).

The Role of the Man and the Woman   In most Asian/Pacific American families, relationship and communication patterns tend to be quite hierarchical, with the father as the identified head of the household. Although many decisions and activities may appear to be determined by the father, other individuals may come into the picture. Generally, if there are grandparents in the household, the father would still act as the spokesperson for the family, but he would consult the grandparents, his wife, and others regarding any major decision. As such, it may be important in any kind of fire service contact that requires a decision and/or choice to allow the parties time to discuss issues in, as much as feasible, a "private" manner. Self-control and keeping things within the family are key values for Asian/Pacific Americans. Fire service officers thus may find that there is more control in a situation by allowing the Asian/Pacific American to come to the same conclusion as he or she with respect to a situation and to exercise his or her own self-choice (which may be the same as what the officer would want the parties to do anyway). For example, the officer can explain a fire safety situation to the father of a family member, and instead of saying directly to the family member that the situation is a violation and must be corrected, the fire service officer can allow the father to suggest that the violation must be corrected. What may appear to be a minor consideration in this case can result in a higher degree of persuasion, control, and cooperation by all parties concerned.

Although there are no clear-cut rules as to whether one goes to the male head of the household or to the female head to make a fire service inquiry, the general rule of thumb is that one would not go too wrong by starting with the father. It should be noted that for most Asian/Pacific American families, the role of the mother in discipline and

FIGURE 5.20 ◆ Asian Indian American children in traditional costumes dance on a float of the India Cultural Society during a parade in New York.

in decision making is very important. While the household may appear to be "ruled" by the father, the mother's role in finances, discipline, education, operations, and decision making is major.

Children, Adolescents, and Youth   Most Asian/Pacific American families involve at least two or more individuals within the same household working outside of the home. Thus, if young children are present, there is a high reliance on either family members or others to help care for them while the parents are at work (Figure 5.20). It is not uncommon for older children to care for younger children within a household. Moreover, latchkey children within an Asian/Pacific American home are common, especially for families that cannot afford external child care. In recent immigrant and refugee families, Asian/Pacific American children have a special role in being the intermediaries between parents and the external community because of the ability of the younger individuals to learn English and the American ways of doing things. Children often serve as translators and interpreters for fire service officers in their communication and relations with Asian/Pacific American families involving recent immigrants and refugees. In such situations, it is suggested that the fire service officer review the role expected of the youthful member of the family, and determine how sensitive an area the translated content is to the different family members and the consequences should the content be incorrectly translated. For example, asking a juvenile to translate to his or her parents who speak no English that the juvenile had been involved in the destruction of public property resulting from playing with matches at the school may result in significant omission and/or changed content. Because of the embarrassment caused to the juvenile and possibly to the parents, the content of the fire service officer's message may be radically altered. In all cases, when a child is acting as a translator, the fire service officer should direct all verbal and nonverbal communication to the parents (as one would normally do without a translator). Otherwise, the parents may view the officer's lack of attention to them as an insult.

VERBAL AND NONVERBAL COMMUNICATION STYLES OF ASIAN/PACIFIC AMERICANS

We do not wish to create any kind of stereotypes, but there are key features of Asian/Pacific American verbal and nonverbal communication styles that necessitate explanation. Misunderstanding resulting from style differences can result in perceptions of poor community services from fire service agencies, conflicts resulting from such misunderstandings, and safety and control issues for fire service officers.

1.   It is important that fire service officers take the time to get information from individuals who may have limitations to their English-speaking abilities. The use of fire service officers who may speak different Asian/Pacific dialects or languages, translators, and language bank resources would greatly enhance this process (Figure 5.21). Often Asian/Pacific Americans have not been helped in public safety situations because officers could not or did not take information from individuals who could not speak English well.

2.   Asian/Pacific Americans tend to hold a more "family" and/or "group" orientation. As such, the lack of the use of "I" statements and/or self-reference should not be evaluated as not being straightforward or as being evasive about oneself or one's relationships. The fire service officer may be concerned because an Asian/Pacific American may wish to use the pronoun *we* when the situation may call for a personal observation involving an "I" statement. For example, in reporting a suspicious fire, the Asian/Pacific American may describe what he or she saw by saying, "We saw. . . ." Using such group statements to convey what the individual saw is consistent with the family and group orientation of Asian/Pacific Americans.

3.   The fire officer must be aware that for many Asian/Pacific Americans, it is considered to be rude, impolite, and to involve a "loss of face" to say "no" directly to an authority figure such

FIGURE 5.21 ◆ Many from the Asian/Pacific American elderly population as well as recent immigrants and refugees require translators and interpreters to communicate with fire services and other public services.

as a fire service officer. Fire service officers need to understand the following possibilities when an answer of "yes" is heard from an Asian/Pacific American. It can mean (1) "Yes, I heard what you said (but I may or may not agree with you)"; (2) "Yes, I understand what you said (but I may or may not do what I understand)"; (3) "Yes, I can see this is important for you (but I may not agree with you on this)"; or (4) "Yes, I agree (and will do what you said)." Because the context of the communication and the nonverbal aspects of the message are equally meaningful, it is vital for fire service officers to be sure of the "yes" answers received, as well as other language nuances from Asian/Pacific Americans. Two examples might be illustrative: (1) If an Asian/Pacific American says that he or she will "try my best to attend," this generally means that he or she will not be there, especially for more voluntary events and situations such as community neighborhood safety meetings. (2) If an Asian/Pacific national says, in response to a question, "It is possible," this generally means do not wait for the event to happen. Such communications, as noted previously, may be more applicable to some Asian/Pacific Americans than others, but sensitivity on the part of fire service officers to these language nuances will facilitate communication. Specific rules for interacting with each Asian/Pacific American group are not necessary, but officers should have a general understanding of language and cultural styles. In a communication situation in which the response of "yes" may be ambiguous, it is suggested that fire service officers rephrase the question so that the requested outcome in action and understanding is demonstrated in the verbal response.

Ambiguous Response
Fire Service Officer: "I need you to remove the fire hazard within 72 hours. Do you understand?"
Asian restaurant owner: "Yes!"
Rephrasing of Question to Show Understanding and Outcome
Fire Service Officer: "What are you going to do within 72 hours?"
Asian restaurant owner: "I have to remove the boxes that are blocking the back door because they are causing a fire hazard in case of an emergency."

4. Asian/Pacific Americans tend to be "high context" in communication style. This means that the fire service officer needs to provide both interpersonal and situational contexts for effective communications. Context for Asian/Pacific Americans means that members of the community know the fire service officers in the community. Community members may have had previous working relationships with the fire service officer (e.g., fire prevention meetings, department open house). Moreover, other members of the community may help to provide information and context for fire service cooperation based on past relationships. Context also means providing explanations and education to members or groups within Asian/Pacific American communities about procedures and fire safety before asking them questions and/or requesting their participation in an activity. By providing background information and by establishing prior relationships with Asian/Pacific American communities, the Asian/Pacific American individual has a context for cooperating with fire departments and officers.

5. Be aware of nonverbal and other cultural nuances that may detract from the effective communication of the fire service officer with a member of the Asian/Pacific American community. Many Asian/Pacific Americans find it uncomfortable and, sometimes, inappropriate to maintain eye contact with authority figures such as uniformed fire service officers. It is considered in many Asian/Pacific American cultures to be disrespectful if there is eye contact with someone who is of higher status, position, importance, or authority. As such, many Asian/Pacific Americans may look down on the ground and/or avert their eyes from gazing at a fire service officer. The fire service officer should not automatically read this nonverbal behavior as indicating a lack of trust or respect, or as a dishonest response. Likewise, for the fire service officer, he or she should be aware of possible nonverbal gestures and actions that may detract from his or her professional roles (e.g., gesturing with the curled index finger for a person to come forward in a manner that might be used only for servants in that person's home culture).

6. Asian/Pacific Americans may not display their emotionality in the same way that the fire service officer expects. The central thesis guiding Asian/Pacific Americans is the Confucian notion of "walking the middle road." This means that extremes, too much or too little of anything, are not good. As such, Asian/Pacific Americans tend to moderate their display of positive and/or negative emotion. Often, in crisis situations, nonverbal displays of emotions are controlled to the point that the affect of the Asian/Pacific American appears "flat." Under such circumstances, the fire service officer needs to understand and interpret such displays of emotion appropriately. For example, just because the Asian homeowner does not appear emotionally shaken by a fire service officer's report that his or her house has burned down does not mean that the person is not experiencing a severe emotional crisis.

SUMMARY OF RECOMMENDATIONS FOR FIRE SERVICE IN THE ASIAN/PACIFIC AMERICAN COMMUNITY

1. As a result of the early immigration laws and other discriminatory treatment received by Asian/Pacific Americans in the United States, the experiences of Asian/Pacific Americans with public safety officials has been fraught with conflicts, difficulties, and mixed messages. Fire service officers should realize that some Asian/Pacific Americans may still remember this history and carry with them stereotypes of fire and other public safety services as something to be feared and avoided. Fire service officials may need to go out of their way to establish trust and to win cooperation in order to accomplish their goals to serve Asian/Pacific Americans effectively.

2. As noted earlier, the label Asian Americans/Pacific Islanders encompasses over 40 very diverse ethnic and cultural groups. Fire service officials need to be aware that great differences exist among the 40 diverse ethnic groups (e.g., of different cultural and language backgrounds) as well as the differences that may result from the individual life experiences of members within any one of the 40 groups (e.g., generational differences). Because a key stereotype of much concern to Asian/Pacific Americans is that they are regarded by mainstream Americans as very much alike, it is important that fire service officers not make such errors in their interactions with Asian/Pacific Americans.

3. There is tremendous diversity among Asian/Pacific Americans, and one way to understand individuals within these communities is to look at some of the motivating forces that might affect decisions by Asian/Pacific American citizens. Earlier in this chapter we provided a seven-part typology that will assist fire service officers in viewing some of these motivational bases. Although there are many ethnicities, cultures, and languages among the 40 or more groups within Asian/Pacific American communities, one way to understand the impact of their immigration and life experiences is by learning the motivational determinants of individuals within different generational and immigrant groups.

4. The self-preferred term for referring to Asian/Pacific Americans varies with contexts, groups, and experiences of the individual. Fire service officials need to be aware of terms that are unacceptable and derogatory and terms that are currently used. When in doubt, fire service officers have to learn to become comfortable in asking Asian/Pacific Americans which terms they prefer. Fire service officers are advised to provide helpful feedback to their peers when offensive terms, labels, and/or actions are used with Asian/Pacific Americans. Such feedback will help reduce the risk of misunderstanding and improve the working relationships between fire service officers and individuals within Asian/Pacific American communities. Moreover, it will help enhance the professional image of the fire department for those communities.

5. Many Asian/Pacific Americans are concerned with their ability to communicate clearly, and this is of particular concern among Asian/Pacific Americans who are immigrants and refugees. Fire service officers need to take the time and be aware that bilingual individuals and nonnative English speakers want to communicate effectively with them. Maintaining contact, providing extra time, using translators, and being patient with speakers will allow Asian/Pacific Americans to communicate their concerns.

6.   Cultural differences in verbal and nonverbal communication often result in misinterpretation of the message and of behaviors. Fire service officers need to be aware of nonverbal aspects of Asian/Pacific Americans in their communication styles, including eye contact, touch, gestures, and affect (show of emotions). Verbal aspects such as accent, limited vocabulary, and incorrect grammar may give fire service officers the impression that an Asian/Pacific American individual is not understanding what is communicated. It is important to remember that the English listening and comprehension skills of Asian/Pacific American immigrants and refugees are usually better than their speaking skills.

7.   Asian/Pacific Americans, because of their past experiences with public safety agencies, along with their own concerns about privacy, self-help, and other factors, are reluctant to report fire hazards and may not seek fire service assistance and help. Fire departments and officials need to build relationships and working partnerships with representative groups from the Asian/Pacific American communities. Relationship building is often helped by outreach efforts such as community open houses, availability of bilingual officers, and participation of fire service officers in community activities.

LATINO/HISPANIC AMERICANS

This section provides specific ethnic and cultural information on Latino/Hispanic Americans. The label Latino/Hispanic Americans encompasses over 25 different ethnic and cultural groups from Central and South America and the Caribbean. We first define this highly diverse group and then provide a historical overview that will contribute to readers' understanding of the relationship between fire service personnel and citizens. We present demographics and elements of diversity among Latino/Hispanic Americans, as well as issues related to ethnic and cultural identity: labels and terms, myths and stereotypes, the extended family and community, gender roles, generational differences, and adolescent and youth issues. We present, in the closing section, recommendations for improved communication and relationships between fire service personnel and members of Latino/Hispanic American communities (Figure 5.22).

Latino/Hispanic Americans are the fastest growing cultural group in the United States in terms of overall numbers of people. Between the 1990 and 2000 censuses, the population increased by more than 50 percent, from 22.4 million in 1990 to 32.8 million in 2000. The Hispanic population reached 41.3 million as of July 1, 2004, according to national estimates by the U.S. Census Bureau (Bernstein, 2005). Growth in all urban and rural areas of the United States has been striking, with the Hispanic population having a sizable presence in virtually all geographic areas within the United States. The population growth can be attributed primarily to higher birthrates (Cohn, 2005) and secondarily to the following three factors: (1) higher immigration from Mexico, Central and South America, and the Caribbean; (2) greater longevity, because this is a relatively young population; and (3) larger numbers of subgroups being incorporated into the Latino/Hispanic American grouping (Korzenny and Korzenny, 2005). Moreover, the U.S. Census estimates for 2004 did not include the number of undocumented immigrants or unauthorized migrants. The Pew Hispanic Institute estimated that the total number of unauthorized migrants in the United States in 2004 was 10.3 million, with 5.9 million from Mexico and another 2.5 million

FIGURE 5.22 ◆ Latino/Hispanic American and other multiethnic children at the Brooklyn Fire Department.

from the rest of Latin America (Passel, 2005). Various other labels have been applied to this group of unauthorized migrants, including illegals, illegal aliens, and illegal immigrants. For the Latino/Hispanic American community, we use the specific term *unauthorized migrant* to mean "a person who resides in the United States, but who is not a U.S. citizen, has not been admitted for permanent residence, and is not included within a set of specific authorized temporary statuses permitting longer-term residence and work" (Passel, Van Hook, and Bean, 2004).

The total number of unauthorized migrants from Mexico and the rest of Latin American is estimated to be 8.3 million, and if these numbers were added to the census estimates for the documented Hispanic population in the United States, the total Hispanic presence would be 49.6 million people. At 49.6 million total Latino/Hispanic people in the United States, that would make the United States the second-largest nation in the world for Latino/Hispanic presence, second only to Mexico, and with a population greater than the third-largest Hispanic nation, Spain, with 40 million people (Central Intelligence Agency, 2006). Although Brazil, with 188 million people, is in South America, the vast majority of the people are of Portuguese descent, and only a minority would be considered to be Latino/Hispanic (Korzenny and Korzenny, 2005). Given the current and anticipated growth of this population, all federal, regional, state, and local fire service departments need to prepare to serve this growing presence in our communities.

LATINO/HISPANIC AMERICANS DEFINED

Hispanic is not a racial group; a person can be white, black, Asian, and so on and still be considered Hispanic. Hispanic is a generic term referring to all Spanish-surname and Spanish-speaking people who reside in the United States and in Puerto Rico (a commonwealth). Latino is generally the preferred label on the West Coast, parts of the East Coast, and the Southeast. The term *Latino* is a Spanish word indicating a person of Latin American origin, and it reflects the gender-specific nature of its Spanish language derivation: Latino, for men, and Latina, for women. Hispanic is generally preferred on the East Coast, primarily by the Puerto Rican, Dominican, and Cuban communities (although individual members within each of these communities may prefer the specific term referring to their country of heritage). Objections to the use of the term *Hispanic* include the following: (1) Hispanic is not derived from any culture or place (i.e., there is no such place as "Hispania"), and (2) the term was primarily invented for use by the U.S. Census Bureau. Sometimes, the term *Spanish speaking/Spanish surnamed* may be used to recognize the fact that one may have a Spanish surname, but may not speak Spanish (which is the case for a large number of Latino/Hispanic Americans). *La Raza* is another term used, primarily on the West Coast and in the Southwest, to refer to all peoples of the Western Hemisphere who share a cultural, historical, political, and social legacy of the Spanish and Portuguese colonists and the Native Indian and African people; it has its origins within the political struggles of this region and the mixing of the races, *el mestizaje*.

Like La Raza, *Chicano* is another term that grew out of the ethnic pride and ethnic studies movement in the late 1960s. Chicano refers specifically to Mexican Americans and is used primarily on the West Coast, in the Southwest, and in the Midwest. It is also commonly utilized in college communities across the United States that provide an ethnic studies curriculum.

In 1976, Congress passed Public Law 94-311, called the Roybal Resolution, which required the inclusion of a self-identification question on Spanish origin or descent in government surveys and censuses. As such, *Hispanic* is the official term used in federal, state, and local governmental writings and for demographic references. Federal standards implemented in 2003 allow the terms *Latino* and *Hispanic* to be used interchangeably (Office of Management and Budget, 1997).

As the Latino/Hispanic American communities grow and develop in the United States, the preferred and specific terms to be used also will evolve and be changed by the members of these communities.

HISTORICAL INFORMATION

The historical background of Latino/Hispanic Americans contains key factors that affect their interactions with and understanding of fire service and public safety personnel. Clearly, this brief historical and sociopolitical overview can only highlight some of the commonalities and diverse cultural experiences of Latino/Hispanic Americans. Our historical review focuses primarily on the larger Latino/Hispanic communities in the United States (those with Mexican, Puerto Rican, and Cuban ethnic and historical roots).

In the 1800s, under the declaration of Manifest Destiny, the United States began the expansionist policy of annexing vast territories to the south, north, and west. As Lopez y Rivas (1973) noted, the United States viewed itself as a people chosen by "Providence" to form a larger union through conquest, purchase, and annexation. With the purchase (or annexation) of the Louisiana Territories in 1803, Florida in 1819, Texas in 1845, and the Northwest Territories (Oregon, Washington, Idaho, Wyoming, and Montana) in 1846, it seemed nearly inevitable that conflict would occur with Mexico. The resulting Mexican-American War ended in 1848 with the signing of the Treaty of Guadalupe Hidalgo, in which Mexico received $15 million from the United States for the land that is now Texas, New Mexico, Arizona, and California, with more than 100,000 Mexican people living in those areas. As is obvious from this portion of history, it makes little sense for many Mexican Americans to be stereotyped as "illegal aliens," especially since more than a million Mexican Americans (some of whom are U.S. citizens and some of whom are not) can trace their ancestry back to families living in the southwestern United States in the mid-1800s (Fernandex, 1970). Moreover, for Latino/Hispanic Americans (especially Mexican Americans), the boundaries between the United States and Mexico are seen as artificial. "The geographic, ecological, and cultural blending of the Southwest with Mexico is perceived as a continuing unity of people whose claim to the Southwest is rooted in the land itself" (Montiel, 1978).

Although one-third of Mexican Americans can trace their ancestry to families living in the United States in the mid-1800s, the majority of this group migrated into the United States after 1910 because of the economic and political changes that occurred as a result of the Mexican Revolution.

Puerto Rico, on the other hand, was under the domination of Spain until 1897, at which time it was allowed the establishment of a local government. The United States invaded Puerto Rico and annexed it as part of the Spanish-American War (along with Cuba, the Philippines, and Guam) in 1898. Although Cuba (in 1902) and the Philippines (in 1949) were given their independence, Puerto Rico remained a territory of the United States. In 1900 the U.S. Congress passed the Foraker Act, which allowed the president to appoint a governor, to provide an Executive Council consisting of 11 presidential appointees (of which only 5 had to be Puerto Rican), and to elect locally a 35-member Chamber of Delegates. In reality, the territory was run by the president-appointed governor and the Executive Council. The Jones Act of 1917 made Puerto Ricans citizens of the United States. It was not until 1948 that Puerto Rico elected its first governor, Luis Munoz Marin. In 1952 Puerto Rico was given commonwealth status, and Spanish was allowed to be the language of instruction in the schools again (with English taught as the second language).

Following World War II, large numbers of Puerto Ricans began migrating into the United States. With citizenship status, Puerto Ricans could travel easily and settled in areas on the East Coast, primarily New York City (in part because of the availability of jobs and affordable apartments). The estimated number of Puerto Ricans in the United States is 2 million on the mainland and 3.8 million on the island (Therrien and Ramirez, 2000).

Cubans immigrated into the United States in three waves. The first wave occurred between 1959 and 1965 and consisted of primarily white middle-class or upper-class Cubans who were relatively well educated and had business and financial resources. The federal government's Cuban Refugee Program, Cuban Student Loan Program, and Cuban Small Business Administration Loan Program were established to help this first wave of Cuban immigrants achieve a successful settlement (Bernal and Estrada, 1985). The second wave of Cuban immigrants occurred between 1965 and 1973. This wave resulted from the opening of the Port of Camarioca, allowing all who wished to leave Cuba to exit. Those who left as part of the second wave were more often of the working class and lower middle class, primarily white adult men and women. The third wave of immigrants leaving Cuba from Mariel occurred from the summer of 1980 to early 1982. This third wave was the largest (about 125,000 were boat-lifted to the United States) and consisted primarily of working-class persons, more reflective of the

Cuban population as a whole than previous waves. Most immigrated into the United States with hopes for better economic opportunities. Within this group of *Marielito* were many antisocial, criminal, and mentally ill persons released by Fidel Castro and included in the boat lift (Gavzer, 1993).

In addition to the three major groups that have immigrated to the United States from Mexico, Puerto Rico, and Cuba are immigrants from 21 countries of South and Central America, as well as the Caribbean. Arrival of these immigrants for political, economic, and social reasons began in the early 1980s and has added to the diversity of Latino/Hispanic American communities in the United States. The total numbers of some groups, such as the Dominicans (a rapidly growing group on the East Coast), are difficult to determine because of their undocumented entry status in the United States.

DEMOGRAPHICS: DIVERSITY AMONG LATINO/HISPANIC AMERICANS

Composed of many different cultural groups, this broad label encompasses significant generational, educational, and socioeconomic differences, varying relocation experiences, and many other varieties of life experience (Figure 5.23). Although the Spanish language may provide a common thread among most Latino/Hispanic Americans, there are cultural and national differences in terms and expressions used, including nonverbal nuances. Moreover, the language of Brazil is Portuguese, not Spanish, and thus the language connection for Brazilian Latino/Hispanic Americans is unique. The Latino/Hispanic American population numbers about 32.8 million and represents 12.0 percent of the U.S. population (not including the 3.9 million people who live in Puerto Rico). About 85 percent of Latino/Hispanic Americans trace their roots to Mexico, Puerto Rico, and Cuba, while the remaining 15 percent are from the other countries of Central and South America, the Caribbean, and Spain. The numerical growth of the Latino/Hispanic American population is the fastest of all American

FIGURE 5.23 ◆ The diversity of the Latino/Hispanic American community is reflected in the crowd at Cinco de Mayo festival.

ethnic groups, with more than 10 million people added to the U.S. population between the 1990 and 2000 censuses (Therrien and Ramirez, 2000).

Latino/Hispanic Americans are concentrated in five states: California (31.1 percent), Texas (18.9 percent), New York (8.1 percent), Florida (7.6 percent), and Illinois (4.3 percent). Some of the key demographics information about this population includes the following:

1. **Age.** The Latino/Hispanic American population tends to be younger than the general U.S. population. The median age is 25.9 years, in contrast to 35.3 for the rest of America.
2. **Size of household.** The average Latino/Hispanic household consists of 3.5 people, which is nearly one person more for every Latino/Hispanic household than that for other U.S. households, which average 2.6 persons per household.
3. **Birthrate.** Latino/Hispanic Americans have a higher birthrate than the general U.S. population. The Latino/Hispanic American birthrate per 1,000 is 104.8, in comparison to 65.4 for the rest of America.
4. **Purchasing power.** The estimated purchasing power of Latino/Hispanic Americans in the United States in 2004 was $700 billion (Korzenny and Korzenny, 2005).
5. **Urban households.** About 88 percent of all Latino/Hispanic Americans live in metropolitan areas, making this group the most highly urbanized population in the United States. For example, Latino/Hispanic Americans constitute notable percentages of the population for the following large cities: El Paso (74 percent), Corpus Christi (57 percent), San Antonio (52 percent), Fresno (41 percent), Seattle-Tacoma (40 percent), Los Angeles (38 percent), Albuquerque (38 percent), Miami–Ft. Lauderdale (37 percent), Tucson (27 percent), San Diego (25 percent), Austin (24 percent), and San Francisco/Oakland (19 percent) (Hornor, 1999).
6. **Language.** Latino/Hispanic American self-identification is most strongly demonstrated in the use and knowledge of Spanish. The Spanish language is oftentimes the single most important cultural aspect retained by Latino/Hispanic Americans. As might be expected, with the large number of Latino/Hispanic Americans born in the United States, English may soon be the dominant language as this population moves into successive future generations.

LABELS AND TERMS

As noted earlier, the term *Latino/Hispanic American* is a convenient summarizing label to achieve some degree of agreement in referring to a very heterogeneous group of people. Similar to the case of Asian/Pacific Americans, the key to understanding which terms to use is based on the principle of self-preference. For those who have origins in the Caribbean (e.g., Puerto Rican, Cuban, Dominican), the term *Latino* may be equally problematic for self-designation and self-identification.

Until 2003, federal and other governmental designations used only *Hispanic*; *Latino* and *Hispanic* are now used interchangeably. The governmental designations are used in laws, programs, and regulations and in most reports and publications. For individuals within any of the groups, often the more specific names for the groups are preferred (e.g., Mexican, Puerto Rican, Cuban, Dominican, Argentinean, Salvadorian). Some individuals may prefer that the term *American* be part of their designation (e.g., Mexican American). For fire service officers, the best term to use in referring to individuals is the term that they prefer to be called. It is perfectly acceptable to ask individuals what they prefer to be called.

The use of slurs such as "wetback," "Mex," "Spic," "Greaser," or other derogatory ethnic slang terms is never acceptable for use by fire service officers, no matter how provoked an officer may be. Other stereotypic terms such as "illegal," "New York Rican," "Macho man," "Latin lover," "Lupe the Virgin," and "low rider" do not convey the kinds of professionalism and respect for community diversity important to fire-fighting and fire service, and need to be avoided in fire service work.

> City officials offered Knoxville Fire Chief Carlos Perez a public apology after the first-ever Hispanic head of the Fire Department filed a formal complaint alleging that racially insensitive comments were made about him. The complaint, filed earlier this month with the city's Civil Service Department, was made against one specific departmental employee who allegedly referred to the chief as "Pedro" in a derogatory manner, city officials confirmed. An investigation of the allegations is complete, city spokeswoman Margie Nichols said Wednesday. "The investigation found no (supporting evidence) that it was used by anyone else," she said. After learning about the incident, City Councilman Steve Hall offered his apology to Perez on behalf of the city at the council's meeting Tuesday night. "We as a government can't condone that kind of behavior from our employees," Hall said after the meeting. "Whether it's aimed at blacks, Hispanics, other minorities or women" (Hickman, 2006, p. B1).

Fire service officers hearing these or other similar words used in their own departments (or with peers or citizens) should provide immediate feedback about the inappropriateness of the use of such. Officers who may, out of habit, routinely use these terms may find themselves or their superiors in the embarrassing situation of explaining to offended citizens and communities why they used the term and how they intended no bias, stereotype, or prejudice.

MYTHS AND STEREOTYPES

Knowledge of and sensitivity to Latino/Hispanic Americans' concerns, diversity, historical background, and life experiences will facilitate the fire service mission with this ethnic group. It is important to have an understanding about some of the myths and stereotypes of Latino/Hispanic Americans that contribute to the prejudice, discrimination, and bias this population encounters. Many fire service officers do not have much experience with the diversity of Latino/Hispanic American groups and learn about these groups only through stereotypes (often perpetuated by movies) and through the very limited contact involved in their fire service duties. Stereotypic views of Latino/Hispanic Americans reduce individuals within this group to simplistic, one-dimensional characters and have led many Americans to lump members of this group into one stereotypic group, "Mexicans." It is important for fire service officers to be aware of the different stereotypes of Latino/Hispanic Americans. The key to effectiveness with any ethnic or racial group is not that we completely eliminate myths and stereotypes about these groups, but that we are aware of these stereotypes and can monitor our thinking and our behaviors when the stereotypes are not true of those persons with whom we are interacting.

Some of the stereotypes that have affected Latino/Hispanic Americans in fire service include the following:

1. **Viewing Latino/Hispanic Americans as "illegal aliens."**   Although many may argue over the number of Latino/Hispanic illegal and undocumented immigrants, this stereotype does not represent the vast majority of Latino/Hispanic Americans in the United States, who are U.S. citizens or legal residents (Figure 5.24). The issues of illegal aliens and undocumented immigrants are complex ones, but cultural awareness and sensitivity will allow fire service officers

FIGURE 5.24 ◆ Mural by Dominican artist depicts Latino/Hispanic community and immigration and travel on a New York street.

the knowledge to avoid an offensive situation by acting on the stereotype of Latino/Hispanic Americans as illegal immigrants.

2. Viewing Latino/Hispanic Americans as lazy and as poor workers. This is a stereotype that has been perpetuated in the workplace. It is difficult to understand why some continue to hold this stereotype or what factors continue to perpetuate it given what we know about the Latino/Hispanic American workforce in the United States and globally. Latino/Hispanic American community advocates make the argument that it is difficult to imagine anyone being labeled as "lazy" or "poor workers" if they are willing to work as laborers from dawn to dusk in the migrant farm fields, day in and day out, year after year. Moreover, comparisons of the labor force participation rate for all persons 16 years old and over showed virtually no differences between Latino/Hispanic Americans (67.9 percent) and all other ethnic and racial groups (67.1 percent) working into today's labor force (Hornor, 1999).

3. Perceiving Latino/Hispanic Americans as uneducated and uninterested in educational pursuits. Prior to 1948 many Latino/Hispanic American children were denied access to the educational system available to others and instead were relegated to "Mexican" schools. A challenge in the U.S. courts on segregated schools allowed Latino/Hispanic children access to the "regular" school system (see *Mendez v. Westminster School District*, 1946, and *Delgado v. Bastrop Independent School District*, 1948). This stereotype of "uneducated" also relates to how Latino/Hispanic officers may be inappropriately stereotyped and seen in terms of their ability to learn and to achieve in fire service and other professional firefighting training.

4. Seeing Latino/Hispanic American young males as gang members and drug dealers. Some hold the stereotype, especially among young males in inner cities, that Latino/Hispanic Americans are commonly involved in gangs and the illegal drug trade. Latino/Hispanic cultures are group oriented, and people from young to old tend to congregate as groups rather than individually or as couples. "Hanging out" as a group tends to be the preferred mode of socialization. However, given the stereotype of Latino/Hispanic American young males as being gang members, it is easy to perceive five young Latino/Hispanic males walking together as constituting "a gang." Such stereotypes have resulted in suspicion, hostility, and prejudice toward Latino/Hispanic Americans and have served as attempts to justify improper treatment (e.g., routine traffic stops) and poor service (e.g., in restaurants and stores).

5. Assuming that all Latino/Hispanic Americans speak Spanish. Twenty-one percent of Latino/Hispanic Americans reported that their families speak predominantly English in the home (Yankelovich Monitor, 2003). Many Latino/Hispanic Americans have been in the United States for six generations or more, and English is the only language that they speak and write. George Perez is a Latino/Hispanic American cultural awareness trainer for many fire service, emergency service, and other public service organizations. As one of the outstanding trainers in his field, he is not surprised by the frequently heard stereotypic comment of "You speak English so well, without an accent. How did you do it?" His reply will usually be, "It's the only language I know! I'm a fifth generation Latino/Hispanic American. I grew up in northern California and received my bachelor's and master's degrees in the field that I teach."

THE LATINO/HISPANIC AMERICAN FAMILY

Understanding the importance of family for Latino/Hispanic Americans might be of significant value in community fire service. Obviously, with over 25 different cultural groups that make up the Latino/Hispanic American category, many differences exist among the groups within this collective with respect to their family experiences. We would like to address some family characteristics that many of the different cultural groups share. *La familia* is perhaps one of the most significant considerations in working and communicating with Latino/Hispanic Americans. (In places where we use the Spanish term, we have done so to indicate the additional cultural meanings encompassed in a term such as *la familia* that are not captured in the English term the *family*.) The Latino/Hispanic American family is most clearly characterized by bonds of interdependence, unity, and loyalty and includes nuclear and extended family members, as well as networks of neighbors, friends, and community members. Primary importance is given to the history of the family, which is firmly rooted in the set of obligations tied to both the past and the future. In considering the different loyalty bonds, the parent–child relationship emerges as primary, with all children owing *respecto* to parents (*respecto* connotes additional cultural meanings than does the English term *respect*). Traditionally, the role of the father has been that of the disciplinarian, decision maker, and head of the household. The father's word is the law, and he is not to be questioned. The father will tend to focus his attention on the economic and well-being issues of the family and less on the social and emotional issues. The mother, on the other hand, is seen as balancing the father's role through her role in providing for the emotional and expressive issues of the family. Extended family members such as grandmothers, aunts and uncles, and godparents (*compadrazgo*) may supplement the mother's emotional support. In the Latino/Hispanic American family, the older son is traditionally the secondary decision maker to the father and the principal inheritor (*primogenito*) of the family. Because of the central nature of the Latino/Hispanic American family, it is common for fire service officers to come into contact with members of nuclear and extended families in the course of working with the Latino/Hispanic American community (Figure 5.25). One key to the success of fire service officers in working with the Latino/Hispanic American extended family network is the knowledge of how best to communicate within the family context and an understanding of with whom to speak for information, observations, and questions.

FIGURE 5.25 ◆ Grandfather and granddaughter who are part of a three-generational, extended Latino/Hispanic American family.

The Role of the Man and the Woman   In many Latino/Hispanic American families, the relationship and communication patterns are hierarchical with the father as the identified head of the household, who is held in high respect. When it comes to family well-being, economic issues, and discipline, the father may appear to be the decision maker; however, many other individuals may come into the picture. Generally, if there are grandparents in the household, the father may consult them, as well as his wife, on major decisions. In the case of fire service or public safety matters, it may be of great importance for officers to provide the father and the family with some privacy, as much as possible, to discuss key issues and situations. With central values such as *respeto* (respect) and *machismo* (see next) in the Latino/Hispanic American culture, it is critical for the father and other family members to demonstrate control in family situations.

Within the Latino/Hispanic American family, the sex roles are clearly defined; boys and girls are taught from childhood two different codes of behavior (Comas-Diaz and Griffith, 1988). Traditional sex roles can be discussed within the context of the two codes of gender-related behaviors: *machismo* and *marianismo*. *Machismo* literally means maleness, manliness, and virility. Within the Latino/Hispanic American cultural context, machismo means that the male is responsible for the well-being and honor of the family and is in the provider role. *Machismo* is also associated with having power over women, as well as responsibility for guarding and protecting them. Boys are seen as strong by nature and do not need the protection that is required by girls, who are seen as weak by nature.

Women are socialized into the role of *marianismo*, based on the beliefs about the Virgin Mary, in which women are considered spiritually superior to men and therefore able to endure all suffering inflicted by men (Stevens, 1973). Women are expected to be self-sacrificing in favor of their husbands and children. Within the context of the Latino/Hispanic American family, the role of the woman is as homemaker and caretaker of the children. In the current U.S. context, the traditional gender roles of women and men in the Latino/Hispanic American community have undergone much change, which has resulted in key conflicts. Since women have begun to work, earn money, and have some of the financial independence men have (e.g., they can go out and socialize with others outside of the family), they have pursued many experiences inconsistent with the traditional Latino/Hispanic female role. Although there are no clear-cut rules as to whether fire service officers should go to the male head of the household or to the female family member, fire service officers would probably be more correct to address the father first in public safety inquiries. Consistent with the cultural values of *machismo* and *marianismo*, the Latino/Hispanic American household appears to be run by the father; however, in actual practice, the mother's role in discipline, education, finance, and decision making is also central.

Children, Adolescents, and Youths   Within the Latino/Hispanic American family, the ideal child is obedient and respectful of his or her parents and other elders. Adults may at times talk in front of the children as if they are not present and as if the children cannot understand the adults' conversations. Children are taught *respecto*, which dictates the appropriate behavior toward all authority figures, older people, parents, relatives, and others. If children are disrespectful, they are punished and scolded. In many traditional families, it is considered appropriate for parents (and for relatives) to discipline a disrespectful and misbehaving child physically.

In Latino/Hispanic American households, there is a high reliance on family members (older children and other adults) to help care for younger children. It is not uncommon for Latino/Hispanic American families to have latchkey children or have children cared for by older children in the neighborhood. In addition, as in other communities in which English is the second language, Latino/Hispanic American children have a special role in being the intermediaries for their parents on external community matters because of the ability of the younger individuals to learn English and the American ways of doing things. Children often serve as translators and interpreters for public safety officers in their communication and relations with Latino/Hispanic American families involved in legal matters, immigration concerns, and community resources. Although the use of children and family members as translators is viewed as professionally and culturally inappropriate, oftentimes it is the only means available to the fire service officer. In such situations, it is suggested that the fire service officer review what role is expected of the youthful member of the family. The fire service officer needs to see how sensitive a topic might be for different family members. In all cases, when a child is acting as a translator for parents, the fire service officer should direct all verbal and nonverbal communication to the parents. Otherwise, the parents may view the officer's lack of attention to them as an insult. Such sensitivity by the fire service officer is particularly important for Latino/Hispanic Americans because of the cultural value of *personalismo*, which emphasizes the importance of the personal quality of any interaction. This cultural concept implies that relationships occur between particular individuals as persons, not as representatives of institutions (e.g., fire department) or merely as individuals performing a role (e.g., as a person who fights fires).

VERBAL AND NONVERBAL COMMUNICATION STYLES OF LATINO/HISPANIC AMERICANS

Although we do not wish to create stereotypes of any kind, key features of Latino/Hispanic American verbal and nonverbal communication styles necessitate explanation.

Misunderstanding resulting from style differences can result in perceptions of poor community services from fire service agencies, conflicts resulting from such misunderstandings, and safety and control issues for the fire service officer.

1. Latino/Hispanic Americans' high cultural value for *la familia* results in a very strong family and group orientation. As such, fire service officers should not view the frequently seen behavior of "eye checking" with other family members before answering and the lack of the use of "I" statements and/or self-reference as Latino/Hispanic Americans not being straightforward about themselves or their relationships. The fire service officer may be concerned because a Latino/Hispanic American interviewee may wish to use the pronoun *we* when the situation may call for a personal observation involving an "I" statement. For example, a Latino/Hispanic American family member who witnessed a store fire may first nonverbally check with other family members before talking and then describing what he or she saw. Such verbal and nonverbal behavior is consistent with the family and group orientation of Latino/Hispanic Americans.

2. Speaking Spanish to others in the presence of a fire service officer, even though the officer had requested responses in English, should not automatically be interpreted as an insult or as an attempt to hide information from the fire service officer (Figure 5.26). In times of stress, those who speak English as a second language will automatically revert to their first and native language. In a fire service situation, many individuals may find themselves under stress and may speak Spanish, which is the more accessible and comfortable language for them. Moreover, speaking Spanish gives the individual a greater range of expression (to discuss and clarify complex issues with other speakers and family members), thus yielding more useful and clearer information to fire service personnel about critical events.

3. It is important for fire service officers to take the time to get information from individuals who may have limited skills in speaking English (the use of fire service officers who speak Spanish, translators, and language bank resources will help). Often Latino/Hispanic Americans have not been helped in fire service and public safety situations because officers could not or did not take information from nonnative English speakers.

FIGURE 5.26 ◆ Hispanic firefighter is interviewed about his work with the Latino/Hispanic community.

4. Although Latino/Hispanic Americans may show respect to fire service officers because of the fire department authority, they do not necessarily trust the fire service officers or the organization. *Respecto* is extended to elders and those who are in authority. This respect is denoted in the Spanish language, by the use of *Usted* (the formal you) rather than *tu* (the informal you). Showing respect, however, does not ensure trust. The cultural value of *confianza* (or trust) takes some time to develop. Like many from ethnic minority communities, Latino/Hispanic Americans, as a **micro culture or minority,** have experienced some degree of prejudice and discrimination from the **macro culture/majority or dominant group** and citizens with such experiences need time to develop trust with public safety officers, who are identified as being a part of the majority community. The following example illustrates how negative images and relationships with fire service officers might be formed:

A man who says he wasn't aware that he needed to pay dues in order to get fire protection for his rural property estimates damage from a fire last week at which firefighters declined to assist were about $30,000. Bivaldo Rueda, 45, said he also suffered burns fighting the fire on his own, and missed a week of work at Tyson Foods because of it. He said although he saw an ambulance at the scene of the Feb. 13 fire, no one offered him medical assistance. His son drove him to a hospital for treatment. Rueda, through a Spanish-speaking interpreter, told *The Springfield News-Leader* on Thursday that shortly after the fire broke out in the garage he owns south of Monett, a member of the Monett Rural Fire Department approached with a hose. But, he said, the firefighter's supervisor ordered him not to use it after learning that Rueda was not a dues-paying member. Rueda said he did not know about the dues requirement, and that he offered to pay at the scene but was refused. According to his account, there were eight firefighters with two fire trucks, along with emergency and Barry County Sheriff's Department personnel at the scene. Some of the firefighters, Rueda said, seemed amused as he scrambled to fill a bucket with water to douse the flames. "I'm struggling to put out the fire and I'm sweating big time," he said. "I felt like I was a major participant in a movie, and the other guys were the audience." Rueda said a few of his neighbors wanted to help, but there was only one bucket. "There was not even a garden hose," he said ("Man Whose Garage Burned Says He Didn't Know About Dues—He'll Pay Now," 2006).

micro culture or minority
Any group or person who differs from the dominant culture. Any group or individual, including second- and third-generation foreigners, who is born in a country different from his or her origin and has adopted or embraced the values and culture of the dominant culture.

macro culture/majority or dominant group
The group within a society that is largest and/or most powerful. This power usually extends to setting cultural norms for the society as a whole. The term *majority* (also *minority*) is falling into disuse because its connotations of group size may be inaccurate.

5. The cultural value of *personalismo* emphasizes the importance of the person involved in any interaction. Latino/Hispanic Americans take into strong consideration not only the content of any communication but also the context and relationship of the communicator. This means that it is important for the fire service officer to provide information about why questions are asked, who is the person asking the question (i.e., information about the officer), and how the information will be used in the context for effective communications. Additionally, context for Latino/Hispanic Americans means taking some time to find out, as well as to self-disclose, some background information (e.g., living in the same neighborhood, having similar concerns about fire safety). Additional contextual elements, such as providing explanations and information to Latino/Hispanic Americans about procedures, regulations, and so forth before asking them questions or requesting their help, will ease the work of the fire service officer. By providing background information and establishing prior relationships with community members, fire service agencies and officers build a context for cooperation with Latino/Hispanic American individuals.

6. Fire service officers should be cognizant of nonverbal and other cultural nuances that may detract from effective communication. Many Latino/Hispanic Americans, especially younger individuals, find it uncomfortable and sometimes inappropriate to maintain eye contact with authority figures such as uniformed fire service officers. Strong eye contact with someone who is of higher position, importance, or authority is considered a lack of

respeto in Latino/Hispanic American cultures. As such, many citizens from this background may deflect their eyes from gazing at fire service officers. It is important that officers not automatically read this nonverbal behavior as indicative of a lack of trust or as a dishonest response.

7.   Latino/Hispanic Americans may exhibit behaviors that appear to be evasive, such as claiming not to have any identification or by saying that they do not speak English. In some of the native countries from which many Latino/Hispanic Americans have emigrated, the public safety agencies are aligned with a politically repressive government. Therefore, many Latino/Hispanic Americans may have similar "fear" reactions to any kind of uniformed officers including fire service in the United States. It is suggested that fire service officers take the time to explain the need for identification and cooperation, and that they acknowledge the importance of comprehension on the part of the Latino/Hispanic American individual(s) involved.

SUMMARY OF RECOMMENDATIONS FOR FIRE SERVICE IN THE LATINO/HISPANIC AMERICAN COMMUNITY

1.   The experience of Latino/Hispanic Americans with public safety officers in the United States has been complicated (1) by the perceptions of Latino/Hispanic Americans regarding the enforcement of immigration laws against illegal aliens and by the discriminatory treatment received by Latino/Hispanic Americans in the United States, and (2) by community conflicts, as well as perceptions of fire service ineffectiveness and unresponsiveness. Fire service officers should realize that some citizens may still remember this history and carry with them stereotypes of public safety services as something to be feared and avoided. Fire service officials need to go out of their way to establish trust, to provide outreach efforts, and to win cooperation in order to effectively accomplish their goals to serve Latino/Hispanic Americans. Building partnerships that are focused upon community collaboration in the fire service is important.

2.   The label *Latino/Hispanic Americans* encompasses over 25 very diverse ethnic, cultural, and regional groups from North, Central, and South America. Fire service officials need to be aware of the differences between the diverse groups (e.g., nationality, native cultural and regional differences and perceptions, and language dialects), as well as the within-group differences that may result from individual life experiences (e.g., sociopolitical turmoil). Because key stereotypes of Latino/Hispanic Americans by mainstream Americans are regarded as more negative than positive, it is important that fire service officers make a special effort to extend respect and dignity to this community of very proud people with a culturally rich heritage.

3.   The preferred term for referring to Latino/Hispanic Americans varies with the contexts, groups, and experiences of Latino/Hispanic American individuals. Fire service officials need to be aware of terms that are unacceptable and derogatory and terms that are currently used. When in doubt, fire service officers have to learn to become comfortable in asking citizens which terms they prefer. Fire service officers are advised to provide helpful feedback to their peers whenever offensive terms, slurs, labels, and/or actions are used with Latino/Hispanic Americans. Such feedback will help reduce the risk of misunderstanding and improve the working relationships of officers within the Latino/Hispanic American communities. Additionally, it will help enhance the professional image of the fire department for those communities.

4.   Many Latino/Hispanic Americans are concerned with their ability to communicate clearly and about possible reprisal from the public safety personnel, as associated with the role of those in uniform in more politically repressive countries. Fire service officers need to take the time required to understand communications and need to be aware that bilingual and nonnative English speakers want to communicate effectively with them. Maintaining contact, providing extra time, using translators, and being patient with speakers encourage citizens to communicate their concerns. Cultural differences in verbal and nonverbal communication often result in misinterpretation of the message and of behaviors. Fire service officers need to be aware of the nonverbal aspects of some Latino/Hispanic American communication styles, such as eye contact, touch, gestures, and emotionality. Verbal aspects such as accent, mixing English with Spanish, limited vocabulary, and incorrect grammar may give the fire service officer the impression that the individual does not understand what is communicated. As in all cases when English is the second language, it is important to remember that listening and comprehension skills with English are usually better than speaking skills.

5. Latino/Hispanic Americans, because of their past experiences with fire service agencies, along with their own concerns about privacy, self-help, and other factors, are reluctant to report hazardous conditions and may not seek fire service assistance and help. It is important for fire departments and officials to build relationships and working partnerships with Latino/Hispanic American communities. This is helped by outreach efforts such as community open houses, bilingual fire service officers, and participation of fire service officers in community activities.

6. Latino/Hispanic Americans tend to hold a severe, punishment-oriented perception of public safety. That is, citizens have strong authoritarian views and an equally strong sense of "rightness" and of punishing the "wrongdoer." Because of this perspective, members from this community may view fire service and public safety personnel as more severe than they really are. It is important for fire departments and officials to be aware of this and to approach Latino/Hispanic Americans with knowledge that they may perceive the fire service as more punitive than it is in actuality.

MIDDLE EASTERN AND ARAB AMERICANS

This section provides specific cultural information on the largest group of Middle Easterners to settle in the United States, Arab Americans. We begin with an explanation of the scope of the term *Middle Eastern* as it is used in this chapter and provide information briefly on non-Arab Middle Eastern groups. We then present a historical overview, demographics and the diversity among Arab Americans, information on basic Arab values and beliefs, as well as myths and stereotypes. A brief presentation of some aspects of the Islamic religion is included within the chapter, as well as a summary of some of the commonalities between Christianity, Judaism, and Islam. Elements of family life are presented, including a discussion of the role of the head of the household and issues related to children and "Americanization." Various cultural practices and characteristics (including greetings, approach, touching, and hospitality) as well as verbal and nonverbal communication issues (such as distance, gestures, emotional expressiveness, and general points about English language usage) are presented (Figure 5.27). The final section describes recommendations for improved communication and relationships between fire service personnel and Arab American communities.

Middle Easterners come to the United States for numerous reasons, such as to gain an education and begin a career, to escape an unstable political situation in their country of origin, and to invest in commercial enterprises with the goal of gaining legal entry into the country. People in fire service and public safety continue to have contact with Middle Easterners from a number of different countries. Fire service officers would benefit from having a rudimentary knowledge about past and present world events related to Middle Easterners and should be aware of stereotypes that others hold of them. Attitudes toward Middle Easterners in this country, as well as geopolitical events in the Middle East, can have implications for fire service.

FIGURE 5.27 ◆ An Arab family of men, women, and children outside their home.

MIDDLE EASTERNERS AND RELATED TERMINOLOGY DEFINED

Among the general population there is considerable confusion as to who Middle Easterners are and, specifically, who Arabs are. Although commonly thought of as Arabs, Iranians and Turks are not Arabs. Many people assume that all Muslims are Arabs and vice versa. In fact, many Arabs are also Christians, and the world's Muslim population is actually comprised of dozens of ethnic groups. (The largest Muslim population is in Asia, not the Middle East.) Nevertheless, the predominant religion among Arabs is Islam, and its followers are called Muslims. They are also referred to as Moslems, but *Muslims* is the preferred term as it is closer to the Arabic pronunciation. People sometimes confuse the words *Arab*, *Arabic*, and *Arabian*.

The following excerpt is from *100 Questions and Answers about Arab Americans: A Journalist's Guide*; It clears up common confusion around terminology related to Arab Americans (Detroit Free Press, 2001).

Should I say Arab, Arabic or Arabian?

Arab is a noun for a person and is used as an adjective as in "Arab country." Arabic is the name of the language and generally is not used as an adjective. Arabian is an adjective that refers to Saudia Arabia and the Arabian Peninsula. When ethnicity or nationality are relevant, it is more precise and accurate to specify the country by using Lebanese, Yemeni or whatever is appropriate.

What all Arabs have in common is the Arabic language, even though spoken Arabic differs from country to country (e.g., Algerian Arabic is different from Jordanian Arabic). The following countries constitute the Middle East and are all Arab countries with the exception of three:

- Aden
- Bahrain

- Egypt
- Iran (non-Arab country)

- Iraq
- Israel (non-Arab country)
- Jordan
- Kuwait
- Lebanon
- Oman
- Palestinian Authority

- Qatar
- Saudi Arabia
- Syria
- Turkey (non-Arab country)
- United Arab Emirates
- Yemen

There are other Arab countries that are not in the Middle East (e.g., Algeria, Tunisia, Morocco, Libya) in which the majority population shares a common language and religion (Islam) with people in the Arabic countries of the Middle East. In this section, we cover primarily information on refugees and immigrants from Arab countries in the Middle East, as they constitute the majority of Middle Eastern newcomers who bring cultural differences and special issues requiring clarification for fire service. The following is a brief description of the populations from the three non-Arab countries.

Iranians and Turks   Iranians use the Arabic script in their writing, but for the most part speak mainly Farsi (Persian), not Arabic. Turks speak Turkish, although there are minority groups in Turkey who speak Kurdish, Arabic, and Greek. More than 99.8 percent of Iranians and Turks are Muslim, Islam being the most common religion among people in many other Middle Eastern countries (Central Intelligence Agency, 2003). However, many Iranians in the United States are Jewish and Bahai, both of which groups are minorities in Iran. Of the Muslim population in Iran, the majority belong to the Shia sect of Islam, the Shia version of Islam being the state religion. Persians are the largest ethnic group in Iran, making up about 50 percent of the population (Central Intelligency Agency, 2003); but there are other ethnic populations, including Kurds, Arabs, Turkmen, Armenians, and Assyrians (among others), most of which can be found in the United States. Iranians and Turks are not Arabs, but some of the cultural values related to the extended family with respect to pride, dignity, and honor are similar to those in the traditional Arab world. Many Iranian Americans and Turkish Americans came to the United States in the 1970s and were from upper-class professional groups such as doctors, lawyers, and engineers. Many of the Jewish Iranians in the United States left Iran after the fall of the Shah. In the United States, there are large Iranian Jewish populations in the San Francisco bay area, Los Angeles, and New York. Also, one can find populations of Muslim Iranians in major U.S. cities, such as New York, Chicago, and Los Angeles.

Israelis   Israel is the only country in the Middle East in which the majority of the population is not Muslim. Approximately 20 percent of the population in Israel is made up of Arabs (both Christian and Muslim). Eighty percent of the Israeli population is Jewish (Central Intelligence Agency, 2003), with the Jewish population divided into two main groups: Ashkenazim and Sephardim. The Ashkenazim are descended from members of the Jewish communities of Central and Eastern Europe. The majority of American Jews are Ashkenazi (Figure 5.28), while currently the majority of Israeli Jews are Sephardim, having come originally from Spain, other Mediterranean countries, and the Arabic countries of the Middle East. Israeli immigrants in the United States may be either Ashkenazi or Sephardic, and their physical appearance will not indicate to an officer what their ethnicity is. An Israeli may look like an American Jew, a Christian, a Muslim Arab, or none of these.

FIGURE 5.28 ◆ Multigenerational Jewish men and women attend an outdoor Jewish wedding in Brooklyn Park.

Most of the Israeli Arabs who live within the borders of Israel are Palestinians whose families stayed in Israel after the Arab–Israeli war in 1948, following the birth of Israel as a nation. The Six-Day War in 1967 resulted in Israel's occupying lands that formerly belonged to Egypt, Syria, and Jordan, but where the majority of the population was Palestinian. Thus until the signing of the Oslo 1993 Peace Accords between Israel and the Palestine Liberation Organization (PLO), Israel occupied territories with a population of approximately 1 million Palestinians. The Palestinian–Israeli situation in the Middle East has created enormous hostility on both sides. In the fall of 2000, failure to reach a negotiated settlement agreement required by the Oslo Peace Accords of 1993 resulted in a period of increased hostilities. The tension continues (as this book goes into press) and can have implications for fire service and public safety officials in the United States, especially in communities where there are large populations of Jews and Arabs, or where Israelis and Palestinians reside in large numbers (e.g., Los Angeles, New York, Chicago).

HISTORICAL INFORMATION

Although many recent Middle Eastern immigrants and refugees in the United States have come for political reasons, not all emigrating Arab Americans left their country of origin due to these concerns. There have been two major waves of Arab immigrants to the United States. The first wave, which came between 1880 and World War I, was largely from Syria and what is known today as Lebanon (at the time these areas were part of the Turkish Ottoman Empire). Of the immigrants who settled during this wave, approximately 90 percent were Christian. Many people came to further themselves economically (thus these were immigrants and not refugees forced to leave their countries); but, in addition, many of the young men wanted to avoid the military in the Ottoman Empire. A substantial percentage of these immigrants were farmers and artisans and became involved in the business of peddling their goods to farmers and moved from town to town.

In sharp contrast to the characteristics and motivation of the first wave of immigrants, the second wave of Arabic immigrants to the United States, beginning after World War II, came in large part as students and professionals because of economic instability and political unrest. As a result, these groups brought a political consciousness that was not previously emphasized by earlier immigrants. The largest group of second-wave immigrants was made up of Palestinians, many of whom came around 1948, the time of the partition of Palestine, which resulted in Israel's independence. In the 1970s, after the Six-Day War between Israel and Egypt, Syria, and Jordan, another large influx of Palestinians came to the United States. In the 1980s a large group of Lebanese came as a result of the civil war in Lebanon. Yemenis (from Yemen) have continued to come throughout the century; Syrians and Iraqis have made the United States their home since the 1950s and 1960s because of political instability in their countries. Thus these second-wave immigrants came largely because of political turmoil and have been instrumental in changing the nature of the Arab American community in the United States.

The most dramatic example of how Arab immigration has affected a U.S. city is the Detroit area in Michigan. There Arabs began to arrive in the late nineteenth century, but the first huge influx was between 1900 and 1924, when the auto industry attracted immigrants from all over the world (Woodruff, 1991). The Detroit–Dearborn area has the largest Arab community in the United States, with Arab Americans constituting about one-fourth of the population in Dearborn (Figure 5.29). A large percentage of the Detroit area's Middle Eastern population is Chaldean, Christian Iraqis who speak the Chaldean language. Although they are from the heart of the Middle East, most do not identify themselves to be Iraqi, and some may be deeply offended if referred to as Arab.

DEMOGRAPHICS: DIVERSITY AMONG MIDDLE EASTERN AND ARAB AMERICANS

Immigrants from all over the Arabic world continue to settle in the United States. For example, in 2000 approximately 12,000 visas (combined total) were issued to immigrants

FIGURE 5.29 ◆ Muslim teacher reading to preschool-aged multiethnic children about the Ramadan holiday.

coming to the United States from Bahrain, Egypt, Iraq, Jordan, Kuwait, Lebanon, Qatar, Saudi Arabia, Syria, the United Arab Emirates, and Yemen (Samhan, 2003).

Approximately 3.5 million Americans of Arabic ancestry are in the United States, but this figure does not include those who have not declared their ancestry to census officials. The communities with the largest Arab American populations are in Los Angeles/Orange County, Detroit, the greater New York area, Chicago, and Washington, DC. California has the largest grouping of Arab American communities.

Differences and Similarities There is great diversity among Arab American groups. Understanding this diversity will assist fire service officers in not categorizing Arabs as one homogeneous group and will encourage people to move away from stereotypical thinking. Arabs from the Middle East come from at least 13 different countries, many of which are vastly different from each other. The governments of the Arabic countries also differ, ranging from monarchies to theocracies to military governments to socialist republics (Central Intelligence Agency, 2003). Arab visiting nationals (Type VI in our typology)—such as foreign students, tourists, business people, and diplomats to the United States from the Gulf States (e.g., Saudi Arabia, Qatar, Oman, Bahrain, United Arab Emirates)—are typically wealthy; but those from countries such as Jordan, Lebanon, and Palestine, for example, do not generally bring wealth to the United States, and, in fact, many tend to be extremely poor. Another area in which one finds differences is in clothing and typical attire. In a number of countries in the Middle East, many older men wear headdresses, but it is less common among men who are younger and/or who have more education. Similarly, younger women in many of the Middle Eastern countries may choose not to wear the head covering and long dress that covers them from head to toe.

There are broad differences among Arab American groups associated with social class and economic status. Although many Arab Americans who come to the United States are educated professionals, there is a percentage that comes from rural areas (e.g., southern Lebanese, West Bank Palestinians, Yemenis). They differ in outlook and receptiveness to modernization. On the other hand, despite traditional values, many newcomers are modern in outlook. Many people have a stereotypical image of the Arab woman, yet not all women of Arabic descent adhere to this image. Certain Arab governments (e.g., Saudi Arabia) place restrictions on women mandating that they do not mix with men, that a woman must always be veiled, and that she not travel alone or drive a car (Ismail, 2003). On the other hand, women from less restrictive Arab countries (e.g., Egypt and Jordan) might exhibit very different behavior from those whose governments grant them fewer freedoms. Nevertheless, women in traditional Muslim families from any country typically have limited contact with men outside their family and wear traditional dress. Some implications of these traditions as they relate to Arab women and male fire service officers in the United States are discussed further in this section.

Basic Arab Values

1. Traditional Arab society upholds honor; the degree to which an Arab can lose face and be shamed publicly is foreign to the average Westerner. Fire service officers recognize that dignity and respect should be shown to all individuals, but citizens from cultures emphasizing shame, loss of face, and honor (e.g., Middle Eastern, Asian, Latin American) may react even more severely to loss of dignity and respect than do other individuals. People will go so far to avoid shame that they will not report hazardous conditions, complain about poor services, and the like. Keep in mind that some cultures emphasizing the concept of shame have sanctioned extreme punishments for loss of face and honor (e.g., death if a woman loses her virginity before marriage).

2.   Loyalty to one's family takes precedence over other personal needs. A person is completely intertwined with his or her family; protection and privacy in a traditional Arab family often override relationships with other people. Members of Arab families tend to avoid disagreements and disputes in front of others, much preferring to resolve issues privately themselves.

3.   Communication should be courteous and hospitable. Harmony between individuals is emphasized. Too much directness and candor can be interpreted as being extremely impolite. From a traditional Arab view, it may not be appropriate for a person to give totally honest responses if they result in a loss of face, especially for self or family members (this may not apply to many established Arab Americans). From this perspective, the higher goals of honor and face-saving are operative. This aspect of cross-cultural communication is not easily understood (and is often criticized) by most Westerners. Certainly, fire service officers will not accept anything but the whole truth, despite arguments rationalized by cultural ideals having to do with face and shame. However, it should not lead a public safety officer to draw explicit attention to a face-saving style. The fire service officer would be well advised to work around the issue of the indirect communication rather than insinuating that the citizen may not be honest.

MYTHS AND STEREOTYPES

The Arab world has long been perceived in the West in terms of negative stereotypes, which have been transferred to Americans of Arab descent. . . . Briefly put, Arabs are nearly universally portrayed as ruthless terrorists, greedy rich sheiks, religious fanatics, belly dancers or in other simplistic and negative images. When these stereotypes are coupled with the growing centrality of the Middle East in world politics and the increased political visibility of Arab-Americans, one result is that our community becomes more susceptible to hate crimes (American-Arab Anti-Discrimination Committee, 1992).

This characterization of discrimination against Arabs still holds true today, over a decade after the publication of the report. The Western media have been continually responsible for representing Arabs in a less than accurate way. When one hears the word *Arab*, several images come to mind: (1) wealthy sheik (despite the class distinctions in the Middle East [as elsewhere] between a wealthy Gulf Arab sheik and a poor Palestinian or Lebanese); (2) violent terrorist (the majority of Arabs worldwide want peace and do not see terrorism as an acceptable means for achieving peace) (Figure 5.30); (3) sensuous harem owner or man with many wives (harems are rare, and for the most part polygamy, or having more than one wife, has been abolished in the Arab world); and (4) ignorant, illiterate, and backward (Arab contributions to civilization have been great in the areas of mathematics, astronomy, medicine, architecture, geography, and language, among others, but this is not widely known in the West) (Macron, 1989). Despite these stereotypes, there are high-profile Arab Americans in all sectors of the professional world. Over 600 Arab Americans are members of the National Arab American Medical Association, and the Michigan Arab American Legal Society has a highly respected and substantial membership as well.

Movies and Television   Perhaps the most offensive form of Arab stereotypes comes from the media; programs and movies routinely portray Arabs as evil womanizers, wealthy oil sheiks wearing turbans, thieves, and terrorists. Even films and programs aimed at children propagate this stereotype of the Arab as a villain. The original lyrics of the opening song of Disney's *Aladdin*, "Arabian Nights," are sung by an Arab who has been portrayed as a stereotype: "Oh I come from a land, from a faraway

FIGURE 5.30 ◆ Many Arab and Middle Eastern Americans have objected to the negative and stereotypic depiction of Arabs in films such as *True Lies*.

place/Where the caravan camels roam/Where they cut off your ear if they don't like your face/It's barbaric, but hey, it's home" (Shaheen, 2001, p. 51).

Because many Americans do not know Arabs personally, media images become embedded in people's minds. Jack Shaheen, author of *Reel Bad Arabs: How Hollywood Vilifies a People,* writes, "by depicting Arabs solely as slimy, shifty, violent creeps, Hollywood has been not only misrepresenting the Arab world, but also creating a climate for hate." For example, the film *The Siege,* released in 1998, continues to receive extremely negative reactions from Islamic groups. This film about Palestinian terrorists was said to have linked Islam to terrorism and showed Muslim Americans being rounded up and placed in internment camps (Shaheen, 2001, pp. 430–33). "Whenever Hollywood productions and TV series portray terrorism or violence, it always has to be an Arab. They seem to think it is our monopoly" (comment by the president of the American-Arab Anti-Discrimination Committee, "Film's Portrayal of Muslims Troubling to Islamic Groups," 1998, p. B10).

ISLAMIC RELIGION

Misunderstanding between Americans and Arabs, or Americans and Arab Americans, can often be traced, in part, to religious differences and a lack of tolerance of these differences. Islam is practiced by the majority of Middle Eastern newcomers to the United States, as well as by many African Americans. Many Arab Americans (especially those from the first wave of Arab immigration) are Christian, however, and prefer that others do not assume they are Muslim simply because they are Arabs.

Remember that Muslims are the people who practice the religion of Islam. By and large, most Americans do not understand what Islam is and, because of stereotyping, wrongly associate Muslims with terrorists or fanatics. Many, but by no means all, Arab Muslims in the United States are religious and have held on to the traditional aspects of their religion, which are also intertwined with their way of life. Islam means submission to the will of God and for traditional, religious Muslims, the will of God (or fate) is a central concept. The religion has been called "Mohammedanism," which is incorrect because it suggests that Muslims worship Mohammed rather than God (Allah). (An alternate spelling is Muhammad, which is closer to the Arabic pronunciation of the name.) It is believed that God's final message to man was revealed to the prophet Mohammed. *Allah* is the shortened Arabic word for the God of Abraham and it is used by both Arab Muslims and Arab Christians.

The Quran (Koran) and the Pillars of Islam   The Quran is the holy text for Muslims and is regarded as the word of God (Allah). The five "Pillars of Islam" or central guidelines form the framework of the religion:

- Profession of faith in Allah (God)
- Prayer five times daily
- Almsgiving (concern for the needy)
- Fasting during the month of Ramadan (sunrise to sunset)
- Pilgrimage to Mecca (in Saudi Arabia) at least once in each person's lifetime

There are several points where fire service officials can respect a Muslim's need to practice his or her religion, such as the need to express one's faith in God and to be respected for it. Normally, people pray together as congregations in mosques, the Islamic equivalent of a church or synagogue. People, however, can pray individually if a congregation is not present. Remember that prayer five times a day is a pillar of Islam and that strict Muslims will want to uphold this "command" no matter where they are. Call to prayer takes place at the following times:

- One hour before sunrise
- At noon
- Midafternoon
- Sunset
- Ninety minutes after sunset

Taboos in the Mosque   A fire service officer will convey respect to a Muslim community if he or she can avoid entering a mosque (Figure 5.31) and interrupting prayers (emergencies may occasionally make this impossible). Religion is so vital in Arab life that fire service officials should always show respect for Islamic customs and beliefs. Thus, other than in emergency situations, officers are advised of the following:

- Avoid entering a mosque, or certainly the prayer room of a mosque, during prayers.
- Never step on a prayer mat or rug with your shoes on.
- Never place the Quran on the floor or put anything on top of it.
- Avoid walking in front of people who are praying.
- Speak softly while people are praying.
- Dress conservatively (both men and women are required to dress conservatively—shorts, for example, are not appropriate).
- Invite people out of a prayer area to talk to them.

Proper protocol in a mosque (also referred to as a *masjid*) requires that people remove their shoes before entering, but this must be left to the fire service officer's discretion. Officer safety, of course, comes before consideration of differences.

FIGURE 5.31 ◆ A mosque is combined with a private elementary and middle school in a brick building on a street in Harlem.

Ramadan: The Holy Month   One of the holiest periods in the Islamic religion is the celebration of Ramadan, which lasts for one month. There is no fixed date because, like the Jewish and Chinese calendars, the Islamic calendar is based on the lunar cycle (related to the phases of the moon) and so dates vary from year to year. During the month of Ramadan, Muslims do not eat, drink, or smoke from sunrise to sunset. The purpose of fasting during Ramadan is to "train one in self-discipline, subdue the passions, and give [people] . . . a sense of unity with all Moslems" (Devine and Braganti, 1991, p. 28). On the twenty-ninth night of Ramadan, when there is a new moon, the holiday is officially over. The final fast is broken, and for up to three days people celebrate with a feast and other activities. Throughout the month of Ramadan, Muslim families tend to pray more often in the mosque than during other parts of the year. For Muslims, Ramadan is as important and holy as Christmas; this fact is appreciated when others who are not Muslim recognize the holiday's importance.

Similarities Between Christianity, Judaism, and Islam   In reading the above tenets of the Muslim faith, notice that certain practices and aspects of Islam are also found in Christianity and Judaism. All three religions are monotheistic; that is, each has a belief in one God. Followers of the religions believe that God is the origin of all and is all-knowing as well as all-powerful. Because God is merciful, it is possible for believers to be absolved of their sins, though the practices for obtaining absolution vary in the different ideologies. All three religions have a central holy book central to the faith: Judaism has the Torah (the first five books of the Old Testament), Christianity the Bible, and Islam the Quran (Koran). All three religions regard their texts to be either the direct word of God or inspired by the word of God. There are similarities between the three books. For example, the concept of the Ten Commandments is present in each. All three contain stories about many of the same people, such as Adam, Noah, Abraham, Moses, David, and Solomon. The Bible and the Quran also both contain stories about Mary, Jesus, and John the Baptist. The oral reading or recitation of each book constitutes part of regular worship. Prophets exist in each tradition and are revered as those who transmit the word of God to the people. Chronologically, Judaism became a religion first, followed by Christianity, and then Islam. Islam builds upon the foundations of the previous faiths, and belief is held in the authenticity of the prophets of earlier books. For example, Islamic belief sees both Moses and Jesus as rightful prophets but only as precursors to Mohammed, whom Muslims believe to be the final prophet of God. As there are different internal interpretations of Christianity and Judaism, so too, followers of Islam interpret their religion in different ways. Differing readings of the Quran can lead to more or less tolerance within the religion. Last, both the Islamic and Judeo-Christian traditions are said to stem from the same lineage; Abraham was father to both Isaac, whose progeny became the people of Israel and through whom the Messiah, Jesus, came, and Ishmael, who started the Arabic lineage.

THE MIDDLE EASTERN AND ARAB AMERICAN FAMILY

Arab Americans typically have close-knit families in which family members have a strong sense of loyalty and fulfill obligations to all members, including extended family (aunts, uncles, cousins, grandparents). Traditionally minded families also believe strongly in the family's honor, and members try to avoid any behavior that will bring shame or disgrace to the family. The operating unit for Arab Americans (and this may be less true for people who have been in the United States for generations) is not the individual but the family. Thus if a person behaves inappropriately, the entire family is disgraced. Similarly, if a family member is harmed (in the Arab world), there would be some type of retribution. For the fire service officer, three characteristics of the Arabic family will affect his or her interaction with family members:

1. Extended family members are often as close as the "nuclear family" (mother, father, children) and are not seen as secondary family members. If there is a public safety issue, fire service officers can expect that many members of the family will become involved in the matter. Although officers might perceive this as interference, from an Arabic cultural perspective, it is merely involvement and concern. The numbers of people involved are not meant to overwhelm an officer.

2. Family loyalty and protection are seen as two of the highest values of family life. Therefore, shaming, ridiculing, insulting, or criticizing family members, especially in public, can have serious consequences such as lost of rapport and relationship with community members.

3. Newer Arab American refugees or immigrants may be reluctant to accept fire service assistance in nonemergency situations. Because families are tightly knit, they can also be closed "units" whereby members prefer to keep private matters or conflicts to themselves. As a result, fire service officers will have to work harder at establishing rapport if they want to gain cooperation.

From the field observations of one of the coauthors (Wong) in training fire service and public service agencies, a fire service officer who is not trained in understanding and responding appropriately and professionally to cultural differences could alienate the family by (1) not respecting the interest and involvement of the family members, and (2) attempting to gain control of the communication in an authoritarian and offensive manner. The consequences may be that he or she would have difficulty establishing the rapport needed to gain information about the conflict at hand and would then not be trusted or respected. To do the job effectively, fire service officials must respect Arab family values, along with communication style differences (the latter will be discussed shortly).

The Role of the Man and the Woman   As in most cultures with a traditional family structure, the man in the Arab home is overtly the head of the household and both his role and his influence are strong. The wife has a great deal of influence, too, but it can often be more "behind the scenes." An Arab woman does not always defer to her husband in private as she would in public (Nydell, 1987). However, as mentioned earlier in the section, there are many women who have broken out of the traditional mold and tend to be more vocal, outspoken, and assertive than their mothers or grandmothers. Traditionally, in many Arab countries, some fathers maintain their status by being strict disciplinarians and demanding absolute respect, thus creating some degree of fear among children and in their wives. Once again, Arab Americans born and raised in this country have, for the most part, adopted middle-class "American" styles of child raising whereby children participate in some of the decision making and are treated in more of an egalitarian manner, just as an adult would be. As with changing roles among all kinds of families in the United States, the father as traditional head of the household and the mother as having "second-class" status are not as prevalent among established Arab Americans in the United States.

In traditional Arab society, men exert influence and power publicly. This power may be seen by Westerners in a negative light, but it is important to caution against misinterpreting a husband's or father's behavior as merely control (Shabbas, 1984). He, and other male figures of importance, can be employed in securing the compliance of the family in important matters. The husband or father can be a natural ally of authorities. Fire service officers would be well advised to work with both the father and the mother. On family matters the woman frequently is the authority, even if she seems to defer to her husband. Communicating with the woman, even if indirectly, while still respecting the father's need to maintain his public status, will win respect from both the man and the woman.

CULTURAL PRACTICES

As with all other immigrant groups, the degree to which people preserve their cultural practices varies. The following descriptions of everyday behavior will not apply equally to all Arab Americans, but they do not necessarily apply only to recent newcomers. Immigrants may preserve traditions and practices long after they come to a new country by conscious choice or sometimes because they are unaware of their cultural behavior (i.e., it is not in their conscious awareness).

Greetings/Approach/Touching   Most recent Arab American newcomers expect to be addressed with a title and their last name (Mr. _____; Miss _____), although in many Arab countries people are addressed formally by Mr. or Mrs. and the first name. Most Arab women do not change their names after they are married or divorced. They, therefore, may not understand the distinction between a "maiden" name and a "married" name. The usual practice is to keep their father's last name for life (Boller, 1992).

Many Arab Americans who have retained their traditional customs shake hands and then place their right hand on their chest near the heart. This is a sign of sincerity and warmth. In the Middle East, Americans are advised to do the same if they observe this gesture (Devine and Braganti, 1991). Fire service officers can decide whether they are comfortable using this gesture—most people would not expect it from an officer, but some might appreciate the gesture as long as the officer were able to convey sincerity. Generally, when Arabs from the Middle East shake hands, they do not shake hands briefly and firmly. Arabs (i.e., not assimilated Arab Americans) tend to hold hands longer than other Americans and shake hands more lightly. Older children are taught to shake hands with adults as a sign of respect. Many Arabs would appreciate an officer shaking hands with their older children. With a recent immigrant or refugee Arab woman, it is generally not appropriate to shake hands unless she extends her hand first. This would definitely apply to women who wear head coverings.

Many Arabs of the same sex greet each other by kissing on the cheek. Two Saudi Arabian men, for example, may greet each other by kissing on both cheeks a number of times. This does not suggest homosexuality, but rather is a common form of greeting. Public touching of the opposite sex is forbidden in the traditional Arabic world, and fire service officers should make every effort not to touch Arabic women, even casually.

Hospitality "Hospitality is a byword among [Arabs], whatever their station in life. As a guest in their homes you will be treated to the kindest and most lavish consideration. When they say, as they often do, 'My home is your home,' they mean it" (Salah Said, as quoted in Nydell, 1987, p. 58). Hospitality in the Arab culture is not an option but more of an obligation or duty. In some parts of the Arab world, if you thank someone for his or her hospitality, the person may answer with a common expression meaning "Don't thank me. It's my duty." (Here the word *duty* has a more positive than negative connotation.) Fire service officers need to understand how deeply engrained the need to be hospitable is and must not misinterpret this behavior for something that it is not. Whether you are entering a home or a business owner's shop or office, an Arab American may very well offer you coffee and something to eat. This is not to be mistaken for a bribe and, from the Arab perspective, carries no negative connotations. The period of time spent socializing and extending one's hospitality gives the person a chance to get to know and see whether he or she can trust the other person. Business is not usually conducted among strangers. Obviously, on an emergency call, there is no time for such hospitality.

VERBAL AND NONVERBAL COMMUNICATION STYLES OF MIDDLE EASTERN AND ARAB AMERICANS

Arabs in general are very warm and expressive people, both verbally and nonverbally, and appreciate it when others extend warmth to them. In some areas of nonverbal communication Americans, without cultural knowledge, have misinterpreted the behavior of Arab Americans simply because of their own ethnocentrism (i.e., the tendency to judge others by one's own cultural standards and norms).

Conversational Distance What is acceptable conversational distance between two people is often related to cultural influences. Fire service officers are very aware of safety issues and keep a certain distance from people when communicating with them. Generally, officers like to stand about an arm's length or farther from citizens to avoid possible assaults on their person. This distance is similar to how far apart "mainstream Americans" stand when in conversation. Cultures subtly influence the permissible distance between two people. When the distance is "violated" a person can feel threatened either consciously or unconsciously. Many, but not all Arabs, especially if they are new to the country, tend to have a closer acceptable conversational distance between each other than do other Americans. In Arab culture, it is not considered offensive to "feel a person's breath." Yet many Americans, unfamiliar with this intimacy in regular conversation, have misinterpreted the closeness. While still conscious of safety, the fire service officer can keep in mind that the closer than "normal" behavior (i.e., "normal" for the officer) does not necessarily constitute a threat.

Gestures   Certain gestures that Arabs from some countries use are distinctly different from those familiar to non-Arab Americans. In a section entitled "Customs and Manners in the Arab World," Devine and Braganti (1991, p. 13) describe some commonly used gestures among Arabs:

- What does it mean? or What are you saying? Hold up the right hand and twist it as if you were screwing in a light bulb one turn.
- Wait a minute. Hold all fingers and thumb touching with the palm up.
- No. This can be signaled in one of three ways: moving the head back slightly and raising the eyebrows, moving the head back and raising the chin, or moving the head back and clicking with the tongue.
- Go away. Hold the right hand out with the palm down, and move it as if pushing something away from you.
- Never. A forceful never is signaled by holding the right forefinger up and moving it from left to right quickly and repeatedly.

As with many other cultural groups, pointing a finger directly at someone is considered rude.

Emotional Expressiveness   Arab women, in particular, are very emotional and fire service personnel sometimes see this emotionalism as losing control. Upon seeing family members in trouble, it would be most usual and natural for an Arab woman to put her hands to her face and say something like, "Oh, my God," frequently and in a loud voice. While other Americans can react this same way, it is worth pointing out that in mainstream American culture, there is a tendency to subdue one's emotions and not to go "out of control." What some Americans consider to be out of control, Arabs (as well as Mexicans, Greeks, Israelis, and Iranians, among other groups) consider to be perfectly normal behavior. In fact, the lack of emotionalism that Arabs observe among mainstream Americans can be misinterpreted as lack of interest or involvement.

Although a communication-style characteristic never applies to all people in one cultural group (and we have seen that there is a great deal of diversity among Arab Americans), certain group traits apply to many people. Arabs, especially the first generation of relatively recent newcomers, tend to display emotions when talking. Unlike many people in Asian cultures (e.g., Japanese and Korean), Arabs have not been taught to believe that the expression of emotion is a sign of immaturity or lack of control. Arabs, as other Mediterranean groups, such as Israelis or Greeks, tend to shout when they are excited or angry and are very animated in their communication. They may repeatedly insert expressions into their speech such as "I swear by God." This is simply a cultural mannerism.

Westerners, however, tend to judge this style negatively. To a Westerner, the emotionalism, repetition, and emphasis on certain statements can give the impression that the person is not telling the truth or is exaggerating for effect. An officer unfamiliar with these cultural mannerisms may feel overwhelmed, especially when involved with an entire group of people. It would be well worth it for the officer to determine the spokesperson for the group, but to refrain from showing impatience or irritation at this culturally different style.

Swearing, the Use of Obscenities, and Insults Fire service officers working in Arab American communities should know that for Arabs, words are extremely powerful. Whether consciously or unconsciously, some believe that words can affect the course of events and can bring misfortune (Nydell, 1987). If a fire service officer displays a lack of professionalism by swearing at an Arab (even words such as *damn*), it will be nearly impossible to repair the damage. Fire service officers who understand professionalism are aware that this type of language and interaction is insulting to all persons. The choice to use obscenities and insults, however, means that fire service officers risk never being able to establish trust with an ethnic community such as that of the Arab Americans. This can translate into not being able to secure cooperation when needed. Even a few officers exhibiting this type of behavior can damage the reputation of an entire department for a long period of time.

English Language Problems If time allows, before asking the question "Do you speak English?" officers should try to assess whether the Arab American is a recent arrival or an established citizen who might react negatively to the question. A heavy accent does not necessarily mean that a person is unable to speak English (although that can be the case). Certain specific communication skills that can be used with limited-English-speaking persons (see Chapter 4) can be applied to Arab Americans. Officers should proceed slowly and nonaggressively with questioning and, whenever possible, ask open-ended questions. A fire service officer's patience and willingness to take extra time will be beneficial in the long run.

SUMMARY OF RECOMMENDATIONS FOR FIRE SERVICE IN THE MIDDLE EASTERN AND ARAB AMERICAN COMMUNITY

1. There are several basic Arab cultural values that fire service officers should keep in mind when interacting with Arab American citizens: (a) A person's dignity, honor, and reputation are of paramount importance and no effort should be spared to protect them, especially one's honor. (b) Loyalty to one's family takes precedence over other needs; thus an individual is completely intertwined with his or her family. (c) Communication should be courteous and hospitable; honor and face-saving govern interpersonal interactions and relationships.

2. Arab Americans have been wrongly characterized and stereotyped by the media, and as with all stereotypes, this has affected people's thinking about Arab Americans. Fire service officers should be aware of stereotypes that may influence their judgments. Common stereotypes of Arabs include (a) illiterate and backward, (b) passive, uneducated woman, (c) thief, and (d) terrorist.

3. Fire service officers can demonstrate to Muslim Arabs a respect for their culture and religion by the following: (a) respecting the times when people pray (five times a day); (b) maintaining courteous behavior in mosques, such as not stepping on prayer mats, not walking in front of people praying, and speaking softly; and (c) working out solutions with community members regarding religious celebrations (e.g., with respect to standby emergency services, building capacity, fire safety).

4. The basic unit for Arab Americans, especially recent arrivals and traditional families, is not the individual but the family (including the extended family). If a family member is involved in a public safety or hazardous condition situation, fire service officers should expect that other family members will become actively involved. The traditional Arabic family is used to working out their conflicts themselves. Fire service officers should not automatically assume that this involvement is an attempt to interfere with public safety affairs.

5. Traditionally and outwardly, the father is the head of the household and much of the conversation should be directed toward him. Fire service officers should, however, keep in mind the following: (a) Many Arab women are outspoken and vocal. Do not dismiss their input because men may appear to be, at least publicly, the ones with the power; and (b) traditional Arab women who do not freely communicate with men may have difficulty expressing themselves to a male fire service officer. In their own families, however, they are often the real decision makers. Consider various ways of getting information (e.g., the use of a female translator, female fire service officer, and indirect and open-ended questions).

6. There are a number of specific cultural practices and taboos that fire service officers should consider when communicating with Arab Americans who have preserved a traditional lifestyle (this does not apply to the majority of Arab Americans who have been in the United States for generations). Avoid even the casual touching of women. Be respectful of the need that some Arab women have to be modest. Never point the sole of one's shoes or feet at a person. Expect people to extend hospitality by offering coffee or food (this is part of their cultural customs and should not be viewed as a bribe). Arab Americans may stand closer to each other than other Americans do when talking. This is not meant to be threatening; it is largely unconscious and reflects a cultural preference for closer interpersonal interaction.

7. There are cultural differences in communication style that can affect the fire service officers' judgments and reactions: (a) Becoming highly emotional (verbally and nonverbally) and speaking loudly are not looked down upon in the Arab world. Fire service officers who may have a different manner of communication should not express irritation at this culturally different style; nor should they necessarily determine that the people involved are being disrespectful. Developing patience with culturally different styles of communication is a key cross-cultural skill. (b) When a person speaks with an accent, it does not necessarily mean that he or she is not fluent in English or is illiterate. Many highly educated Arab Americans speak English fluently but with an accent, and would be insulted if they were treated as if they were not educated.

8. During times of crises in the Middle East, Arab Americans become targets of prejudice and racism. Fire service departments need to monitor communities and keep informed of world events so that Arab American communities have more protection during times when they may be vulnerable to property destruction, community riots, arson, and hate crimes.

9. As community members desire to be involved in decision-making processes that affect their lives, fire service officials are strongly urged to communicate personally with Arab American leaders. If fire chiefs and other public safety executives reach out to communities, showing their desire to foster good relations, they are more likely to have positive interactions with the community, while at the same time increasing the willingness of the community to provide tips crucial to keeping the peace.

NATIVE AMERICANS/AMERICAN INDIANS

This section provides specific cultural information on Native Americans/American Indians, including aspects of their history that both directly and indirectly can affect the relationship with fire service officials. It presents information about Native American identity and group identification terms (Figure 5.32) as well as explains briefly

FIGURE 5.32 ◆ Native American student looks over a few unrolled posters featuring American Indians and Native Alaskans amid the assorted merchandise of an American Indian museum gift shop.

demographics, the tribal system, reservations, Native American mobility, and family structure. This section also addresses the diversity that exists among Native American groups by describing cultural similarities and differences found among various Indian groups. The final section outlines recommendations for improved communication and relationships between fire service personnel and members of Native American/ American Indian communities.

HISTORICAL INFORMATION AND BACKGROUND

Recorded history disputes the origins of the first "Indians" in America. Some researchers claim that they arrived from Asia more than 40,000 years ago; others assert that they did not arrive from anywhere else. In either case, despite their long history in North America and the fact that they were the first "Americans," traditional U.S. history books did not recognize their existence until the European conquests, beginning with Christopher Columbus in 1492. The perception of the native peoples as either nonexistent or simply insignificant reflects an ethnocentric view of history. Even the word *Indian* is not a term that native Americans originally used to designate their tribes or communities. Because Columbus did not know that North and South America existed, he thought he had reached the Indies, which then included India, China, the East Indies, and Japan. In fact, he arrived in what is now called the West Indies (in the Caribbean) and he called the people he met "Indians" (*Los Indios*). Eventually, this became the name for all the indigenous peoples in the Americas. However, before the white settlers came to North and South America, almost every "Indian" tribe had its own name, and despite some shared cultural values, native Americans did not see themselves as one collective group or call themselves "Indians." Most tribes refer to themselves in their own languages as "The People," "The Allies," or "The Friends." Some of the terms that whites use for various tribes are not even authentic names for that tribe. For example, the label *Sioux*, which means enemy or snake, was originally a term given to that group by an enemy tribe and then adopted by French traders. In many cases, a tribe's real name is not necessarily the name commonly used.

Today in public schools across the country, some educators are only beginning to discuss the nature of much of the contact with Native Americans in early American history. Traditionally, rather than being presented as part of the American people's common legacy, Native American cultural heritage had often been presented as bits of colorful "exotica." The reality is that Euro American and Indian relations have been characterized by hostility, contempt, and brutality. The native peoples have generally been treated by Euro Americans as less than human or as "savages," and the rich Native American cultures have been ignored or crushed. For this reason, many culturally identifying American Indians do not share in the celebrations of Thanksgiving or Christopher Columbus Day. To say that Columbus discovered America implies that Native Americans were not considered "human enough" to be of significance. Ignoring the existence of the Native Americans before 1492 constitutes only one aspect of ethnocentrism. American Indians' experience with the "white man" has largely been one of exploitation, violence, and forced relocation. It is this historical background that has shaped many Native American views of Euro Americans and their culture. Although the majority of people in the United States have a sense that Native Americans were not treated with dignity in U.S. history, many are not aware of the extent of the current abuse toward them. This may be due to the fact that this group is a small and traditionally "forgotten minority" in the United States, constituting approximately 1.5 percent of the overall population (including Alaska Natives; U.S. Census Bureau, 2002a).

It would not be accurate to say that no progress at all has been made in the United States with respect to the awareness and rights of our nation's first Americans. On November 6, 2000, President Bill Clinton renewed his commitment to tribal sovereignty by issuing an executive order on consultation with tribal governments.

> The purpose of the Order is to establish meaningful consultation and collaboration with tribal officials in the development of federal policies having tribal implications, to strengthen the administration's government-to-government relationship to tribes, and to reduce the imposition of unfounded mandates by ensuring that all executive departments and agencies consult with tribes and respect tribal sovereignty as they develop policies on issues that impact Indian communities (Legix, 2000).

President George W. Bush, in 2002, announced the following in a proclamation in honor of National American Indian Heritage Month:

> My administration is working to increase employment and expand economic opportunity for all Native Americans. Several federal agencies recently participated in the National Summit on Emerging Tribal Economies to accomplish this goal. In order to build upon this effort, my Administration will work to promote cooperation and coordination among Federal agencies for the purpose of fostering greater economic development of tribal communities. By working together on important economic initiatives, we will strengthen America . . . with hope and promise for all Native Americans (White House Government News Release, 2002).

FIGURE 5.33 ◆ A group of Native American leaders and United States government officials pose outside.

Despite progress that has been made on the records in Washington, DC, many Native Americans do not see that the spirit behind the foregoing sentiments have actually changed their lives. American Indians have among the highest school dropout rates and unemployment rates of all the ethnic and racial groups. As for protection from the federal government, American Indians continue to fight legal battles over the retention of Indian lands and other rights previously guaranteed by U.S. treaties. They still feel abused by a system of government that has committed many treaty violations against tribes and individuals. The U.S. government has often not acted in good faith toward its Native American citizens by seriously and repeatedly disregarding Indian rights that have been guaranteed in the form of binding treaties (Figure 5.33).

Consequently, individuals and tribes are reluctant to trust the words of the government or people representing "the system" because of the breach of many treaties. Whether they are aware of it or not, fire service agents are perceived this way and carry this "baggage" into encounters with Native Americans. Historically, the "uniformed officer" from outside the reservation has been a symbol of rigid and authoritarian governmental control that has affected nearly every aspect of an Indian's life, especially on reservations. Often, firefighter officers (like most citizens) have only a limited understanding of how the government, including the fire service system, caused massive suffering by not allowing American Indians to preserve their cultures, identities, languages, sacred sites, rituals, and lands. Because of this history, fire service officers have a responsibility to educate themselves about the history of the treatment of Native Americans in order to deal with them effectively and fairly today. Fire service officers must understand American Indian communities and put forth extra efforts to establish rapport. This will increase the possibility of success at winning cooperation and respect from people who have never before had any reason to trust any representative of government.

For some American Indians, an additional phenomenon aggravates the repeated breach of trust by the federal government. There is a very proud tradition of American Indians serving in the military that has gone largely unrecognized. For more than 200 years, they have participated with distinction in United States military actions. American military leaders, beginning with George Washington in 1778, recognized American Indians to be courageous, determined, and as having a "fighting spirit" (CEHIP, 1996). American Indians had already contributed to the military in the 1800s and did so on an even larger scale beginning in the 1900s. In World War I, it is estimated that 12,000 American Indians served in the military.

> More than 44,000 American Indians, out of a total Native American population of less than 350,000, served with distinction between 1941 and 1945 in both the European and Pacific theaters of war. Native American men and women on the home front also showed an intense desire to serve their country and were an integral part of the war effort. More than 40,000 Indian people left their reservations to work in ordinance depots, factories, and other war industries. American Indians also invested more than $50 million in war bonds and contributed generously to the Red Cross and the Army and Navy Relief societies. The Native American's strong sense of patriotism and courage emerged once again during the Vietnam era. More than 42,000 Native Americans, more than 90 percent of them volunteers, fought in Vietnam (CEHIP, 1996).

In Desert Storm, Bosnia, and the war in Afghanistan and Iraq (Figure 5.34), Native Americans continue to serve with honor in the United States armed forces

FIGURE 5.34 ◆ American Indian women Marines in uniform have served side-by-side with other Native American soldiers in the U.S. military.

with an estimated number exceeding 12,000 American Indians (Indian Country Today, 2003).

In being the first woman soldier to die in the Iraqi War and the first Native American woman to die in combat, Private First Class Lori Piestewa of the United States Army and from the Hoppi tribe served with devotion to duty. The U.S. Army reported that PFC Piestewa died fighting the enemy when her convoy was ambushed on March 23, 2003, near Nasiriyah, Iraq. PFC Piestewa died in battle along with eight other soldiers as PFC Jessica Lynch was taken prisoner and later rescued by U.S. troops on April 1, 2003. PFC Piestewa was 23 years old, a single mother of a four-year-old son and a three-year-old daughter. She carried on the tradition of her father, a Vietnam veteran, and her grandfather, a World War II veteran. Her determination shone through in a local television interview before being sent to Iraq from her hometown of Tuba, Arizona: "I am ready to go," she said, looking straight at the camera. "I learned to work with people. It's very important to me to know that my family is going to be taken care of." After her death, thanks to donations from the public, a scholarship fund was set up for Piestewa's children. Two Phoenix landmarks, Squaw Peak and Squaw Peak Freeway, have been renamed Piestewa Peak and Piestewa Freeway—a decision hailed by American Indian groups who consider the word "squaw" offensive (Legon, 2005).

NATIVE AMERICAN IDENTITY

Fire service officials may find themselves confused as to who is an American Indian. Individuals may claim to have "Indian blood," but tribes have their own criteria for determining tribal membership. Because the determination of tribal membership is a fundamental attribute of tribal sovereignty, the federal government generally defers to tribes' own determinations when establishing eligibility criteria under special Indian entitlement programs. However, a number of Indian tribes, for historical and political reasons, are not currently "federally recognized" tribes. Members of such tribes, although Indian, are not necessarily eligible for special benefits under federal Indian programs. On the other hand, fraud in this area is quite rampant, whereby people falsely claim Indian ancestry to take unfair advantage of governmental benefits and other perceived opportunities.

Fire service and public safety officers may find themselves in situations in which an individual claims to be Indian when he or she is not. If officers have any doubt, they should inquire as to what tribe the person belongs and then contact the tribal headquarters to verify that person's identity. Every tribe has its own administration and authority (Figure 5.35), the members of which will be able to answer questions of this nature. By verifying information with tribal authorities rather than making personal determinations of "Indianness," fire service officers will help create a good rapport between tribal members and fire service and public safety officials.

According to the 2000 census, approximately 2.5 million (or 0.9 percent of the entire U.S. population) identified themselves as American Indian or Alaska Native (U.S. Census Bureau, 2002a). However, there is a great deal of numerical misinformation because of the lack of a method for verifying the accuracy of people's claims of being Native American. The term *Native American* came into popularity in the 1960s and referred to groups served by the Bureau of Indian Affairs (BIA), including American Indians and natives of Alaska. *Native American*, however, has rarely been accepted by all

FIGURE 5.35 ◆ Indian Health Service is part of the US Public Health Service Hospital, such as this one in Lexington, Kentucky, or the one at Fort Defiance Indian Hospital in Fort Defiance, Arizona. This is one of many hospitals on the Navajo Reservation at which Indian voters get to decide whether to run their own medical services rather than rely on federally controlled facilities.

American Indian groups and is seldom used on reservations. Alaska natives, such as Eskimos and Aleuts, are separate groups and prefer the term *Alaska Native*. To know by what terms individuals or tribes prefer to be called, fire service officers should listen to the names they use for themselves rather than try to guess which one is correct.

In the area of mislabeling, some Native Americans have Spanish first or last names (because of intermarriage) and may "look" Hispanic or Latino (e.g., the Hopis). Identification can be difficult for the fire service officers, so they should not assume that the person is Latino just because of the name or appearance. Many Native Americans do not want to be grouped with Latinos because (1) they are not Latinos; (2) they may resent the fact that some Latinos deny their Indian ancestry and, instead, identify with only the Spanish part of their heritage; and (3) many tribes have a history of warfare with the *mestizo* populations of Mexico. As an aside, the majority population in Mexico, Central America, and South America is of "Indian" ancestry. However, they adopted or were given Spanish names by the *conquistadores* (conquerors). Many Hispanics in U.S. border communities are really of Indian, not Spanish, heritage, or they may be a mixture of the two.

NATIVE AMERICAN POPULATION, TRIBES, AND RESERVATIONS

The 2000 census revealed that of all cities in the United States with a population of 100,000 or more, New York and Los Angeles had the largest American Indian populations. In general, the West has the largest population (43 percent of Native American population), the South has 31 percent, with 17 percent in the Midwest and 9 percent in the Northeast (U.S. Census Bureau, 2000a). Cherokee, Navajo, and Latin

FIGURE 5.36 ◆ American Indian tribal council holds business meeting at the tribal headquarters.

American Indians comprised the most populated tribal groupings. See Figure 5.37 for a listing of Native American tribes and the corresponding regions in which they live in North America.

According to BIA figures in 2003, there were over 562 "federally recognized" tribal governments in the United States (the word *tribe* or *nation* may also be used), and these include native groups of Alaskans such as Aleuts (Figure 5.36). The federally recognized tribes each have a distinct history and culture and often a separate language. Federal recognition means that a legal relationship exists between the tribe and the federal government. Many tribes still do not benefit from federally recognized status. Some may be state recognized and may be in the process of seeking federal recognition, whereas others may not seek recognition at all. The issue of the increasing rivalry among some Indian groups seeking recognition is reflected in the sentiment that some Indian tribes are pitted against each other over government benefits and resources.

An Indian reservation is land that a tribe has reserved for its exclusive use through the course of treaty making. It may be on ancestral lands or simply the only land available when tribes were forced to give up their own territories through federal treaties. A reservation is also land that the federal government holds in trust for the use of an Indian tribe. The Bureau of Indian Affairs (BIA) administers and manages 55.7 million acres of land that, as of 2003, have been held in trust for American Indians, Indian tribes, and Alaska Natives. The 2003 estimate held that about 1.2 million American Indians (BIA, 2003) were living on reservations and that there were approximately 300 reservations (exact figures are difficult to obtain). The largest of the reservations is the Navajo reservation and trust lands, which extend into three states. Since reservations are self-governing, most have tribal fire and police services, and there are issues of jurisdiction discussed later in this section. American Indians are not forced to stay on reservations, but many who leave have a strong desire to remain in touch with and be nourished by their culture on the reservations. For this reason and because of culture shock experienced in urban life, many return to reservations.

In general, the Indian population is characterized by constant movement between the reservation and the city, and sometimes relocation from city to city. In urban areas, when fire service officers contact an Indian, they will not necessarily know how acculturated to city life that individual is. In rural areas it is easier for officers to get to know the culture of a particular tribe. In the city, the tribal background may be less important than the fact that the person is an American Indian.

Since the early 1980s, more than half of the population of Native Americans have been living outside of reservation communities. Many have left to pursue educational and employment opportunities, as life on some of the reservations can be very bleak. Although a large number return home to the reservations to participate in family activities and tribal ceremonies, many attempt to remake their lives in urban areas. A percentage of Native Americans do make adjustments to mainstream educational and occupational life, but the numbers are still disproportionately low.

DEMOGRAPHICS: DIVERSITY AMONG NATIVE AMERICANS/AMERICAN INDIANS

As with other culturally or ethnically defined categories of people (e.g., Asian, African American), it would be a mistake to lump all Native Americans together and to assume that they are homogeneous. For example, in Arizona alone are a number of different tribes with varying traditions: there are Hopis in the northeast, Pimas and Papagos in the south, Apache in the North Central region, and Yuman groups in the west. All of these descend from people who came to what is now called Arizona. The relative "newcomers" are the Navajos and Apaches, who arrived about 1,000 years ago. These six tribes represent differences in culture, with each group having its own history and life experiences.

Broadly speaking, in the United States, distinct cultural groups exist among Native Americans in Alaska, Arizona, California, the Central Plains (Kansas and Nebraska), the Dakotas, the Eastern Seaboard, the Great Lakes area, the Gulf Coast states (Florida, Alabama, Mississippi, Louisiana, Texas), the Lower Plateau (Nevada, Utah, Colorado), Montana, Wyoming, New Mexico, North Carolina, Oklahoma, and the Northwest (Washington, Oregon, and Idaho). Every tribe has evolved its own sets of traditions and beliefs and each sees itself as distinct from other tribes, despite some significant broad similarities. It is beyond the scope of this book to delve deeply into differences among tribes; however, Figure 5.37 lists many of the tribes in North America (Saltzman, 2003). The list itself will impress upon the reader the variety and number of tribes, each sharing similarities and differences with other tribes.

Similarities Among Native Americans   It is possible to talk about general characteristics of Native American groups without negating the fact of their diversity. The cultural characteristics that are described in the following section do not apply to all such Americans, but rather to many who are traditionally "Indian" in their orientation to life. While being aware of tribal differences, the fire service officer should also understand that there is a strong cultural link between the many worlds and tribes of Native Americans and their Indian counterparts throughout the American continent.

Philosophy Toward the Earth and the Universe

The most striking difference between . . . Indian and Western man is the manner in which each views his role in the universe. The prevailing non-Indian view is that man is superior to all other forms of life and that the universe is

Northeast
Abenaki, Algonkin, Beothuk, Delaware, Erie, Fox, Huron, Illinois, Iroquois, Kickapoo, Mahican, Mascouten, Massachuset, Mattabesic, Menominee, Metoac, Miami, Micmac, Mohegan, Montagnais, Narragansett, Nauset, Neutrals, Niantic, Nipissing, Nipmuc, Ojibwe, Ottawa, Pennacook, Pequot, Pocumtuck, Potawatomi, Sauk, Shawnee, Susquehannock, Tionontati, Wampanoag, Wappinger, Wenro, Winnebago.

Southeast
Acolapissa, Alibamu, Apalachee, Asis, Atakapa, Bayougoula, Biloxi, Calusa, Catawba, Chakchiuma, Cherokee, Chesapeake Algonquin, Chickasaw, Chitamacha, Choctaw, Coushatta, Creek, Cusabo, Gaucata, Guale, Hitchiti, Houma, Jeags, Karankawa, Lumbee, Miccosukee, Mobile, Napochi, Nappissa, Natchez, Ofo, Powhatan, Quapaw, Seminole, Southeastern Siouan, Tekesta, Tidewater Algonquin, Timucua, Tunica, Tuscarora, Yamasee, Yuchi.

Plains
Arapaho, Arikara, Assiniboine, Bidai, Blackfoot, Caddo, Cheyenne, Comanche, Cree, Crow, Dakota (Sioux), Gros Ventre, Hidatsa, Iowa, Kansas, Kiowa, Kiowa-Apache, Kitsai, Lakota (Sioux), Mandan, Metis, Missouri, Nakota (Sioux), Omaha, Osage, Otoe, Pawnee, Ponca, Sarsi, Sutai, Tonkawa, Wichita.

Great Basin
Bannock, Paiute (Northern), Paiute (Southern), Sheepeater, Shoshone (Northern), Shoshone (Western), Ute, Washo.

Plateau
Carrier, Cayuse, Coeur D'Alene, Colville, Dock-Spus, Eneeshur, Flathead, Kalispel, Kawachkin, Kittitas, Klamath, Klickitat, Kosith, Kutenai, Lakes, Lillooet, Methow, Modoc, Nez Perce, Okanogan, Palouse, Sanpoil, Shushwap, Sinkiuse, Spokane, Tenino, Thompson, Tyigh, Umatilla, Wallawalla, Wasco, Wauyukma, Wenatchee, Wishram, Wyampum, Yakima.

Southwest
Apache (Eastern), Apache (Western), Chemehuevi, Coahuiltec, Hopi, Jano, Manso, Maricopa, Mohave, Navaho, Pai, Papago, Pima, Pueblo, Yaqui, Yavapai, Yuman, Zuni, Pueblo could potentially be defined further: Acoma, Cochiti, Isleta, Jemez, Laguna, Nambe, Picuris, Pojoaque, Sandia, San Felipe, San Ildefonso, San Juan, Santa Ana, Santa Clara, Santo Domingo, Taos, Tesuque, Zia.

Northwest
Calapuya, Cathlamet, Chehalis, Chemakum, Chetco, Chilluckkittequaw, Chinook, Clackamas, Clatskani, Clatsop, Cowich, Cowlitz, Haida, Hoh, Klallam, Kwalhioqua, Lushootseed, Makah, Molala, Multomah, Oynut, Ozette, Queets, Quileute, Quinault, Rogue River, Siletz, Taidhapam, Tillamook, Tutuni, Yakonan.

California
Achomawi, Atsugewi, Cahuilla, Chimariko, Chumash, Costanoan, Esselen, Hupa, Karuk, Kawaiisu, Maidu, Mission Indians, Miwok, Mono, Patwin, Pomo, Serrano, Shasta, Tolowa, Tubatulabal, Wailaki, Wintu, Wiyot, Yaha, Yokuts, Yuki, Yuman (California).

FIGURE 5.37 ◆ Tribal Groupings and Corresponding Regions in North America
Source: Aaron Olson

his to be used as he sees fit . . . an attitude justified as the mastery of nature for the benefit of man [characterizes Western man's philosophy] (Bahti, 1982).

Through this contrast with Western philosophy (i.e., that people have the capacity to alter nature), the reader can gain insight into the values and philosophies common to virtually all *identifying* Native Americans. While acknowledging the character of each Indian tribe or "nation," there is a common set of values and beliefs involving the earth and the universe, resulting in a deep respect for nature and "mother earth." According to American Indian philosophy, the earth is sacred and is a living entity. Through a spiritual involvement with the earth, nature, and the universe, individuals bind themselves to their environment. American Indians do not see themselves as superior to all else (e.g., animals, plants, etc.), but rather as part of all of creation. Through religious ceremonies and rituals, the Indian is able to transcend himself such that he is in harmony with the universe and connected to nature.

The inclination of some people who do not understand this philosophy would be to dismiss it as primitive and even backward. The costumes, the rituals, the ceremonies, the dances, are often thought of as colorful, but strange. Yet from a Native American perspective:

It is a tragedy indeed that Western man in his headlong quest for Holy Progress could not have paused long enough to learn this basic truth [of one's connectedness to nature]—one which he is now being forced to recognize (with the spoilage of the earth), much to his surprise and dismay. Ever anxious to teach "backward" people, he is ever reluctant to learn from them (Bahti, 1982).

Many non-Indians now embrace certain Native American beliefs regarding the environment; what people once thought of as "primitive" they now see as essential in the preservation of our environment.

When fire service officers make contact with people who are celebrating or praying, whether on reservations or in communities, it is vitally important to be as respectful as possible. They must refrain from conveying an air of superiority and ethnocentrism, conveying an attitude that "those rituals" are primitive. Native American prayers, rituals, and ceremonies represent ancient beliefs and philosophies, many of which have to do with the preservation of and harmony with the earth. An officer should, at all costs, try to avoid interrupting prayers and sacred ceremonies, just as one would want to avoid interrupting church services. (As an aside, fire service officers should also be aware that taking photographs during ceremonies would constitute an interruption and is forbidden. In general, fire service officers should seek permission before taking photos of American Indians; this is true for many of the Indian tribes. Visitors to some reservations may be told that their cameras will be confiscated if they take pictures.)

Acculturation to Mainstream Society   Significant differences exist among the cultures, languages, histories, and socioeconomic status groupings of Native American tribes, communities, and individuals. Nevertheless, in studies on suicide and ethnicity in the United States, much has been written about patterns of what can be described as self-destructive behavior that have been generalized to many Indian groups. According to the Friends Committee on National Legislation (FCNL), the suicide rate for American Indians and Alaska Natives is 72 percent greater than that for all races in the United States. The suicide rate for males between the ages of 15 and 34 is double that of the national average (FCNL, 2001).

Mortality rates attributed to alcohol consumption are nearly seven times as high for American Indians and Alaska Natives as for other races (FCNL, 2001). Alcoholism is the leading health and social problem of American Indians; 75 percent of the deaths for people under 45 years most often follow alcohol use (e.g., unintentional injury) (Indian Health Services [IHS], 2003). However, it cannot be stressed enough that the origins of the psychosocial problems that some American Indians experience in mainstream society are not a result of their own weaknesses or deficiencies. A major cause of the problems date back to the way government has handled and regulated Indian life. The dominant society in no way affirmed the cultural identity of Native Americans; thus many have internalized the oppression that they experienced from the outside world. Furthermore, many young people feel the stresses of living between two cultural worlds. They are not fully part of the traditional Indian world as celebrated on the reservation or in a community that honors traditions, and they are not fully adapted to the dominant American culture. People who are caught between two cultures and are successful in neither run the risk of contributing to family breakdown, often becoming depressed, alcoholic, and suicidal, as just noted. Comparing the suicide rate between white and Native American youths, the most recent statistics from the Pan American Health Organization (2006) indicate that, again, for Indian males, 15 to 24 years old, it was one-third higher. Interestingly, according to the Committee on Cultural Psychiatry (1989), Indian groups that have remained tightly identified with their culture because of isolation from the mainstream culture and because of remaining on indigenous lands, do not exhibit the type of behavior just described. At least at the time of the committee's study, tribes exemplifying healthier attitudes toward their identities and a lower suicide rate include the southwestern Pueblos and the Navajo.

Despite the persistence of many social problems, progress has been made with respect to education and political participation. Fire service officials must not hold on to the stereotype of American Indians as being uneducated. There is a growing population attending colleges and rising to high positions in education, entertainment, sports, and industry.

The National Congress of American Indians (NCAI) is the oldest, largest, and most representative Indian organization devoted to promoting and protecting the rights of American Indians and Alaska Natives as a whole group (Figure 5.38). As an example of

FIGURE 5.38 ◆ Replica of the Congressional Gold Medal awarded to the Navajo Code Talkers of WWII. The obverse of the medal features two Marine Navajo Code Talkers communicating a radio message. Centered along the top of the medal is the inscription "NAVAJO CODE TALKERS." Centered along the bottom is "BY ACT OF CONGRESS 2000."

a specific organizational action, in 1999, the NCAI condemned the use of sports team "mascots" using Native American and native cultural terms (e.g., "Redskins").

Language and Communication   It is possible to make some generalizations about the way a group of people communicate, even when there is great diversity within the group. The following contains information about nonverbal and verbal aspects of communication as well as tips for the fire service officer interacting with Native Americans. Recognizing the differences in cultural language and communications, the fire service has begun to develop culturally appropriate fire prevention education materials:

> Forty fire service professionals from 36 organizations responsible for fire safety and fire protection on Native American reservations met in Oklahoma City June 17 and 18 to receive training in a culturally sensitive fire prevention program. The program, "Wisdom of the Fire," is patterned after the NFPA's successful "Learn Not to Burn©" fire prevention program for elementary school children. Wisdom of the Fire (WOF) was developed through a partnership between NFPA, the Assembly of First Nations (Canada) and the Aboriginal Fire Fighters of Canada. Terry Diabo, director of operations of the Kahnawake (Ontario) Fire Brigade and former president of the Aboriginal Fire Fighters, conducted the two-day train-the-trainer program. . . . The presentation was so well received that the entire task force voted to recommend seeking funding to conduct a train-the-trainer program for Native Americans and Alaska Natives. The consensus was that Wisdom of the Fire was exactly what Native American fire safety educators had been looking for to convey fire safety messages with impact. Wisdom of the Fire takes the customary fire safety messages and delivers them in language and stories that are familiar to members of the First Nations. Diabo emphasized that the key to success for Native American tribes would include building in local culture to make the program more appealing and relative to its audience. The Wisdom of the Fire materials given to each attendee would provide the template for building a successful local program because it was already written in a manner that was familiar to Native peoples. A significant component of the program incorporates roles for tribal leaders and elders to deliver the fire safety messages (Tippett, 2004).

The paragraphs that follow describe patterns of communication and behavior as exhibited by some American Indians who are traditional in their outlook. No description of communication traits, however, would ever apply to everyone within a group, especially one that is so diverse.

• **Openness and self-disclosure:**   Many Native Americans, in early encounters, will approach people and respond with caution. Too much openness is to be avoided, as is disclosing personal and family problems. This often means that the officer has to work harder at establishing rapport and gaining trust. In the American mainstream culture, appearing friendly and open is highly valued, especially in certain regions, such as the West Coast and the South. Because different modes of behavior are expected and accepted, the non-Indian may view the Indian as aloof and reserved. The Indian perception can be that the Euro American person, because of excessive openness, is superficial and thus untrustworthy. Mainstream American culture encourages speaking out and open expression of opinions, whereas American Indian culture does not.

- **Silence and interruptions:**　The ability to remain quiet, to be still, and to observe is highly valued in Native American culture; consequently, silence is truly a virtue. American Indians are taught to study and assess situations and to act or participate only when properly prepared. American Indians tend not to act impulsively for fear of appearing foolish or bringing shame to themselves or their family. When fire service officials contact Native Americans, they may mistake this reticence to talk as sullenness or lack of cooperation. The behavior must not be misinterpreted or taken personally. A cultural trait must be understood as just that (i.e., a behavior, action, or attitude that is not intended to be a personal insult). The officer must also consider that interrupting an Indian when he or she speaks is seen as very aggressive and should be avoided whenever possible.

- **Talking and questions:**　Talking just to fill the silence is not seen as important. The small talk observed in mainstream society ("Hi. How are you? How was your weekend?" and so on) is traditionally not required by Native Americans. Words are considered powerful and are therefore chosen carefully. This may result in a situation in which Native Americans retreat and appear to be withdrawn if someone is dominating the conversation. When fire service officials question Native Americans who exhibit these tendencies (i.e., not being prone to talkativeness), the officer should not press aggressively for answers. Aggressive behavior, both verbal and physical, is traditionally looked down upon. Questions should be open ended, with the officer being willing to respect the silence and time it may take to find out the needed information.

- **Nonverbal communication: Eye contact/touching:**　With respect to American Indian cultures, people often make the statement that American Indians avoid making direct eye contact. Although this is true for some tribes, it does not hold true for all. To know whether this applies or not, an officer can simply watch for this signal (i.e., avoidance of eye contact) and follow the lead of the citizen. In this section we explain the phenomenon of avoidance of eye contact from the perspective of groups who adhere to this behavior. Some Indian tribes have the belief that looking directly into one's eyes for a prolonged period of time is disrespectful, just as pointing at someone is considered impolite. Lakota tribe members, for example, generally believe that direct eye contact is an affront or an invasion of privacy (Mehl, 1990). Navajo tribe members have a tendency to stare at each other when they want to direct their anger at someone. An Indian who adheres to the unspoken rules about eye contact may appear to be shifty and evasive. Officers and other fire service officials must not automatically judge a person as guilty or suspicious simply because that person is not maintaining direct eye contact. To put a person at ease, the officer can decrease eye contact if it appears to be inhibiting the Native American citizen. Avoidance of eye contact with the officer can also convey the message that the officer is using an approach that is too forceful and demanding. Where such norms about avoidance of eye contact apply, and if an officer has to look at a person's eyes, it would help to forewarn the person (i.e., "I'm going to have to check your eyes").

　　With regard to their sense of space, most Native Americans are not comfortable being touched by strangers, whether a pat on the back or an arm around the shoulder. Either no touching is appropriate or it should be limited to a brief handshake. Married couples do not tend to show affection in public. Additionally, people should avoid crowding or standing too close. Keep in mind that many Indian relations with strangers are more formal than those of the mainstream culture; therefore, an officer may be viewed as overly aggressive if he or she does not maintain a proper distance. Officers, therefore, who are going to pat down or search a Native American should first explain the process.

LANGUAGE

Some Native Americans speak one or more Indian languages. English, for many, is a second language. Those who do not speak English well may be inhibited from speaking for fear of "losing face" because of their lack of language ability. In addition, because of a tendency to speak quietly and less forcefully, fire service officers will need patience and extra time; interaction must not be rushed. The Native American who is not strong in English must spend more time translating from his or her own language to English when formulating a response (this is true of all second-language speakers who are not yet fluent). As with all other languages, English words or concepts do not always translate exactly into the various Native American languages. These languages are rich and express concepts reflecting views of the world that do not have direct English equivalents. It is mandatory that the utmost respect be shown when Native Americans speak their own language. Remember that the American Indians have a long history of forced assimilation into the Anglo society, in which, among other things, many were denied the right to speak their native languages.

OFFENSIVE TERMS, LABELS, AND STEREOTYPES

Use of racial slurs toward any group is never acceptable in firefighting and fire service, no matter how provoked an officer may be. Officers hearing disrespectful terms and stereotypes about Native Americans should educate fellow officers as to the lack of professionalism and respect for community diversity that such terms convey.

A number of words are offensive to Native Americans: "chief" (a leader who has reached this rank is highly honored), "squaw" (extremely offensive), "buck," "redskin," "Indian brave," and "skins" (some young Native Americans may refer to themselves using the latter term, but would be offended by others using the term). In addition, the use of Indian tribal names or references as mascots for sports teams has been highly objectionable. Other terms used to refer to Native Americans are "apple" (a slightly dated term referring to a highly assimilated Indian, "red" on the outside; "white" on the inside) and "the people" (more commonly used by some groups of American Indians to refer to themselves). In some regions of the country, a reservation is called a "rez" by American Indians, and in Oklahoma (where there is only one reservation), the term *reservation* is negative and the term *community* is used. It is also patronizing when non-Indians use certain kinship terms, such as *grandfather* when talking to an older man, even though other American Indians may be using those terms themselves.

Other terms and expressions that can be offensive and are used commonly include the following: "sitting Indian style," to refer to sitting in a cross-legged position on the floor; "Indian giver" to characterize someone who takes back a present or an offer; the children acted like "wild Indians," to describe uninhibited and unorganized energetic behaviors; let's have a "pow wow," meaning a discussion; "bottom of the totem pole," meaning lowest ranking; and he is on the "warpath," to describe an aggressive or retaliatory behavior pattern. In addition, it is worth noting that some American Indians deliberately choose not to reveal their ethnic identity in the workplace because of concerns about stereotypes about American Indians. Coworkers may make comments about American Indians or use offensive expressions because they do not "see one in the room." People can be deeply offended and hurt by "unintentional" references to American Indians.

Many people growing up in the United States can remember the stereotypical picture of an Indian as a wild, savage, and primitive person. In older textbooks, including history books, recounting Native American history, American Indians were said to "massacre" whites, whereas whites simply "fought" or "battled" the American Indians (Harris and Moran, 1991). Other common stereotypes or stereotypical statements that are highly resented are: "All Indians are drunks" (despite the fact that there is a large percentage of alcohol-related arrests, not all American Indians have a problem with liquor; furthermore, the argument has been put forth that the white man introduced "fire water" or alcohol to the Indian as a means of weakening him); "You can't trust an Indian"; "Those damn Indians" (as if they are simply a nuisance); and "The only good Indian is a dead one" (a remark that can be traced back to a statement made by a U.S. general in 1869) (Harris and Moran, 1991).

> Native Americans find it offensive when non-Indians make claims which may or may not be true about their Indian ancestry: "I'm part Indian—My great-grand-father was Cherokee (for example). . . ." Although this may be an attempt to es-tablish rapport, it rings of "Some of my best friends are American Indians . . ." (i.e., to "prove" that one does not have any prejudice). [People] should not as-sume affinity [with American Indians] based on novels, movies, a vacation trip, or an interest in silver and turquoise jewelry. These are among the most offen-sive, commonly made errors when non-Indians first encounter an American Indian person or family. Another is a confidential revelation that there is an In-dian "Princess" in the family tree—tribe unknown, identity unclear, but a bit of glamour in the family myths. The intent may be to show positive bonding. . . , but to the Indian they reveal stereotypical thinking (Attneave, 1982).

RESPECT FOR ELDERS

Unlike mainstream American culture, American Indians value aging because of the respect they have for wisdom and experience. People do not feel that they have to cover up or counteract signs of aging, as this phase of life is highly revered. The elders of a tribe or the older people in Native American communities must be shown the ut-most respect by people in fire service. This includes acknowledging their presence in a home visit, even if their involvement is not directly related to the fire service or pub-lic safety matter at hand. In some tribes (e.g., the Cherokee) the grandmother often has the most power in the household and is the primary decision maker. It is advis-able for people in fire service to include the elders in discussions in which they can give their advice or perspectives on a situation. The elders are generally respected for their ability to enforce good behavior within the family and tribe (Figure 5.39). It should also be noted, however, that because of assimilation or personal preference among some Native Americans, the elders in any given household may tend to avoid interfering with a married couple's problems. And although the elders are respected to a higher degree than in mainstream culture, they may withdraw in some situations in which there is fire service contact, letting the younger family members deal with the problem. If in doubt, it is advisable to begin the contact more formally, deferring to the elders. Then officers can observe how the elders participate and if the younger family members include them.

EXTENDED FAMILY AND KINSHIP TIES

Within mainstream society, people usually think of and see themselves first as indi-viduals, and after that they may or may not identify with their families or various communities and groups with which they are affiliated. In traditional Native Ameri-can culture, a person's primary identity is related to his or her family and tribe. Some fire service officers may be in positions to make referrals when there is a problem with

FIGURE 5.39 ◆ Native American elders are highly respected, such as this Navajo woman who provides wise counsel and is wearing a traditional stone necklace.

an individual (e.g., an adolescent) in a family. A referral for counseling for that person individually (and without the family) may be culturally alienating. Individual Western-style counseling or therapy is a foreign way to treat problems. Additionally, Native American culture is highly group oriented.

Today, some family and tribal cohesiveness has lessened because of forced assimilation, extreme levels of poverty, and lack of education and employment. However, for many Native Americans, there are still large networks of relatives who are in close proximity with each other (Figure 5.40). It is not uncommon for children to be raised by someone other than their father or mother (e.g., grandmother, aunts). When fire service officials enter an Indian's home and, for example, ask to speak to the parents of a child, they may actually end up talking to someone who is not the biological mother or father. Officers must understand that various other relatives can function exactly as a mother or father would in mainstream culture. It should not be assumed that American Indian "natural" parents are being lazy about their child-rearing duties, even when the child is physically living with another relative (and may be raised by several relatives throughout childhood). The intensely close family and tribal bonds allow for this type of child raising. The officer must not assume that something is abnormal with this type of arrangement or that the parents are neglecting their children.

Children and Separation from Parents It is crucial that fire service officers understand the importance of not separating children from family members, if at all possible. Many families in urban areas and on reservations have memories of or have heard stories from elder family members that involved the federal government's systematic removal of Indian children from their homes; in many cases, children were

FIGURE 5.40 ◆ Two Navajo women with their children as they appeared at the Inter-Tribal Indian Ceremonial at Gallup, New Mexico.

placed in boarding schools that were often hundreds of miles away. This phenomenon, including education for the children that stripped them of their language and culture, began in the late nineteenth century. Reports exist that say: "Until 1974, the Bureau of Indian Affairs (BIA) was operating 75 boarding schools with more than 30,000 children enrolled" (Ogawa, 1990). Although for many families, the severe trauma of children being forcibly separated from parents took place years ago, the aftereffects still linger. In the early twentieth century there was a famous case in which Hopi Indian fathers were sentenced to years in high-security prisons and were subject to the fullest persecution. Their crime was hiding their children from BIA officials because they did not want the children to be taken by the BIA boarding schools. By hiding the children, the Hopi fathers violated federal law. The severe injustice and the legal and psychological ramifications of this case continue to remain and to be discussed today. The memory of a "uniform coming to take away a child" is an image that can be conjured up easily by some American Indians. It is this "baggage" that the fire service officer today brings into encounters; he or she may be totally unaware of the power of Native Americans' memories of these deplorable actions. Because Native American parents can be very protective of their children, an officer would be well advised to let the parents know about any action that needs to be taken with regard to the child. Fire service officials can establish a good rapport with Indian families if they treat the children well.

SUMMARY OF RECOMMENDATIONS FOR FIRE SERVICE IN THE NATIVE AMERICAN/AMERICAN INDIAN COMMUNITY

1.   Those who are entrusted with fire service in rural areas, in cities, or on Native American reservations should exhibit respect and professionalism when interacting with peoples who have traditionally been disrespected by governmental authority. It is important to remember

that the U.S. government has violated many treaties with Americans Indians and that their basic rights as Americans have repeatedly been denied.

- ◆ Understand the initial resistance to your efforts to establish rapport and goodwill, and do not take it personally.
- ◆ Make an effort to get to know the community in your particular area. Make positive contact with American Indian organizations and individuals. This behavior on your part will be unexpected and will result in more cooperation.

2.   The younger, more environmentally conscious generation of Americans has adopted much valuable ideology from the culture of America's original peoples. Convey a respect for Native American values. They are not alienating or "un-American," and many believe that those very values of preservation are necessary for our environmental survival.

3.   Native Americans have been victims of forced assimilation whereby their languages, religions, and cultures have been suppressed. The negative effects have stayed with generations of American Indians. The point of contact between a fire service professional and an American Indian can often involve issues related to poor adjustment to urban life. While the law must be upheld, consider the conditions that led some Native Americans toward, for example, alcoholism and unemployment. Having empathy for the conditions that lead to a person's circumstances need not make one any less effective in his or her line of duty.

4.   Preferred mainstream American ways of communication often run counter to American Indian styles of communication. Keep the following in mind when trying to build rapport with these citizens: (a) do not take advantage of the American Indian just because he or she may be silent, appear passive, or not be fluent in English; and (b) do not interrupt American Indian people when they are speaking; it is seen as aggressive and rude.

5.   Many Native Americans who favor traditional styles of communication will tend toward: (a) closed behavior and slow rapport building with strangers (this does not mean that the person is aloof or hostile; rather, this can be a cultural trait); (b) silent and highly observant behavior (this is not necessarily an indication that an individual does not want to cooperate); (c) withdrawal if the method of questioning is too aggressive (remember to use time, patience, and silence, which will assist you in getting the response you need); and (d) indirect eye contact for members of some tribes, but not all (your penetrating or intense eye contact may result in intimidating the person and, consequently, in his or her withdrawal).

6.   The words "chief," "redskin," "buck," "squaw," "braves," and "skins" are offensive when used by a non-Indian (sometimes younger American Indians may use some of the terms themselves [e.g., "skins"], but this does not make it acceptable for others to use the terms). If you need to refer to the cultural group, ask the person with whom you are in contact whether he or she prefers the term *Indian*, *Native American*, or another tribal name. Your sensitivity to these labels can contribute to establishing a good rapport. If asking seems inappropriate, listen carefully to how the individuals refer to each other.

7.   The extended family is close-knit and interdependent among Native American peoples. Keep in mind the following:

- ◆ Be respectful and deferential to elders.
- ◆ The elders should be asked for their opinion or even advice, where applicable, as they are often major decision makers in the family.
- ◆ If there are problems with a child, consider other adults, besides the mother and father, who may also be responsible for child rearing.
- ◆ Whenever possible, do not separate children from parents. This can bring back memories of times when children were forcibly taken from their parents and sent to selected mission schools or government boarding schools far from their homes.

8.   From a Native American perspective, many people feel that they are abused by a system of government that is neither honest nor respectful of their culture. They believe that the

government degrades the land upon which all people depend. To a large extent, Indian rights are still ignored because members of the dominant society do not always uphold the laws that were made to protect American Indians. This background makes it especially difficult for people in fire service vis-à-vis their relationships and interactions with Native Americans. For this reason, fire service officials need to go out of their way to demonstrate that they are fair, given the complexities of history and current law. In addition, chief executives and command staff of fire service departments have a special responsibility to provide an accurate education to officers on American Indian cultural groups within their service domains, with an emphasis on government–tribal relations, and to address the special needs and concerns of the American Indian peoples.

EUROPEAN AMERICANS

We are concluding this chapter on a public safety approach to specific cultures with some additional information about European Americans who are recent immigrants and refugees. Although it has been taken for granted that the majority of the European Americans in the United States had developed this country and are fully assimilated into the mainstream, majority culture, this is certainly not true for those European Americans who are recent immigrants and refugees. As such, this section provides specific cultural information on these European Americans, including aspects of their background that both directly and indirectly can affect the relationship with fire service officials.

One of the biggest myths about European Americans is that "they are all alike." It is true that the majority of people in the United States are of European descent, but the majority of Europeans are not of the same ethnicity, nationality, nor do they even have the same physical characteristics. Europe is a continent that is divided into four regions (east, west, north, and south) and has a population of 730 million people (Population Reference Bureau, 2005).

Europe has 45 different countries and each country is unique, with its own customs, religion, government, and language. To illustrate Europe's diversity and heterogeneity, the European Union has 20 different official languages for its European Parliament, compared to the United Nations, which has six official languages. Moreover, it is the world's largest translation operation and has 60 interpreters in use when its 25 member state parliament is in session (Owen, 2005).

European regional differences (Figure 5.41) are noticeable in physical characteristics too. People from southern Europe (i.e., the countries of Italy, Spain, and Greece) have a darker, olive color skin complexion compared to the other regions of Europe (i.e., the countries of Finland, Russia, and Great Britain), where the majority of people have a lighter beige color skin complexion. Figure 5.42 provides a list of the 45 countries that constitute Europe (Population Reference Bureau, 2005).

The typology provided at the beginning of this chapter suggests that as fire service and public safety organizations prepare and train their personnel to work with European Americans who are recent immigrants and refugees, a focus on the key differences within each of the typological groups would be most effective, especially those within the first two categories: (1) Type I: recently arrived immigrant or refugee (less than five years in the United States, with major life experiences *outside* of the United States), and (2) Type II: immigrant or refugee (five or more years in the United States, with major life experiences *outside* of the United States). We have provided this convenient framework for application with European Americans who are recent

FIGURE 5.41 ◆ Globe image of Europe shows Europe (upper right) separated from northern Africa by the Mediterranean Sea and the Arabian Peninsula.

1. Albania	24. Lithuania
2. Andorra	25. Luxembourg
3. Austria	26. Macedonia, The Former Yugoslav Republic of
4. Belarus	
5. Belgium	27. Malta
6. Bosnia and Herzegovina	28. Moldova
7. Bulgaria	29. Monaco
8. Croatia	30. Netherlands
9. Cyprus	31. Norway
10. Czech Republic	32. Poland
11. Denmark	33. Portugal
12. Estonia	34. Romania
13. Finland	35. Russia
14. France	36. San Marino
15. Germany	37. Serbia and Montenegro
16. Greece	38. Slovakia
17. Holy City (Vatican City)	39. Slovenia
18. Hungary	40. Spain
19. Iceland	41. Sweden
20. Ireland	42. Switzerland
21. Italy	43. Turkey
22. Latvia	44. Ukraine
23. Liechtenstein	45. United kingdom

FIGURE 5.42 ◆ The Countries of Europe

Source: Aaron Olson, adapted from Population Reference Bureau (2005).

Figure 5.43 ◆ A Russian restaurant that caters to the numerous recent Russian immigrants and refugee in Brighton Beach, Brooklyn.

immigrants and refugees with regard to their motivational components within each of the groupings in a fire service situation involving European American immigrants and refugees (Figure 5.43).

SUMMARY

The background, history, and other cultural highlights have been presented about African American, Asian/Pacific, Latino/Hispanic, Middle Eastern, and Native American groups with regard to the key issues important for cultural sensitivity for fire service officers and first responder representatives. These specific groups were focused upon because (1) the group is a relatively large ethnic or racial group within the United States; (2) the traditional culture of the group differs widely from that of mainstream American culture; and/or (3) typically or historically there have been problems between the particular group and fire service departments and officers. We have also included a section on recent immigrant and refugee groups from European countries because these groups have similar community issues and concerns with fire service and public safety agencies as do many of the minority communities within the United States. In these culture-specific sections, general information is presented in the following areas: demographics, diversity within the cultural group, group identification terms, offensive labels, stereotypes, family structure, and communication styles (both verbal and nonverbal). The chapter summarizes the key concerns for fire officers with respect to each of the particular cultural groups and provides recommendations for fire service officials and other public safety agencies.

■■

Discussion Questions and Issues

1. **Historical Implications.** Under the Historical Information headings in this chapter, the authors highlighted information and background that might be barriers and challenges for fire service and public safety officers and agencies in their work with African American, Asian/Pacific, Latino/Hispanic, Middle Eastern American, and Native American/American Indians. Discuss the following issues for each of the specific groups:

   a. The fact of African Americans having been slaves in this country has created great psychological and social problems for blacks and whites for generations to come. How is this true for both races? What are the implications for fire service?

   b. Many anti-Asian/Pacific American laws and events may leave Asian/Pacific Americans with the view that fire service agencies are not user friendly. What are the implications of this view for fire service? What are ways to improve such possible negative points of view?

   c. Many associations are made about illegal immigrants and undocumented aliens that may leave Latino/Hispanic Americans with the view that public safety agencies may not be sensitive and responsive. What are ways to improve such possible negative points of view for fire service?

   d. Many Middle Eastern and Arab Americans are aware of how they might be viewed as possible terrorists or supporters of terrorist activities. What are ways for fire service and public safety personnel to work with Middle Eastern and Arab American communities given these perceptions?

   e. The famous Lakota chief, Sitting Bull, spoke on behalf of many American Indians when he said of white Americans: "They made us many promises . . . but they never kept but one: They promised to take our land, and they took it." Given the sensitivity of Native Americans/American Indians to the ownership of their land or trespassers on their land, what are ways for fire service personnel to show sensitivity to these issues when working with Native American/American Indian communities?

2. **Offensive Terms.** The authors advise refraining from using any offensive terms for African American, Asian/Pacific, Latino/Hispanic, Middle Eastern American, and Native American/American Indians (or any other group), even where there is no one from any of the specific groups present. Give two practical reasons for this.

3. **Diversity Among African American, Asian/Pacific, Latino/Hispanic, Middle Eastern American, and Native American/American Indians.** Which groups are you most likely to encounter in fire service in your area? Which groups do you anticipate encountering in your future work?

4. **Similarities of European American Recent Immigrants and Refugees.** Why do you think that many European Americans who are recent immigrants and refugees (e.g., recent East European immigrants from Bosnia, Bulgaria, Serbia, Ukraine) keep to their own communities? In what ways might they be similar to other minority communities? How can fire service agencies be of greater service to these recent immigrant and refugee communities? What kind of outreach and educational fire service would be most effective with these communities?

5. **Effects of Myths and Stereotypes.** Myths and stereotypes about African American, Asian/Pacific, Latino/Hispanic, Middle Eastern American, and Native American/American Indians have greatly affected these groups. What are some of the stereotypes that you have heard of or have encountered regarding these groups? What effects would these stereotypes have on fire service in your area? What are ways to

manage these stereotypes in fire service? How might your awareness of these stereotypes be helpful in working with these communities on homeland security issues?

6. **Verbal and Nonverbal Variations Among Cultures.** How do you think that the information in this chapter about verbal and nonverbal communication styles can help fire service officers in their approaches to African American, Asian/Pacific, Latino/Hispanic, Middle

Eastern American, and Native American/American Indian citizens? When you can understand the cultural components of the styles and behaviors, does this help you to become more sensitive and objective about your reactions? Provide some examples of rephrasing questions in such a way that they elicit responses that show understanding and the intended actions on the part of African American, Asian/Pacific, Latino/Hispanic, Middle Eastern American, and Native American/American Indians.

Website Resources

Visit these websites for additional information about topics covered in Chapter 5:

African American Web Connection: http://www.aawc.com/aawc.html
This website is devoted to providing the African American community with valuable resources on the Web. It includes such topics as arts and poetry, businesses, churches, organizations, and topics of concern, to name a few.

American-Arab Anti-Discrimination Committee (ADC): http://www.adc.org
This website provides information about the ADC's civil rights efforts and useful summary data about cases and complaints regarding discrimination and hate crimes involving Arab Americans.

American Civil Liberties Union (ACLU): http://www.aclu.org
This website contains multiple locations for information about many issues including racial and cultural group discrimination.

Arab American Institute Foundation (AAI): http://www.aaiusa.org
This website is dedicated to the civic and political empowerment of Americans of Arab descent. AAI provides policy, research, and public affairs services to support a broad range of community activities. In addition, AAI also is a census information center on demographics of Arab Americans. On the website, you can access a PDF file entitled, "Healing the Nation: The Arab American Experience After September 11."

Asian American Legal Defense and Education Fund (AALDEF): http://www.aaldef.org
This website provides information about civil rights issues with Asian/Pacific Americans and highlights issues of immigration, family law, government benefits, anti-Asian violence and public safety misconduct, employment discrimination, labor rights, and workplace issues.

Asian American Network: http://www.asianamerican.net
This website provides a national listing of many of the networks of Asian/Pacific American community-based organizations in United States (as well as some in Asia).

Bureau of Indian Affairs (BIA): http://www.doi.gov/bureau-indian-affairs.html
As of the final preparation of this book, this website was "temporarily shut down as part of the Cobell Litigation accusing the U.S. government of improperly managing Indian assets entrusted to it" (see http://www.washingtonpost.com/ac2/wp-dyn/A25625-2002Apr21?); the page still references a wide variety of links to governmental and nongovernmental organizations related to Native Americans.

Connecting Cultures: http://www.connecting-cultures.net
This website introduces Connecting Cultures, a consulting organization that designs and implements workshops and seminars on diversity, religion, and cross-cultural commu-

nication. It provides resources for understanding how culture and religion impact the way we communicate and work with one another. Connecting Cultures' specialty is Arab and Muslim Americans, as well as Arab culture, the Middle East, and the Muslim world.

European Association for American Studies: http://www.eaas.info/national.htm
This website highlights the work of the European Association for American Studies, which is a confederation of national and joint national associations of American Studies in Europe. The aims of the association are to encourage the study of and research in all areas of American culture and society and to promote cooperation and intercommunication between European scholars of the United States from all parts of Europe and from various disciplines.

League of United Latin American Citizens (LULAC): http://www.lulac.org
This website provides information on education, training, scholarships, and services to underprivileged and unrepresented Latino/Hispanic Americans. LULAC is the largest and oldest Latino/Hispanic organization in the United States with over 115,000 members.

Middle East Institute (MEI):
http://www.mideasti.org/
This website provides a summary of information about Middle Eastern nations and American policy makers and organizations. MEI has worked to increase knowledge of the Middle East and to promote understanding between the peoples of the Middle East and America.

The Middle Eastern Network Information Center (MENIC): http://menic.utexas.edu/menic.html
The website for the Middle Eastern Network Information Center at the University of Texas offers a wide range of teaching materials on various levels and on specific Middle Eastern countries.

National Asian Pacific American Legal Consortium: http://www.napalc.org
This website provides a national network of information about legal and civil rights issues affecting Asian/Pacific Americans in terms of litigation, advocacy, public education, and public policy.

National Congress of American Indians (NCAI): http://www.ncai.org
The website for the national organization contains information on issues, events, and other information relevant to the organization and to Native Americans.

National Council of La Raza (NCLR): http:// www.latino.sscnet.ucla.edu/community/nclr.html
This website provides information about capacity-building assistance to support and strengthen Hispanic community-based programs and information regarding applied research, policy analysis, and advocacy for Latino/Hispanic Americans.

Native American Nations: http://www.nativeculturelinks.com/nations.html
This website contains links by tribal names in alphabetical order to web pages that either have been set up by the Native American nations themselves or are devoted to a particular nation. Included are both recognized and unrecognized tribes.

VERA Institute of Justice:
http://www.vera.org
The website provides information on the research projects and activities of the VERA Institute of Justice. This organization works closely with leaders in government and civil society to improve the services people rely on for safety and justice. VERA develops innovative, affordable programs that often grow into self-sustaining organizations, studies social problems and current responses, and provides practical advice and assistance to government officials in New York and around the world.

■ ■

References

American-Arab Anti-Discrimination Committee. (1992). *1991 Report on Anti-Arab Hate Crimes: Political and Hate Violence Against Arab-Americans*, Washington, DC: ADC Research Institute.

Aquino, K., M. M. Stewart, and A. Reed. (2005). "How Social Dominance Orientation and Job Status Influence Perceptions of African-American Affirmative Action Beneficiaries." *Personnel Psychology*, 58(3): 703–44.

Arab American Institute. (2000). *Arab-Americans: Issues, Attitudes, and Views.* Washington, DC: Zogby, International/Arab American Institute Study.

Attneave, C. (1982). "American Indians and Alaska Native Families: Emigrants in Their Own Homeland." In M. McGoldrick et al. (Eds.), *Ethnicity and Family Therapy.* New York: Guilford Press.

Bahti, T. (1982). *Southwestern Indian Ceremonials.* Las Vegas, NV: KC Publications.

Bakari, A. (2004, February 29). "5 Tragic Stereotypes, Part I." *Global Black Network.*

Barnes, J. S., and C. E. Bennett. (2002). *The Asian Population: 2000.* Washington, DC: U.S. Census Bureau.

Bennett, C. E. (2002). *A Profile of the Nation's Foreign-Born Population from Asia* (2000 Update). Washington, DC: U.S. Census Bureau, Tables P23–206.

Bennett, L., Jr. (1989, November). "The 10 Biggest Myths about the Black Family." *Ebony,* pp. 1, 2.

Bernal, G., and A. Estrada. (1985). "Cuban Refugee and Minority Experiences: A Book Review." *Hispanic Journal of Behavioral Sciences,* 7: 105–28.

Bernal, G., and M. Gutierrez. (1988). "Cubans." In L. Comas-Diaz and E. E. H. Griffith (Eds.), *Cross-Cultural Mental Health.* New York: John Wiley, pp. 233–61.

Bernstein, R. (2005, June 9). *Hispanic Population Passes 40 Million, Census Bureau Reports.* Washington, DC: U.S. Census Bureau News.

Boller, P. J., Jr. (1992, March). "A Name Is Just a Name—or Is It?" *FBI Law Enforcement Bulletin,* p. 6.

Bureau of Indian Affairs (BIA). (2003). See link at http://www.doi.gov/bureau-indian-affairs.html

CEHIP. (1996). "20th Century Warriors: Native American Participation in the United States Military." Prepared for the United States Department of Defense by CEHIP, Inc., Washington, DC, in partnership with Native American advisers, Rodger Bucholz, William Fields, Ursula P. Roach. Washington, DC: Department of Defense. See link at http://www.history.navy.mil/faqs/faq61-1.htm

Central Intelligence Agency. (2003). *World Fact Book.* See link at http://cia.gov/cia/publications/factbook/index.html

Central Intelligence Agency. (2006). *World Fact Book.* Washington, DC: Central Intelligence Agency.

Cohn, D. (2005, June 9), "Hispanic Growth Surge Fueled by Births in U.S." *Washington Post,* p. A1.

Comas-Diaz, L., and E. E. H. Griffith. (Eds.) (1988). *Cross-Cultural Mental Health.* New York: John Wiley.

Committee on Cultural Psychiatry, Group for the Advancement of Psychiatry. (1989). *Suicide and Ethnicity on the United States.* New York: Brunner/Mazel.

Council of Economic Advisors. (1998). "Changing America: Indicators of Social and Economic Well-Being by Race and Hispanic Origin." *Crime and Criminal Justice,* Chapter 7. Washington, DC: Author.

Council of Economic Advisors for the President's Initiative on Race. (1998, September 1998). *Changing America: Indicators of Social and Economic Well-Being by Race and Hispanic Origin.* Washington, DC: National Center for Health Statistics, Bureau of Justice Statistics.

CSIS. (2003). *Consumer Services Information Systems Project.* Refer to information in "Glossary" at http://csisweb.aers.psu.edu

Daniels, R. (1988). *Asian America: Chinese and Japanese in the United States since 1850.* Seattle: University of Washington Press.

Davis, R. C., et. al. (2002, August 6). "Turning Necessity into Virtue: Pittsburgh's Experience with a Federal Consent Decree." *Vera Institute of Justice.* See link at http://www.cops.usdoj.gov/Default.asp?Item=565

Delk, J. D. (1995). *Fires & Furies: The Los Angeles Riots of 1992.* Palm Springs, CA: ETC Publications.

Delgado v. Bastrop Independent School District, 388 W. D. Texas (1948).

Detroit Free Press. (2001). *100 Questions and Answers about Arab Americans: A Journalist's Guide.* Detroit, Michigan: Author.

Devine, E., and N. L. Braganti. (1991). *The Traveler's Guide to the Middle Eastern and North African Customs and Manners.* New York: St. Martin's Press.

"Documentation, Language Issues Arise for Hispanic Firefighters." (2006, June 17). *Daily Herald (Provo, Utah),* p. B5.

Enriquez, D. (2006, March 10). "Man Sues in Hate-Crime Case: Lawsuit Compares Firefighters' Actions to Ku Klux Klan's." *Milwaukee Journal Sentinel (Wisconsin).*

Federal Emergency Management Agency (FEMA) and U.S. Fire Administration (1994). *Report of the Joint Fire/Police Task Force on Civil Unrest: Recommendations for Organization and Operations During Civil Disturbance.* Washington, DC: FEMA.

Fernandex, L. F. (1970). *A Forgotten American.* New York: B'nai B'rith.

"Film's Portrayal of Muslims Troubling to Islamic Groups." (1998, November 8). *Boston Globe,* p. B10.

Frazier, J. B. (2006, June 17). "Documentation, Language Issues Arise for Hispanic Firefighters." *Daily Herald* (*Provo, Utah*), p. B5.

Frey, W. H. (1998, June–July). "New Demographic Divide in the US: Immigrant and Domestic 'Migrant Magnets.'" *The Public Perspective,* 9(4): 25–40.

Friends Committee on National Legislation (FCNL). (2001). See link at http://www.fcnl.org/issues/nat/sup/indians_healthfacts_41701.htm

Gavzer, G. (1993, March 21). "Held without Hope." *Parade.*

Gilbert, J., N. Carr-Ruffino, J. M. Ivancevich, and M. Lownes-Jackson. (2003). "An Empirical Examination of Inter-Ethnic Stereotypes: Comparing Asian American and African American Employees." *Public Personnel Management,* 32(2): 251–66.

Glasse, C., and H. Smith. (2003). *The New Encyclopedia of Islam,* rev. ed. Walnut Creek, CA: Alta Mira Press.

Goleman, D. (1992, April 21). "Black Scientists Study the 'Pose' of the Inner City." *New York Times.*

Gracia, J. J. E. (2000). *Hispanic/Latino Identity: A Philosophical Perspective.* Malden, MA: Blackwell Publications.

"Great 1906 San Francisco Earthquake and Fire." (2006, April). *American History.*

Harris, P. R., and R. T. Moran. (1991). *Managing Cultural Differences: High Performance Strategies for a New World of Business.* Houston, TX: Gulf Publishing.

Hickman, H. (2006, January 19). "Apology Offered to Fire Chief; Perez Says Subordinate Referred to Him as 'Pedro' in Derogatory Manner." *Knoxville News-Sentinel (Tennessee),* p. B1.

Hines, P. M., and N. B. Franklin. (1982). "Black Families," in M. McGoldrick et al. (Eds.), *Ethnicity and Family Therapy.* New York: Guilford Press.

Hoge, P. (2005, November 12). "2nd Fire in 8 Years Guts Market in Chinatown; No Injuries Reported; Early Investigation Finds No Foul Play." *San Francisco Chronicle,* p. B1.

Hornor, L. L. (Ed.). (1999). *Hispanic Americans: A Statistical Sourcebook—1999 Edition.* Palo Alto, CA: Information Publications.

Indian Country Today Staff Reports. (2003). "First American Female and Native Soldier Killed in Iraq War Is Remembered." See link at http://www.indiancountry.com/content.cfm?id=1050072510

Indian Health Services (IHS) (2003). Quoted in Democratic Policy Committee (DPC) Reports, September 27, 2004. Available: http//democrats.senate.gov/dpc/dpc-now.efm?doc_name=sr-108-2-259.

Ismail, L. (2003, October). President, Connecting Cultures, Washington, D.C. personal communications.

Korzenny, F., and B. A. Korzenny. (2005). *Hispanic Marketing: A Cultural Perspective.* Burlington, MA: Elsevier Butterworth Heinmann.

Lee, L. C., and N. W. S. Zane. (1998). *Handbook of Asian American Psychology.* Thousand Oaks, CA: Sage Publications.

Lee, S. J. (1996). *Unraveling the "Model Minority" Stereotype: Listening to Asian American Youths.* New York: Teachers College Press.

Legix. (2000, November 9). *Social and General Update.* (a Native American lobbying firm specializing in issues involving housing, tribal justice, sovereignty, education, economic development, and gaming) See link at http://www.legix.com

Legon, J. (2005). "War in Iraq: Heroes of War." See link at http://edition.cnn.com/SPECIALS/2003/iraq/heroes/piestewa.html

Lohman, D. L. (1977). "Race Tension and Conflict." In N. A. Watson (Ed.), *Police and the Changing Community.* Washington, DC: International Association of Chiefs of Police.

Lopez y Rivas, G. (1973). *The Chicanos.* New York: Monthly Review Press.

Macron, M. (1989). "Arab Contributions to Civilization." *ADC Issue* 6. Washington, DC: American-Arab Anti-Discrimination Committee.

"Man Whose Garage Burned Says He Didn't Know About Dues— He'll Pay Now." (2006, February 25). *Associated Press.*

McAdoo, H. P. (1992). "Upward Mobility and Parenting in Middle Income Families." In Burlers, A. K., Banks, W. C., McAdoo, H. P., and Azibo, D. A. Y. (Eds.), *African American Psychology: Theory, Research, and Practice.* Newbury Park, CA: Sage Publications.

Mehl, L. (1990). "Creativity and Madness." Presentation sponsored by the American Institute of Medical Education, Santa Fe, New Mexico.

Mendez v. Westminster, 64 F. Supp. 544 (D.C. Cal. 1946), aff.d, 161 F. 2d 774 (9th Cir. 1947).

Michalak, L. (1988). "Cruel and Unusual: Negative Images of Arabs in American Popular Culture." *ADC Issue 15.* Washington, DC: American-Arab Anti-Discrimination Committee.

Montiel, M. (1978). *Hispanic Families: Critical Issues for Policy and Programs in Human Services.* Washington, DC: COSSMHO, Coalition of Spanish Speaking Mental Health Organizations.

National Center for Health Statistics: Series 21, Number 53, 1995. *Report of Final Natality Statistics,* 1996; Monthly Vital Statistics Report Vol. 46, No. 11, National Center for Health Statistics, 1998; National Vital Statistics Report June 25, 2003.

National Congress of American Indians. (2003). http://www.ncai.org Resolution [#PHX-03-024] May 25, 1993, by Senator Inouye (D-HI); eventually passed as the American Indian Religious Freedom Act Amendments of 1994.

Native American Free Exercise of Religion Act of 1993. Senate Bill 1021 Introduced to 103rd Congress.

Niderost, E. (2006, April). "The Great 1906 San Francisco Earthquake and Fire," *American History.*

Nydell, M. K. (1987). *Understanding Arabs: A Guide for Westerners.* Yarmouth, ME: Intercultural Press.

Office of Management and Budget. (1997, October 30). "Revisions to the Standards for the Classification of Federal Data on Race and Ethnicity." *Federal Register,* 62(280): 58,782–90.

Ogawa, B. (1990). *Color of Justice: Culturally Sensitive Treatment of Minority Crime Victims.* Sacramento: Office of the Governor, State of California, Office of Criminal Justice Planning.

Owen, J. (2005, February 22). "With 20 Official Languages, Is EU Lost in Translation?" *National Geographic News.*

Pan American Health Organization (2006). *PAHO Basic Health Indicator Data Base.* Washington, DC: Author.

Passel, J. S. (2005). *Unauthorized Migrants: Numbers and Characteristics.* Washington, DC: Pew Hispanic Center.

Passel, J. S., J. Van Hook, and F. D. Bean. (2004). *Estimates of Legal and Unauthorized Foreign Born Population for the United States and Selected States, Based on Census 2000.* Washington, DC: Report to the Census Bureau, Urban Institute.

Population Reference Bureau. (2005). *2005 World Population Data Sheet.* Accessed from http://www.prb.org/pdf05/05WorldDataSheet_Eng.pdf

Prashad, V. (2001). *The Karma of Brown Folk.* Minneapolis: University of Minnesota Press.

President's Commission on Mental Health. (1978). *Report of the Special Population Subpanel on Mental Health of Asian/Pacific Americans,* Vol. 3. Washington, DC: U.S. Government Printing Office.

Rahall, N. J. (U.S. Representative, 2003). House Resources Committee, Native American Sacred Lands Act, June 11, 2003.

Redhorse, J., A. Shattuck, and F. Hoffman. (Eds.). (1981). *The American Indian Family: Strengths and Stresses.* Isleta, NM: American Indian Research and Development and Associates.

Saltzman, L. (2003). *Compact History Geographic Overview.* See link at http://www.dickshovel.com/up/html

Samhan, H. (2003). Executive Director, Arab American Institute, Washington, DC, personal communication.

Shabbas, A. (1984). *Cultural Clues for Social Service Case Workers and Special Educators* (unpublished monograph). Berkeley, CA: Arab World and Islamic Resources and School Services.

Shaheen, J. G. (2001). *Reel Bad Arabs: How Hollywood Vilifies a People.* New York:

Olive Branch Press. (p. 51, song from the movie, *Aladdin*)

Smith, H. (1991). *The World's Greatest Religions*. San Francisco: HarperCollins

Special Services for Groups. (1983). *Bridging Cultures: Southeast Asian Refugees in America*. Los Angeles: Special Services for Groups.

Spielman, F. (2006, January 5). "Firefighter Who Aired Slur Denied Return to Austin: Arbitrator 'Strikes a Balance,' Sends Him to Hispanic Neighborhood." *Chicago Sun Times*, p. 40.

Stevens, E. (1973). "Machismo and Marianismo." *Transaction-Society*, 10(6): 57–63.

Suleiman, M. W. (Ed.). (2000). *Arabs in American: Building a New Future*. Philadelphia: Temple University Press.

Takaki, R. (1989). *Strangers from a Different Shore: A History of Asian Americans*. Boston: Little, Brown.

Therrien, M., and R. R. Ramirez. (2000). "The Hispanic Population in the United States." *Current Population Reports*, P20-535. Washington, DC: U.S. Census Bureau.

Tippett, J. (2004, August 15). "Native American Fire Service Professionals Train on New Fire Safety Program." *IFCA Newsletter*. Fairfax, VA: International Fire Chiefs Association.

U.S. Census Bureau. (2002a). *Census 2000 Brief Reports*. Washington, DC: U.S. Government Printing Office.

U.S. Census Bureau. (2002b). http://www.cwn-sus.gov/prod/2002pubs/c2kbr11o1-15.pdf

U.S. Census Bureau. (2003, April). "The Black Population in the United States: March 2002." *Current Population Reports*.

U.S. Bureau of the Census (2003b). Population by Race and Hispanic Origin, 2000.

Weber, S. N. (1991). "The Need to Be: The Socio-Cultural Significance of Black Language." In L. Samovar and R. Porter (Eds.), *Intercultural Communication: A Reader,* 6th ed. Belmont, CA: Wadsworth Press.

White House Government News Release. (2002). See link at http://www.whitehouse.gov/news/releases/2002/11/20021101-7-html

Woodruff, D. (1991, February 4), "Letter from Detroit: Where the Mideast Meets the Midwest—Uneasily." *Business Week,* p. 30A.

Yankelovich Monitor (2003). *2003 Yankelovich Monitor Multicultural Marketing Study.* Chapel Hill, N.C.: Author.

Yoder J. D., and L. L. Berendsen. (2001). "'Outsider Within' the Firehouse: African American and White Women Firefighters." *Psychology of Women Quarterly,* 25(1): 27–36.

Zogby International Study. (2000). "Arab Americans: Protecting Rights at Home and Promoting a Just Peace Abroad." A publication of the Arab American Institute. See link at http://www.aaiusa.org

Firefighter Image and Cultural Sensitivity

6 CHAPTER

Key Terms

pluralistic, p. 249   **subculture, p. 245**

Overview

This chapter considers the changing role and image of firefighters that will result in increased cultural awareness and effectiveness in the fire service in the twenty-first century. Given the impact of greater diversity on the fire service in both the community and within fire departments, we explore how cultural understanding can be translated into more effective fire service (Figure 6.1). To increase cultural awareness, readers are provided with a simple model for quick analysis of differences in various cultures, ethnic groups, and generations. The content strengthens the case that improved fire service performance and professionalism are dependent on cross-cultural skills and sensitivity among firefighters.

Objectives

After completing this chapter, participants should be able to:

- Identify the changing role image of fire service officers.
- Describe the need for firefighter officers to have a positive self-image and image as a fire service *professional*.
- Explain the importance of being culturally aware in outlook and in dealings with citizens, especially in multicultural communities.
- Summarize why those who serve the public are held to higher standards than ordinary citizens with respect to the need for tolerance and acceptance of diversity.
- Highlight how greater diversity in fire departments can be translated into more effective community fire service.

FIGURE 6.1 ◆ Multicultural and diversity common goals of fire service are seen in the response by a Venezuelan firefighter spraying water at a smoky blazing fire from a cherrypicker in Maracaibo, Venezuela.

LEARNING TASKS

Knowing the images of the fire service for the ethnic, racial, and cultural groups in one's service area is critical toward understanding the multicultural challenges of your area. Talk with your agency and find out how it is involved in the fire service to diverse communities. Be able to:

- Find out the changing roles for the fire service that are highlighted for your area.
- Identify the top three reasons for firefighter officers to have a positive self-image.
- Determine the existing challenges the agency has in dealing with citizens in multicultural communities.
- Find out the leadership expectations and agency standards for communicating and responding to citizens in your area.
- Ascertain how greater diversity in fire departments can be translated into more effective community fire service in your city and county fire departments.

PERCEPTIONS

The following quotes exemplify the diverse images of the fire service, the importance of professionalism, and the necessity that today's firefighters have people and leadership knowledge and skills, in addition to firefighting technical skills:

Have you ever heard the expression that the fire service is one hundred years of tradition unhampered by progress [Figure 6.2]. While this may be humorous, it also has some amount of truth associated with it. While we may have made great

FIGURE 6.2 ◆ Firefighting traditions and skills have a lengthy history as illustrated by the multistory tenement buildings saved by firefighters on East 35th Street in New York City.

strides in some areas, we are still doing business the way we always have. Look at the firefighter line of duty death rate and note that it is being maintained even with the many improvements in safety that have been made (White, 2002, University of Maryland Fire and Rescue Institute [MFRI] Field Instructor).

For years, the Fire Department has faced criticism and lawsuits over the lack of women, blacks and Hispanics in its ranks. Now, as the department seeks a more diverse crop of recruits to take the forthcoming firefighter exam, it has decided to do a little rebranding, with the help of a SoHo marketing agency. Gone are the images from the last recruiting campaign, in 2002, which portrayed firefighters as noble and heroic. In their place are kinder, gentler images aimed at promoting the benefits and flexibility of a firefighter's lifestyle—a firefighter in civilian dress spending time with her daughter in a park, lieutenants at a backyard barbecue and firefighters playing basketball in a gym. "The point is that life in the Fire Department is not just fighting fires," said Peter Arnell of the Arnell Group, the SoHo agency that has donated its services for fire-safety and minority-recruiting campaigns. "It's a great career with incredible opportunities for advancement" (Chan, 2006).

The management of the Boston Fire Department is in need of a complete overhaul, including the appointment of civilian leaders to tackle extraordinary problems ranging from racial division on the force to costly and dangerous personnel practices, according to a report released Thursday. The 58-page report by a special commission appointed 11 months ago by Boston Mayor Thomas M. Menino calls for radical changes to the command structure,

FIGURE 6.3 ◆ Newspaper clippings with photographs of news stories highlight the courageous and hazardous work of firefighters in their public service roles.

promotional system, and department culture. The panel described a department suffering from inept management and a blind adherence to traditions that no longer make sense for a modern firefighting force. While the commission praised the performance of firefighters in the field, it warned that virtually every other area needed improvement and that morale among the department's 1,600 employees was low. "Throughout the interviews and evident in the survey responses are a significant number of firefighters, officers and civilians who express a lack of confidence in the department and its direction," the commission wrote. "This lack of confidence is pervasive" (Armstrong 2000).

These perceptions underscore a common reality in which the image of the fire service projected to the public by firefighters, their organizations, and the media has multiple impacts involving personal characteristics and organizational affiliations on professional image construction (Roberts, 2005). In the first quotation, a seasoned field instructor has emphasized the importance of the public image created by the fire service and its personnel in maintaining tradition and culture at the expense of progress and innovation. The second quote underscores the critical role of fire service image and firefighter identity in the recruitment of fire service personnel in today's work environment. The third quote highlights the emerging and changing roles involving people skills, leadership, and management that firefighter officers must have in the fire service to be effective in communities in the twenty-first century (Figure 6.3).

INTRODUCTION

People create images, both accurate and inaccurate, of themselves and others, as well as images of their roles. Our behavior is then influenced by these mental pictures, and others respond to what we project, which may not necessarily reflect reality. The most powerful of these images is the one we have of ourselves, followed by those formed about our multiple roles. Similarly, we also create images of our organizations, our nation, and the fire service system and its activities.

IMPACT OF IMAGES ON HUMAN BEHAVIOR

Firefighters today are both knowledge and service workers who must focus on effective performance, leadership, and management skills, according to the Report of the National Fire Service Research Agenda Symposium, June 1–3, 2005:

> Fire service members at the lower- and mid-levels are not receiving sufficient training and education in the areas of leadership and management skills. In most cases this type of decision-making and empowerment training is only offered to upper level fire command officers through the EFO program or college classes. Selection processes do not allow many firefighters and officers to obtain this valuable training in their early career development. In addition to adequately preparing them for advancement, they would be better equipped to make appropriate field decisions that could contribute to improved personnel safety—and positively influence the overall fire service culture (National Fallen Firefighters Foundation, 2005, p. 21).

As noted in the quote by Chief Crawford in Chapter 3 and by the report above, fire departments must become both learning organizations and community service agencies (Figure 6.4). The former have the responsibility to provide fire service officers with new knowledge, skills, and behaviors to be effective on the job.

Such an image of firefighters may explain why so many fire service agencies are mandating a two- or four-year college degree for entering recruits. Brainpower, not brawn, is what will make for effectiveness in the fire service in the vastly changing

FIGURE 6.4 ◆ Firefighter trainees practice and prepare rescue equipment on a fire truck.

Traditional Work Culture	*Evolving Work Culture*
Attitude	
Reactive; preserve status quo	Proactive; anticipate the future
Orientation	
Fire service as dispensers of public safety	Fire service as helping professionals
Organization	
Paramilitary with top-down command system; intractable departments and divisions; centralization and specialization	Transitional toward more fluid, participative arrangements and open communication system; decentralization, task forces, and team management
Expectations	
Loyalty to your superior, organization and buddies, then duty and public service; conformity and dependency	Loyalty to public service and duty, and one's personal and professional development; demonstrate leadership competence and interpersonal skills
Requirements	
Political appointment, limited civil service; education—high school or less	Must meet civil service standards; education—college and beyond, lifelong learning
Personnel	
Largely white males, military background and sworn only; all alike, so structure workload for equal shares in static, sharply defined slots	Multicultural without regard to sex or sexual orientation; competence norm for sworn and nonsworn personnel; all different, so capitalize on particular abilities, characteristics, potential
Performance	
Obey rules, follow orders, work as though everything depended on your own hard efforts toward gaining the pension and retirement	Work effectively; obey reasonable and responsible requests; interdependence means cooperate and collaborate with others; ensure financial/career future
Environment	
Relatively stable society where problems were somewhat routine and predictable, and authority was respected	Complex, fast-changing, multicultural society with unpredictable problems, often global in scope, and less emphasis upon authority, with importance of leadership and community input

FIGURE 6.5 ◆ Fire Service Role Transitions
Source: Herbert Wong, based on Harris, Moran, and Moran (2004)

world of this new millennium. Figure 6.5 differentiates between perceptions of the disappearing fire service work culture and the emerging culture. Contrasting the two views helps to explain why, within the firefighting field, a different image of the fire service must be created by firefighter officers and then projected to the public. This is but a synopsis of the transition underway in the world of work and the fire service, which in turn alters our views of various community service roles. The shift and trends

FIGURE 6.6 ◆ Changing requirements in fighting terrorism such as firefighters working as search-and-rescue specialists continue to search for survivors among the wreckage at the World Trade Center.

seem to be away from the traditional approaches—from reactive to proactive. That is, the shift is from solely effective responding and firefighting toward prevention and proactive fire service (Figure 6.6). The shift is toward preserving the public well-being and community service, and toward protecting life and property, and from just public-sector firefighting to synergistic fire services by both public and private sectors working in cooperation (Barr and Eversole, 2003).

Within this larger context, comprehending the significance and power of image and its projection becomes critical for the would-be firefighter. Image is not a matter of illusion or mere public manipulation of appearances. Behavioral scientists have long demonstrated the vital connection between image and identity, between the way we see ourselves and our actual behavior (Roberts, 2005). Realists among us have always known that life's "losers," many of them convicted felons, lack an adequate sense of identity and suffer from feelings of poor self-worth. Some people raised in dysfunctional families go so far as to engage not only in self-depreciation and abuse but also in self-mutilation. Aware of these factors, one state (California) established a special commission to promote self-esteem among school youth, so as to curb crime and delinquency (Williams, 1990). The impact of self-image upon the firefighters' role and work behaviors is illustrated in Figure 6.7. Behavioral communication may be understood in terms of senders and receivers of messages (as noted in Chapter 4).

Figure 6.7 suggests that we view the multiple images we form in terms of concentric or overlapping circles, centered on the all-important self-image at the core. As we see ourselves, we project an image to which people respond. Normally, we are greatly in control of the reactions and treatment we receive through the image we project. Our self-concept has been long in development, the outcome not only of life experience but also of others' input to us. For example, if a child continuously receives negative feedback from parents and teachers, he or she begins to believe in personal worthlessness. Usually, these young people fail to achieve unless an intervention calls into question their distorted concept of self. Such persons are likely to project weak

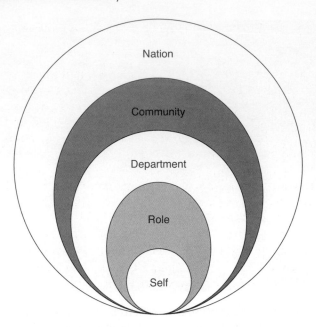

FIGURE 6.7 ◆ Multiple Images and Self-Image at the Core
Source: Herbert Wong

images that often prompt others to take advantage of them. On the other hand, if they have self-confidence and project a positive image, both verbally and nonverbally, an acceptable response from others is probable. (The exception would be responses from bigoted persons who are reflecting an inner bias and refusal to accept people different from themselves.) The best antidote, in most circumstances, to underachievement or even racism is for a person to have a healthy self-image.

Furthermore, if a person is confused about identity, as is often the situation with adolescents, people tend to respond in an uncertain way toward that person. Therefore, in the selection of fire service candidates, firefighter officials seek those with a strong self-image and self-appreciation. Fire service academy training should reinforce, not undermine, this self-confidence. Fire service supervisors are advised to provide regular positive reinforcement of their officers' positive sense of self.

As illustrated in Figure 6.7, our positive (as well as our negative) images of the "self" also have parallels to our images of our role, department, community, and nation. Individuals who have positive images of themselves tend also to have positive images of their role, department, community, and nation (which extends in the obverse, or opposite direction, for those who have negative images of themselves). As such, creating positive images of the fire service and the firefighting profession would be paramount for diverse communities and for effective public services.

TWENTY-FIRST-CENTURY FIREFIGHTER IMAGES

We play many roles in the course of life—male or female, married or single, child or parent, student or mentor. However, our concern here is specifically on work and professional roles within the fire service system and the changing images projected by its practitioners. In terms of behavioral communication theory, fire service officers project an image to the public based on how they see themselves, and citizens usually respond to that image (Schuler, 2004). Two examples—one old, one new—illustrate this point:

In 1913 the city purchased an American LaFrance motorized fire engine and in 1915, they created Chemical Truck No.1 on a Buick touring car chassis. In 1919, the use of horses was discontinued and another truck was purchased. This was an American LaFrance seventy-five foot aerial ladder truck. In this same year, the position of fire chief became a full time position and Ben E. Bangerter was hired. There were also eight other paid firemen, partially paid first and second chiefs and sixty-six volunteers. By 1926, the paid force had grown to eleven men. The members of the paid force spent nearly all of their time at the fire station since it was difficult to notify them of a fire. As a result, the station was a place to go to exchange gossip, play cards or play snooker pool. This gave them a bad public image. In 1937, eight new men were hired and a two platoon system was adopted. Ten men platoons served alternating twenty-four hour shifts. The men from the other shift were allowed to leave the station, but they were still on call. This helped to combat the negative image (Schrader, 1990) [Figure 6.8].

Since then, the role of the fire service has evolved as society has changed its perceptions of values and priorities, community and public services, first responders, and fire and emergency services. However, new immigrants to the United States and Canada, whether through legal or illegal entry, come with their own cultural perceptions and perspectives from their homeland. These can include negative images of fire, emergency, police, and other public services there, which may consist of images of oppressors and bribe takers. For these individuals, professional fire and public services in the new homeland may need to go out of their way to help these new immigrants overcome the mental barriers they carry due to their cultural background.

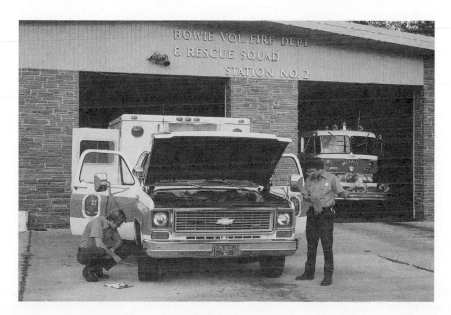

FIGURE 6.8 ◆ Firefighters have long shifts observable by the neighborhood at their fire stations in which their local communities may build positive as well as negative images about their work.

Modern mass media have a powerful influence on how the public perceives the fire service. It begins with fictional fire service and first responders as depicted in novels and emergency service stories, expands to reports of firefighting activities in newspapers and magazines, and extends to radio and television broadcasts about firefighting and first responder services. Today U.S. movies and television fire service and first responder shows are viewed internationally. The range of media impact on images of the fire service is extensive.

With the coming of the information age and its new communication technologies, local firehouse and nonemergency events may suddenly take on negative images internationally (Figure 6.9). With the help of the Internet, satellites, television, and computers, the inhabitants of our planet may react as a global village to what may seem to be very local issues involving the fire service or fire department conflicts. For example, the following news story appearing in the *Los Angeles Times* could appear instantly in the global village to create negative images of fire service leadership and the firefighting profession:

> When tales of racist behavior, hazing and harassment in the Los Angeles Fire Department are recounted, the Tennie Pierce story usually comes first. Some of Pierce's fellow firefighters at Station 5 in Westchester sneaked dog food into his spaghetti dinner and watched him eat it—an incident that officials have called a prime illustration of how harassment is entrenched within the department. A recent audit of the Fire Department, widely considered one of the best in the nation, found rampant racism, sexual harassment and hazing. Fire officials, including Chief William Bamattre, acknowledge that the personnel problems are real. But they say that what lies behind them are major shifts in mission and makeup the agency is undergoing, rather than inherent

FIGURE 6.9 ◆ Movies and television programs such as the *Third Watch* tell the stories of the members of the 55 Squad—the police, firefighters, and paramedics who work the "third watch" between 4 p.m. and 12 a.m.—the finest and the bravest who protect and serve the citizens of New York.

bigotry. To Pierce, the only African American firefighter working that shift at the station, the 2004 incident was clearly racist. "I think it was done to demoralize me," said Pierce, who stands 6 feet 5. "How could 10 people sit up there and let a man eat dog food and not stop him? Not one person?" Last November, Pierce, 50, filed suit against the department and three firefighters: a captain who allegedly purchased the dog food, another captain who reportedly knew about the dog food but did nothing to stop Pierce from eating it and the firefighter who is accused of mixing it into his pasta and serving it to him (Richardson, 2006).

The negative images created in this example can quickly outweigh the good performed by this fire department. The principal lesson to be learned from any actions that result in negative images is that the organizational image of a reasonable and effective agency can be undermined by the unprofessional and illegal actions of a relatively small percentage of fire service officers or their leadership. It is not enough for a firefighter officer to be a competent professional; he or she must also assist by preventing or reporting wrongdoing by colleagues.

INFLUENCING THE PUBLIC POSITIVELY

Psychologists remind us that the image we have of our role affects not only our behavior but also the behavior of those to whom we project this image. The U.S. Marine Corps, for example, is currently undergoing a change of image from warriors to peacekeepers and humanitarians (Figure 6.10). What if all of those described previously under the definition of fire service officers actually saw themselves as such "service"

FIGURE 6.10 ◆ American Marines have changed parts of their image from soldiers to community service and global helpers.

officers, thus projecting that image to the public? A change in role image and actions among community providers of the fire service might prompt the majority of the community to perceive them more positively. Neighborhood fire service officers who interact with the community are more likely to improve public perception of fire prevention and the fire service. Clearly, fire service leadership and communication skills will facilitate superior fire service, as shown in the following example in Ohio:

> The city of Bellbrook has a fire department to be proud of, Bellbrook Fire Chief Scott Hall told the City Council on March 28. In his annual report to the city, Hall said that the 40-member department, which includes five full-time and 11 part-time employees, and 24 volunteer members, responded to 642 fire calls in 2004 and 538 emergency medical services calls, with a fire loss of just $300. The department logged 6,409 staff hours of training out of a total of 39,118 personnel hours, he said. Hall said most calls for service were on Sundays (192), with the most frequent time for a call from noon to 6 p.m. The most runs (31 percent) were for calls in the area north of Ohio 725 between Bledsoe Drive on the west and Little Sugarcreek Road on the east. The department has 10 paramedics, two intermediate medical technicians, 24 emergency medical technicians, two liaisons to the Regional Hazardous Materials Response Team, five members on the Regional Critical Incident Stress Management Team, and five members on the Federal Urban Search and Rescue Task Force. One member is a certified fire explosion investigator and another member serves as the city representative on the Greene County board of health, he said. Hall gave credit for the department's excellence to his leadership team and to platoon leaders and their five- to seven-member platoons, which include special operations, health and safety, fire operations and EMS (Ullmer, 2005).

CHANGING THE FIRE SERVICE CULTURES

In the article "Panel Urges Wide Change in Fire Department: Hub Report Faults Command, Promotional System, Culture," a special commission report noted:

> "The management skills need modernizing," the mayor said in an interview. "We are changing to fit the times." . . . The commission also called for a fundamental change in the culture of the department. It said many minority and female firefighters, who do not have longtime links to the department, often feel alienated. The panel said part of the problem was an emphasis on tradition that is out of balance with the desire to modernize the agency. "Tradition is an anchor around our necks," the report quotes a firefighter as telling the commission. "The [fire department] has too many historians and not enough visionaries. Our fear of change is killing us." Many in the department also complained of an "old-boy network" that created the perception that only certain people received plum assignments, or that others were disciplined unfairly because they were not part of the "in" crowd. The panel said much of the racial conflict dates back 25 years to a court order directing the department to hire more minorities. "Many other organizations that faced the same problems have moved beyond the decades-old issues and it is long past time for the Boston Fire Department to do the same," the commission wrote. "This failure by the department to address properly such important issues as race and diversity cannot be allowed to continue" (Armstrong, 2000).

Traditional fire department cultures need to transition away from the more rigid military and bureaucratic cultures of the past to more flexible, corporate-like, proactive, and high-tech cultures. Fire service leaders have become change agents in this twenty-first century process. To facilitate the changeover, many fire service executives and managers have applied methods of cultural analysis to their own unique work environment.

WAYS OF ANALYZING CULTURES

There are many ways to study a culture, such as a systems approach that might analyze it in terms of kinship, religion, education, economics, politics, associations, health, and pastimes. However, Harris, Moran, and Moran (2004) provide a simple model that will enable those in the fire service to get a grasp of a culture, whether on a large or small scale, irrespective of these individual groupings. These 10 benchmarks of their model can be used to better comprehend a foreign organizational or local culture, especially that of a minority or ethnic group. Figure 6.11 summarizes the characteristics and steps for cultural analysis, first in terms of cultures in general and then specifically in the context of fire service culture.

---

1. **Sense of self and space.**   Examine how people in this cultural group perceive themselves and distance themselves from others. Culture not only helps to confirm one's identity but also provides a sense of space, both physical and psychological. Such cultural conditioning of behavior may dictate a humble bearing in one culture or macho posturing in another. Some cultures support rugged individualism and independent action, whereas others teach that self-worth is attained through group conformity and cooperation. In some cultures, such as American, the sense of space dictates more physical distance between individuals, whereas in Latin/Asian cultures, less distance is desirable. Some cultures are very structured and formal, whereas others are more flexible and informal. Some cultures are very closed and determine one's place narrowly and precisely, whereas others are more open and fluid. Each culture validates self in unique ways.

   Fire service culture, for example, expects firefighter officers to project a sense of authority and assertiveness, to be respected and in control, to act in emergency situations with a sense of duty, to save lives, and to serve the community and the common good. Sense of space is experienced in terms of a firehouse, a service area, a neighborhood in community services, a district, or a city area.

2. **Communication and language.**   Examine the communication system in the culture, both verbal and nonverbal. Apart from the national or regional language that might be spoken, study the dialects, accents, slang, jargon, graffiti, and other such variations. For example, there are differences in the way the English language is spoken in England and within the British Isles and Commonwealth nations; in North America, between the Canadians and the Americans (and within the latter between regions and groups, such as black English). Levine and Adelman remind us to go beyond language to comprehend the full communication by seeking the meanings given to body language, gestures, and signals (Levine and Adelman, 1992).

   The fire service culture has its own jargon and code system for communicating rapidly within the field of firefighting; organizational communications dictate a formal system for reporting, for exchanges with superior ranks, and for dealing with public officials and the media (Figure 6.12). A code of silence may exist about speaking to outsiders concerning fire service business and personnel.

FIGURE 6.11 ◆ Characteristics for Cultural Analysis

FIGURE 6.12 ◆ Fire service organizational communications.

3. Dress and appearance. Examine the cultural uniqueness relative to equipment, outward garments, adornments, and decorations or lack thereof; the dress or distinctive clothing demanded for different occasions (Figure 6.13) (e.g., business, sports, weddings or funerals); the use of color and cosmetics; hair and beard styles, or lack thereof; and body markings.

In fire service culture, policy, regulations, and even custom may determine a uniform with patches and insignia of rank, plus certain equipment to be worn and even the length of

FIGURE 6.13 ◆ Firefighter/emergency medical technician places equipment, outward garments, and other tools securely in their proper storage place in the back of a fire and rescue vehicle.

Figure 6.13 (*continued*)

hair permissible. Exceptions may be permitted for those in administration, inspection, education, or prevention work.

4. Food and eating habits (to include alcohol consumption). Examine the manner in which food is selected, prepared, presented, and eaten. According to the culture, meat, such as beef or pork, may be prized or prescribed, or in contrast forbidden all together—one person's pet may be another's food (e.g., alligators, snakes, eels, etc.). Sample national dishes, diverse diets, and condiments for tastes vary by culture. It is important to realize that some cultural groups can be conditioned to accept various foods or seasoning that your own body would not tolerate without reactions. Feeding habits may range from the use of hands or chopsticks to the use of utensils or cutlery and may vary even in the use of the same utensils (e.g., Americans and Europeans do not hold and use the fork in the same manner). Subcultures can be studied from this perspective (as in soldier's mess, executive dining rooms, vegetarian restaurants, prescriptions for females, etc.). Even drinking alcohol differs by culture—in Italy, it is more associated with eating a meal, whereas in Japan and Korea it is ritualized in the business culture as part of evening entertainment and strengthening business relations, sometimes done to excess.

In fire service culture, the emphasis has been on fast foods, hearty meals, and firehouse cooking; customs also have included off-duty relaxation and camaraderie as in a "firefighter's drinking hole," sometimes marked by too much alcoholic consumption. The new generation of firefighters is more concerned about healthy foods, keeping physically fit, and stress management—including diet, exercise, no smoking, and no substance abuse.

5. Time and time consciousness. Examine the time sense as to whether it is exact or relative, precise or casual. In some cultures, promptness is determined by age or status (e.g., at meetings—subordinates arrive first, the boss or elders last). Is the time system based on 12 or 24 hours? In tribal and rural cultures, tracking hours and minutes is unnecessary, because timing is based on sunrise and sunset, as well as the seasons, which also vary by culture (e.g., rainy or dry seasons vs. fall/winter/spring/summer).

Schedules in the postindustrial work culture are not necessarily eight hours; businesses may operate on a 24-hour basis because of the emergency nature of the tasks such as in the fire service (Figure 6.14) or the global telecommunications and electronic mail nature of international business. Chronobiologists are concerned about the body's internal clock and performance under such circumstances, so analyze body temperature and composition relative to sleepiness, fatigue, and peak periods (e.g., as with jet fatigue when passing through time zones, or firefighters involved in natural disasters, etc.).

The fire service culture operates on a 24-hour schedule with sliding work shifts that do affect performance; some departments adopt the military timekeeping system of 24 hours. Normally, promptness is valued and rewarded.

6. Relationships. Examine how the culture fixes human and organizational relationships by age, sex, status, and degree of kindred, as well as by wealth, power, and acquired wisdom. In many cultures, marriage and the family unit are the most common means for establishing relations between the sexes and among parents, children, and other relatives. In the Far East, this is accomplished through an extended family that may involve aunts, uncles, and cousins living in the same household. Many cultures also operate with the male head of household as the authority figure, and extend this out from home to community to nation, explaining the tendency in some countries to have male dictators. In traditional cultures, custom sets strict guidelines about male–female relations prior to marriage, about the treatment of the elderly (in some cultures, they are honored, in others ignored), and about female behavior (wearing

Figure 6.13 (*continued*)

FIGURE 6.14 ◆ Emergency response by three firefighters in surrounding a volatile mental health patient in front of a fire truck.

veils and appearing deferential to males in contrast to free expression in dress and being considered equal to males). "Underworld" or criminal cultures sometimes adopt a family pattern with the "godfather" as the head and various titles to distinguish roles.

In fire service culture, organizational relations are determined by rank and protocol, as well as by assignment to different departmental units and responsibilities (Figure 6.15). Although policy may dictate that all fellow officers and citizens be treated equally, unwritten

FIGURE 6.15 ◆ Fire executives obtain and provide information important to the community.

Figure 6.15 (*continued*)

practices may differ. The partnership system usually entails close relations and trust between individuals whose life and welfare is dependent upon the other.

7. **Values and norms.** Examine how the culture determines need satisfaction and procedures, how it sets priorities, and how it sets values with some behavior lauded and other behavior decried. Thus cultures living on a survival level (e.g., the homeless) function differently from cultures living in affluence. In some Pacific Island cultures, for instance, the more affluent one becomes, the more one is expected to share with the group. On the other hand, cultural groupings with high security needs may often value material things (e.g., money and property), as well as law and order. In the context of the group's value system, the culture sets norms of behavior within that society or organization. Acting upon a unique set of premises, standards of membership are established affecting individual behavior—for example, the conventions may require total honesty with members of one's own group but less disclosure with those from other groups. Other standards may be expressed in gift-giving customs; rituals for birth/marriage/death; and guidelines for showing respect, privacy, and good manners. The culture determines what is legal or illegal behavior through a codified system or custom; what may be legal in one culture may be illegal in another.

In fire service culture, for instance, subordinates are expected to show respect for officers of superior rank, whereas the reverse may be tolerated for those who have broken the law. Publicly and by departmental regulations, a code of ethics is in place in which illegal activities, bribery, and corruption are punishable offenses. This culture also espouses traditional American values (e.g., duty, loyalty, patriotism).

8. **Beliefs and attitudes.** Examine the major belief themes of a people and how these affect their behavior and relations among themselves, toward others, and what happens in "their world." A near cultural "universal" seems to be a concern for the supernatural, evident in religious adherence and practices, which are often dissimilar by group. "Primitive" or tribal cultures are described as "animists" because they experience the supernatural in nature (e.g., American Indians or Native Americans)—a belief with which some modern environmentalists may resonate. Other religious traditions hold varying beliefs with regard to the supernatural. The differences are apparent in the Western cultures with Judeo-Christian traditions, as well as Islamic, in contrast to Eastern cultures, dominated by Buddhism, Confucianism, Taoism, and Hinduism. Religion, spiritualism, or secularism expresses the philosophy of a people about important realities of life's experiences. Some cultural groups are more fundamentalist and rigid in their religious beliefs, whereas others are more open in interpreting religious doctrine. A people's belief system may be intertwined with a cultural stage of human development—hunting, farming, industrial, or postindustrial. For example, some people in advanced technological societies may substitute a belief in science or cosmic consciousness for more traditional beliefs.

In fire service culture, for example, there has been a strong belief in group loyalty, teamwork (Figure 6.16), pragmatism, bravery, and public service. God and religion have been acknowledged in oaths, in religious societies of fire service officers, by appointment of department chaplains, and during burial ceremonies of officers who die in the line of duty. Until recently, the fire service culture tended to be perceived as "chauvinistic," but that is changing with the introduction of more female officers and the education of the workforce on diversity issues.

9. **Mental processes and learning.** Examine how some cultures emphasize one aspect of brain, knowledge, and skill development over another, thus causing striking differences in the way their adherents think and learn. Anthropologist Edward Hall suggests that the

Figure 6.15 (*continued*)

FIGURE 6.16 ◆ In fire service culture, there has been a strong belief in group loyalty and teamwork, as shown here with firefighters working with other rescuers in an open, narrow Coast Guard motorboat on floodwaters rising toward the rooftops of a barn during a flood in the Midwest.

mind is internalized culture and involves how a people organize and process information. Life in a particular group or locale defines the rewards and punishments for learning or not learning certain information or in a certain way. In some cultures, the emphasis is on analytical learning—abstract thinking and conceptualization—whereas in others it is on rote learning or memorization; some cultures value logic, whereas others reject it; some cultures restrict formal education to only males or the wealthy, whereas others espouse equal education for all. Although reasoning and learning are cultural universals, each culture has a distinctive approach. However, the emergence of the computer and telecommunications as learning tools is furthering unity through the globalization of education.

In fire service culture, recruit academies and other forms of in-service training may differ by locality as to content, instructional emphasis, and method. Anti-intellectualism among some firefighters is being countered through professional development and standards established by the federal/state governments and their credentialing processes, as well as by fire service curricula in higher education. As a result, modern fire service is moving from more reactive, pragmatic, action-oriented behaviors based on feelings and experiences toward proactive, thoughtful, analytical, and informed response behaviors. Professional competence is judged now by high performance and level of learning, not just by years on the job and connections.

10. Work habits and practices. Another dimension for examining a group's culture is through members' attitudes toward work. These attitudes would be assessed through examining the types of work, divisions of work, dominant work habits and procedures, and the work rewards and recognitions that are provided. Some cultures adopt a work ethic stating it is desirable to engage in worthwhile activities—even sports and the arts—whereas others limit prescribed activities to labor for income. Work worthiness is meas-

Figure 6.16 (*continued*)

ured differently by cultures along domains such as income produced, job status, or service to the community. In Japan, for example, cultural loyalty is transferred from the family to the organization, which is dependent upon the quality of individual performance. Classification of vocational activity is somewhat dependent upon the culture's stage of development—a people can be characterized primarily as hunters, farmers, or factory or knowledge/service workers, with the trend away from physical labor toward use of mental energy aided by new technologies. The nature of work, as well as the policies, procedures, and customs related to it, can be in transition. In the postindustrial culture, there is more emphasis on the use of advanced technologies, such as automation and robotics, as well as upon quality of working life—from compensation and benefits to stress management or enhancement of one's potential on the job. In conjunction with work, a culture differs in the manner and mode of offering praise for good and brave deeds, outstanding performance, length of service, or other types of accomplishment. Promotions, perks, and testimonials are all manifestations.

In fire service culture, a hierarchical structure has organized work into specializations, divisions, and other such operational units engaged primarily in firefighting. Rewards and recognitions in the past have been largely commendations, advancement in rank, and retirement dinners, but now are being expanded to include assignments for professional development, interagency exchanges, and even sabbatical leaves for educational advancement.

Figure 6.16 (*continued*)

Source: Excerpted with permission from Philip R. Harris, Robert T. Moran, and Sarah V. Moran (2004) *Managing Cultural Differences,* 6th ed. Burlington, MA: Elsevier Butterworth-Heinemann.

The 10 general classifications for cultural analysis provide a model for looking at a group's characteristic modes of beliefs and behaviors. A firefighter in emergency situations is obviously not going to have the time to do an in-depth cultural analysis of the impacted communities. However, especially in community fire service types of situations, the officer can undertake some preparation before beginning an assignment in a given area where citizens will largely be from unfamiliar ethnic groups. Most people have preconceived notions or only a stereotypical understanding of cultural groups with which they have not previously had contact. The more information an officer or employee can gather about the cultural patterns of a distinctive community group, the less likely he or she will be to make false assumptions about people. Knowledge of these cultural patterns would help the officer understand the behavior and actions of individual groups and would contribute to an overall ability to establish rapport and gain trust.

Admittedly Figure 6.11 offers only 10 general classifications for cultural analysis, whether the culture consists of a nation, an organization, a profession, a group, a generation (e.g., youth or seniors), or an ethnic or racial group. There are other dimensions of cultural classification, but the categories provided here specifically help firefighter officers more quickly and systematically comprehend individual cultures (such as those described in Chapter 5 of this book). Further, during international travel, whether on duty or vacation, these same major characteristics can be observed to make the intercultural experience more meaningful. When fire service officers from other countries visit, these guidelines are useful in encouraging them to talk about their national or fire service cultures. All of the aspects of culture noted in these 10 general classifications are interrelated; there is a danger in trying to compartmentalize this complex concept of group culture and miss the sense of the whole.

MULTICULTURAL FIRST RESPONDER IMAGE ISSUES

If the representatives of the fire service are more culturally aware and sensitive, they are more likely to project to the public a positive image as public servants. The following are examples of actual newspaper headlines affecting the image of the department involved. Although the media does not always accurately report what happened, we do know that there are still instances of fire service personnel insensitivity and that even inaccurate news reports leave deep impressions in the minds of viewers. Some negative headlines have included statements such as the following:

- "S.F. Fire Chief Finds Trouble Smoldering: The Department Is Fighting On-the-Job Alcohol Use That Has Spawned Three Lawsuits" (*Sacramento Bee,* May 4, 2005)
- "Four Women File Sex-Bias Suit" (*The Oregonian,* March 11, 2005)
- "Grand Jury Will Hear Evidence in Fire Station Noose Incident Soon; The Firefighters Who Found the Nooses Have Since Been Assigned to Other Stations" (*Florida Times-Union—Jacksonville*, May 26, 2006)
- "Ex-Fire Captain Testifies of Picking Up Women in Engines" (*McClatchy Newspapers,* August 18, 2005)
- "Jury Awards Black Firefighters in Discrimination Case" (*Associated Press,* July 5, 2005)

However, there are certainly numerous reports of fire service departments engaged in positive interactions with the communities they seek to protect and that reflect positively gender diversity (Figure 6.17), such as the following:

When the crew from Fire Engine Company 22 raced off at 7:50 a.m. the other day for the first call of their 24-hour shift, a woman reporting chest pains,

FIGURE 6.17 ◆ Gender diversity in the fire service necessitates wanting more women to consider a career as a frontline firefighter.

their big red rig was primed for action but missing a typical feature: a man. The four members of Engine 22, Division A, a captain, an engineer, a fire-fighter-paramedic and a firefighter, protect the Point Loma neighborhood of San Diego, an affluent peninsula on the Pacific Ocean. They are one of the few crews in the nation made up entirely of women, winding up together last October, as the captain, Joi Evans, said, because of "the way the cards fell." Together they work, cook, shop, train and sleep in small dorm rooms in the station house, around the clock for 10 days a month, at a time when women are making some inroads into the fire service nationwide but are still only a sliver of the front line in one of the most physically grueling and male-dominated professions. With women accounting for about 8 percent of the 880 uniformed firefighters assigned to its station houses, compared with the national average of 2.5 percent, the San Diego Fire-Rescue Department, which has a female assistant chief, is considered one of the best departments for women to work, according to Women in the Fire Service, an advocacy group based in Madison, Wis (Kershaw, 2006).

Throughout this book the underlying message to the fire service has been that working and communicating on the basis of one's own cultural background may result in distorted perceptions. Instead, fire service officers should try to better understand the unique world of the community being served. In illustration, gestures have different meanings in varied cultures. The gesture of rolling the head from side to side (which is similar to the "no" head movement in the West), for example, signifies "I'm listening" or "I'm in agreement" in southern India. Now transpose this gesture into a North American urban area where a native of southern India is questioned by a fire service or EMS officer with respect to whether a 9-1-1 emergency request was placed. The fire service officer asks, "Did you call 9-1-1?" If not yet acculturated to the ways of the United States or Canada, the response from the immigrant may be a head roll from side to side, a gesture that means "yes" in India. This response would likely be interpreted as "no" by the fire service officer. The misunderstanding inherent with this interaction could severely hamper rescue efforts.

In addition to ethnic or racial cultures, the fire service institution has a unique organizational culture. Hence, there is a distinct fire service **subculture** throughout the world evident in agencies and departments, as noted in the following quote regarding the fire service in the United Kingdom:

> The new Fire Service will need excellent leaders, able to tackle poor management and to inspire ambitious performance. In turn, they will need to be given good support. The service of the future needs to make the best possible use of new technology and to adopt the streamlined structure, flexible skills and sophisticated management systems found elsewhere in the public and private sectors. . . . New leadership and management styles are required. Brigades in general need to improve their human resources (HR) procedures and practices. External advice will probably be needed in some cases. The Inspectorate in its new form and the College will have critical roles to play. A start could be made by these central bodies identifying a set of core values for the Service as a whole and a programme to develop methods of encouraging ownership of them (Independent Review of Fire Service, 2002).

Similarities and dissimilarities in various fire service cultures are important to consider when seeking interagency cooperation on a call or project; although all

subculture

A group with distinct, discernible, and consistent cultural traits existing within and participating in a larger cultural grouping.

may be in the field of fire and emergency services, each entity has a unique organizational culture that can encourage or discourage collaboration. It is also possible to analyze the cultural differences of specialization within the fire service and first responder fields. Although the cultures of fire departments, fire marshals, forest service smoke jumpers, and other emergency services have much in common, there are also distinct differences caused by the nature of their fire service duties. Finally, the workplaces of some agencies are locked into a traditional industrial or military model, whereas others are attuned to a more innovative, team-oriented, postindustrial work culture.

CULTURALLY SENSITIVE FIRE PROTECTION

Nowhere is a fire department's sense of professionalism and ethics challenged more than in the manner in which we treat diverse ethnic and racial group members, whether within the department or outside citizens and visitors. Apart from the indigenous Native Americans, immigrants from the fifteenth through the nineteenth centuries were mainly from Europe and Africa, and were gradually assimilated into the mainstream population. But in the twentieth century, the newcomers were primarily from Asia, South and Central America, the Middle East, and the countries of the former Soviet Union; for them, the process of acculturation is still ongoing.

One strategy for improving relations is to bring more diverse members into the fire service (drawing from minority populations) (Figure 6.18), a phenomenon that is taking place slowly in North America (for example, see Assistance to Firefighters Grant Program Staffing for Adequate Fire and Emergency Response [SAFER] Grants, a part of the Assistance to Firefighters Grant Program, Department of Homeland Security, 2006). However, what constitutes the term *minority group* is becoming increasingly debatable. For example, in some states (e.g., California)

FIGURE 6.18 ◆ Extended families of Hispanic firefighters are often part of the team to battle wildfires throughout the United States.

FIGURE 6.19 ◆ Customers line up at the window of an outdoor cafeteria at the Infanta Restaurant in the Little Havana district of Miami, Florida.

and cities (e.g., Miami, Florida), the Hispanic or Latino "minority" has become or is becoming the majority, while Caucasians will become the minority group in such areas.

Any new group integrating into the United States will present new fire service challenges that require changes in firefighter tactics and behavior in the twenty-first century, as noted in Figure 6.19.

Although most immigrants to the United States concentrate on becoming Americans (Figure 6.19), there are new groups who have come as refugees, fleeing persecution, strife, and distant wars. Their minds and efforts are directed toward their native countries and their suffering peoples. These include the Tamils of Sri Lanka, the Eritreans and Ethiopians of eastern Africa, the Congolese and Sierra Leoneans of western Africa, the Kurds of Iraq and Turkey, the Hmong tribesmen of Laos, the Muslims from the Middle East, and even the followers of the Falun Gong sect from China, along with thousands of others, such as Serbs and Kosovars from Yugoslavia. These people usually arrive poor and live together in less desirable urban centers of the United States, striving to preserve their traditional cultures, especially their languages and music. Often such desperate peoples are at the agricultural or rural stage of cultural development. Normally, they have little knowledge of English but are hardworking and take jobs that more prosperous Americans no longer want. Their common concern is for relatives and friends in their homeland. Thus, they frequently send back a large portion of their very modest earnings, take out bank loans, buy bonds, or engage in all kinds of voluntary relief work to assist the citizens of their native countries. They seek information about those left behind through newsletters and newspapers in their native language, or even by using the Internet and maintaining websites.

Those left behind in the old country are very dependent on the generosity of these new immigrants; the economies of some of their Third World countries are entirely sustained on gifts from these exiles. For example, 35 groups support the destitute Sri Lankan Tamils, who are totally dependent on their expatriates abroad. One Tamil bank takes in $350 billion a year in this way, about half as much as the Sri Lankan government's war budget ("There Is Another Country," 2000).

The downside is that some of these American funds go to support divisions left behind from past conflicts, including civil wars. Of the 30 groups designated by the U.S. State Department as terrorists, most are doing their fund-raising in this country. In July 2000, 200 police officers in Charlotte, North Carolina, arrested a man suspected of raising money for the Hezbullah extremists in Lebanon and Palestine. Furthermore, some of the old-country rivalries and feuds have been exported to the United States, often on the basis of clan loyalties—to further promote (even online) their specific causes with conferences, petitions, and protests are organized. All this contributes to furthering ethnic tensions and sometimes to illegal activities. For instance, the large Afghan community in Flushing, New York, is split between the supporters of the Taliban back home and its opponents. The 70,000 rural-mountain-type Hmongs in Minnesota fought with the Americans in the Vietnam War. They have thriving farmers' markets, but cooperation among themselves is undermined by clan divisions, along with arguments between moderates and radicals over military action in Laos against the still existing communist government.

While most of these contemporary immigrants are law-abiding, hardworking, family oriented, and dedicated to education for their children, the realities described have great implications for the fire service. Such upwardly mobile peoples are vulnerable to victimization by U.S. criminals, as well as by the lawbreakers among their own. The illegal activities of the Russian–American Mafia in Bensonhurst, Brooklyn, confirm this unpleasant factor (adapted from "There Is Another Country," 2000, pp. 26–27).

ORGANIZATIONAL TRENDS IN THE FIRE SERVICE

Change must come from within the department before all personnel can demonstrate effectiveness and sensitivity in the community. In concluding this section, it is useful to present two organizational trends that provide increasingly important changes to affect fire service agencies.

Trend 1: Transforming the Organizational Culture The culture of fire service agencies must internalize changes and accept diversity within both the department and the community. In the past this culture typically was dominated by white males, primarily of Irish or Italian backgrounds, who were often military veterans. With the influx of a multicultural workforce into the firefighting fields, system change is necessary if the organizational culture is to reflect the needs and concerns of this new generation consisting of a culturally diverse society.

One of the nation's leading diversity consultants, Elsie Cross, confirmed the reality of organizations' need to change. Throughout 20 years of working with many large corporate clients, this African American consultant recognized that most corporate cultures have actually been white male cultures, in which the majority controlled the power in the organization by dominating meetings, making all key decisions, establishing exclusive information loops, and choosing their own successors. To compensate for the past discrimination and to increase organizational effectiveness, Elsie

Cross Associates recommends the following strategies for managing diversity and valuing differences (White, 1992):

Introduce culture change to overhaul policies and procedures by means of focus groups and workshops that define basic assumptions, written codes, and rituals; incorporate this feedback into a revised mission statement that becomes a proclamation of the "new organizational culture vision and values."

- Identify the barriers that block individual or group diversity goals; identify the champions capable of building broad, internal coalitions around cultural awareness and group action against discrimination and for planned change.
- Require that the top command make at least a five-year commitment of resources to redefining the agency culture toward more effective management of diversity.
- Hold managers accountable for implementing the diversity changes through regular evaluation of their efforts.
- Confront the pain of racism and sexism at the personal and group level in the organization.

This management model has proven successful, and Cross credits the National Training Laboratories for Applied Behavioral Sciences for helping her to develop it. The strategy is invaluable to fire service leaders in a multiracial and multicultural society.

Trend 2: Organizations Must Move Beyond Cultural Awareness to Meet Today's Multicultural Challenges Fire service leaders should ask themselves this question: Does our agency provide adequate training for officers in cultural awareness? When a fire service agency has provided the basic training in cultural awareness and inaugurated effective equal employment and affirmative action policies, it is time to adopt another strategic plan that moves the fire service agency toward cultural competence. George F. Simons, a consultant on diversity, believes we must move beyond mere avoidance of ethnic, racial, and gender discrimination and sexual harassment. Firefighter supervisors and managers should not have to "walk on eggshells" while at work trying not to offend women and minorities. For managing or training in a multicultural environment, this expert counsels that advanced diversity training should be conducted that teaches people the skills to communicate and collaborate effectively with each other as colleagues, especially in mixed-gender, mixed-ethnic, and mixed-racial groups. Simons and Abramms (1999) offer the following advice:

- Organizational policy that supports productive partnerships and creative sharing with diverse membership from majority and minority groups should be developed.
- Executives and other leaders should participate in special coaching programs that include assessment, mind management, and communications training that enable them to model behavior and values that inspire their associates to work ingeniously with differences.

To further promote firefighter professionalism and first responder strategies in a diverse society, officers must be able to look up to leaders who are not ethnocentric and biased in their manner of leading their agency and responding to the communities they serve. For a firefighter officer to reflect cultural awareness and understanding of all peoples (as well as to apply the principle of fair treatment to every fire service action and communication), he or she must have role models at all levels of the chain of command (Figure 6.20), especially at the level of the fire chief. The fire chief should explicitly demonstrate **pluralistic** leadership and, ideally, be "an ambassador

pluralistic
The existence within a nation or society of groups distinctive in ethnic origin, cultural patterns, religion, or the like. A policy of favoring the preservation of such groups within a given nation or society.

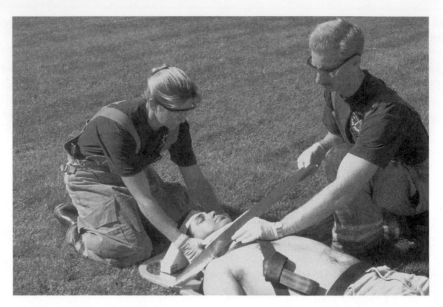

FIGURE 6.20 ◆ Woman firefighter instructor demonstrates how to secure the head of a supine patient to a long board with straps.

for diversity" within his or her department. Early in this chapter, we discussed fire-fighters who employ effective leadership skills with diverse populations. Attaining this outlook cannot happen consistently without direction from the top. For this rea-son, Figure 6.21 contains a list of characteristics and attributes of a pluralistic leader developed by MGH Consulting (1993).

Pluralistic Leaders

1. Look for ways to serve as catalysts for changing the work environment and community to actively welcome and value diverse people.

2. Are committed to eliminating all of the various "isms" (racism, sexism, etc.) that exist in their immediate work environment or neighborhood and speak to the greater concerns for equality, fairness, and other democratic ideals.

3. Help diverse people to be seen fairly and to be valued in the work environment and broader community.

4. Accept feedback about how to improve their relationships with those who are different by remaining open to change and growth.

5. Acknowledge their own prejudices or stereotypes and see the limitations that these will bring to their work.

6. Take time to assess their individual progress toward achieving the qualities and characteristics necessary to lead a diverse workforce.

7. Mentor others who need to gain sensitivity toward diversity.

8. Value differences among people and cultures as one of the great treasures of the human family and global community.

FIGURE 6.21 ◆ Characteristics and Attributes of a Pluralistic Leader

Source: MGH Consulting, reprinted with permission of MGH Consulting, 2454 Cameron Drive, Union City, CA 94587

For professionals in any facet of the fire service system, change begins first with modeling the leadership role and second with the practitioners' image of their role, which is then projected to the public. In the evolving new work culture, adhering to the image of the fire service and first responder is appropriate and requires that those in public service become more culturally sensitive in their outlook and approach. Effectively dealing with cultural issues within the field of the fire service then becomes a means for exercising leadership that demonstrates that sensitive firefighting respects cultural uniqueness. Professional fire service leaders must take the initiative to guide their departments into becoming ones that are culturally competent. Also, leaders must capitalize on the diversity of people within the organization and community, establish synergy, and seek to develop human potential for the betterment of both. The challenges for the fire service, in particular, are to recognize and appreciate diversity within both the community and the workforce, while using such insights advantageously. Human diversity must become a source of renewal rather than tolerable legislated requirements within our agencies, communities, and society. To accomplish this goal within the fire service and emergency service systems, multicultural awareness and skills training must become an integral part of the human resource development of first responders. The type of firefighting advocated in this chapter also requires competent supervision and management with fire administration. It is also important that departments, regionally or statewide, work together to enhance fire safety in the community at the same time promoting the common good and civility. There must be an interagency collaboration and synergy in regard to not only firefighting but also training, information exchange, and recruitment within a cultural and diversity context.

■ ■

Discussion Questions and Issues

1. **Cultural Impact.** Consider the long-term influence of the institution of slavery on the African American family in general and its male youth in particular. What is the implication of this negative heritage for today's African American citizens as a whole and its influence within the subcultures of fire service officers and community public services? Identify hopeful trends that repair or redress the damage of slavery and discrimination for this valuable segment of the American population.

2. **Extended Families and the Law.** Many Asian, Near Eastern, and Middle Eastern cultures function with an extended rather than a nuclear family. When such immigrants come to North America, they attempt to carry on this larger family tradition (e.g., food preparation of large meals, working hard in a family-owned business, remodeling and enlarging houses or estates, acquiring several automobiles or a fleet, intermarriage with relatives). Consider how such customs may cause the new arrivals to violate local laws and regulations relative to fire codes, child labor, multiple-family dwellings and occupancy, multiple-car parking on public streets, and so forth. Have you observed anything in this regard with respect to Southeast Asian, Cuban, Haitian, Asian, Indian, Pakistani, or other immigrants? What can a fire service officer do to help these newcomers from violating safety codes and laws of which they are unaware?

3. **Culture and Community Fire Service.** Every group—whether ethnic, racial, vocational, or professional—has characteristics defining its culture. The fire service organizational culture is no exception. People who do not belong to a particular in-group often develop myths and stereotypes about another group in an attempt to categorize behavior that they may not understand. Make a list of myths and stereotypes about fire service officers that exist among community members. After you have compiled and shared your list with others, discuss the following:

How did these myths or stereotypes arise?

Is there any truth to the myths or stereotypes?

Do they in any way affect communication with citizens or prevent you from doing your job well?

Do any of the stereotypes or myths create special difficulties or challenges with people from diverse backgrounds? If so, how?

What can a modern, professional fire department do to counter or lessen the influence of such myths and stereotypes in the community?

Website Resources

Visit these websites for additional information about image and cultural sensitivity issues:

Asian Firefighters Association (AFA): http://www.asianfire.org
This website provides information about the Asian Firefighters Association, with the mission to create and maintain equality of opportunity for Asians and other minorities within the fire service throughout the San Francisco community.

Fire and Aviation Management: http://www.nps.gov/applications/fire/
This website provides information about Fire and Aviation Management within the National Park Service of the U.S. Department of the Interior.

International Association of Black Professional Fire Fighters (IABPFF): http://www.iabpff.org
This website features fact sheets, news articles, and publications for cultivating and maintaining professional competence among firefighters, establishing unity, keeping alive the interest among retired members for the purpose of improving the social status of African Americans, and increasing professional efficiency.

International Association of Fire Chiefs (IAFC): http://www.iafc.org
It is an online resource for fire service executive and leadership issues and publications developed by the IAFC.

International Association of Fire Fighters (IAFF): http://www.iaff.org
It is a website for information about the range of issues confronting firefighters and their departments.

National Association of Hispanic Firefighters (NAHF): http://www.nahf.org
It is a website for information about the recruitment, retention, and advancement of the Hispanic firefighter by developing and conducting national unbiased culture awareness programs in these areas.

National Fire Academy: http://www.usfa.fema.gov/nfa
It is a website for information about training, careers, jobs, publications, and research for the fire service.

National Interagency Fire Center (NIFC): http://www.nifc.gov
It is a website for information concerning key newsworthy issues concerning the fire service in the United States.

U.S. Fire Administration: http://www.usfa.fema.gov
A resource of fire service resources, programs, and statistics collected and published by the Federal Emergency Management Administration (FEMA).

U.S. Forest Service: http://www.fs.fed.us/
A website of the U.S. Forest Service fire service resources, programs, and statistics

collected and published by the U.S. Department of Agriculture (USDA).

Women Chief Fire Officers Association (WCFOA): http://www.womenfireofficers.org
It is a website for information to provide a proactive network that supports, mentors, and educates current and future women chief officers.

Women in the Fire Service, Inc. (WFS):
http://www.wfsi.org
It is a website for information about training conferences, careers, jobs, publications, and research pertinent to networking women in today's firefighting world.

References

Armstrong, D. (2000, January 21). "Panel Urges Wide Change in Fire Department: Hub Report Faults Command, Promotional System, Culture." *Boston Globe*, p. A1.

Assistance to Firefighters Grant Program Staffing for Adequate Fire and Emergency Response (SAFER) Grants (2006). A part of the Assistance to Firefighters Grant Program, Department of Homeland Security, see http://www.firegrantsupport.com/safer/

Barr, R. C., and J. M. Eversole. (2003). *The Fire Chief's Handbook,* 6th ed. Tulsa, OK: PenWell.

Bowman, J. S. (Ed.). (1991). *Ethical Frontiers in Public Management*. San Francisco: Jossey-Bass.

Chan, S. (2006, May 9). "Fire Department Tries a Softer, Gentler Approach." *New York Times.*, p. 3.

Harris, P. R. (1994). *High Performance Leadership: HRD Strategies for the New Work Culture*. Amherst, MA.: Human Resource Development Press.

Harris, P. R. (1998). *The New Work Culture: HRD Transformational Management Strategies*. Amherst, MA.: Human Resource Development Press.

Harris, P. R., and R. T. Moran. (2004). *Managing Cultural Differences Instructor's Guide*. Burlington, MA: Elsevier Butterworth-Heinemann.

Harris, P. R., R. T. Moran, and S. V. Moran. (2004). *Managing Cultural Differences,* 6th ed. Burlington, MA: Elsevier Butterworth-Heinemann.

Hart, L. B. (1999). *Learning from Conflict*. Amherst, MA: Human Resource Development Press.

Independent Review of Fire Service. (2002). *The Future of the Fire Service—Reducing Risk, Saving Lives*. United Kingdom: Author.

Jewett, C. (2005, August 18). "Ex-Fire Captain Testifies of Picking Up Women in Engines." *McClatchy Newspapers Sacramento Bee,* p. B1.

"Jury Awards Black Firefighters in Discrimination Case." (2005, July 5). *Associated Press.*

Justice, T., and D. Jamieson. (1999). *The Complete Guide to Facilitation: Enabling Groups to Succeed*. Amherst, MA: Human Resource Development Press.

Kershaw, S. (2006, January 23). "Answering the Fire Bell in the Company of Women." *New York Times,* p. 1.

Levine, D., and Adelman, M. (1992). *Beyond Language: Cross-Cultural Communication*. Englewood Cliffs, NJ: Prentice Hall.

Lewis, C. W. (1991). *The Ethics Challenge in Public Service*. San Francisco: Jossey-Bass.

Lynn Learning Labs. (1998). *Mentoring: Passing the Torch*. Amherst, MA: Human Resource Development Press.

McGregor, E. B. (1991). *Strategic Management of Human Knowledge, Skills, and Abilities*. San Francisco: Jossey-Bass.

Moran, R. T., P. R. Harris, and W. G. Stripp. (1993). *Developing Global Organizations: Strategies for Human Resource Professionals*. Boston: Gulf Publications Series/Butterworth-Heinemann.

Murphy, B. (2006, May 26). "Grand Jury Will Hear Evidence in Fire Station Noose Incident Soon; The Firefighters Who Found the Nooses Have Since Been Assigned to Other Stations." *Florida Times-Union—Jacksonville,* p. B1.

National Fallen Firefighters Foundation. (2005). *National Fire Service Research Agenda Symposium*. Emmitsburg,

MD: U.S. Fire Administration and Center for Fire Research.

Owen, W. (2005, March 11). "Four Women File Sex-Bias Suit." *The Oregonian.*

Richardson, L. (2006, March 11). "Fire Station Pranks or Harassment? Practical Jokes Are Part of LAFD's Culture, but a Black Firefighter Who Ate Dog Food Sees Racism." *Los Angeles Times,* p. A1.

Roberts, L. M. (2005). "Changing Faces: Professional Image Construction in Diverse Organizational Settings." *Academy of Management Review,* 30(4): 685–711.

Sample, H. A. (2005, May 4). "S.F. Fire Chief Finds Trouble Smoldering: The Department Is Fighting On-the-Job Alcohol Use That Has Spawned Three Lawsuits." *Sacramento Bee,* p. A4.

Schrader, J. A. (1990) *The Heritage of Blue Earth County, Minnesota.* Mankato, MN: Minnesota Heritage Publishing.

Schuler, M. (2004). "Management of the Organizational Image: A Method for Organizational Image Configuration." *Corporate Reputation Review,* 7(1): 37–53.

Simons, G. F., and B. Abramms. (1999). *The Questions of Diversity: Reproducible Assessment Tools for Organizations and Individuals.* Amherst, MA: Human Resource Development Press.

Simons, G. F., C. Vazquez, and P. R. Harris. (1993). *Transcultural Leadership: Empowering the Diverse Workforce.* Boston: Gulf Publications Series/Butterworth-Heinemann.

Sullivan, S. (2000). *Mastering Leadership* [audiovisual workshop]. Amherst MA: Human Resource Development Press.

"There Is Another Country." (2000, August 19). *Economist,* pp. 26–27.

Ullmer, K. (2005, April 7). "Chief Praises Bellbrook Fire Department; Team's Leadership Cited in Annual Report to Council." *Dayton Daily News* (*Ohio*), pp. Z2–11.

White, C. (2002). *Changing the Fire Service Culture.* MFRI Field Instructor Curriculum for the Maryland Fire and Rescue Institute (MFRI) of the University of Maryland, College Park, Maryland.

White, J. P. (1992, August 9). "Elsie Cross vs. the Suits: One Black Woman Is Teaching White Corporate America to Do the Right Thing." *Los Angeles Times Magazine.* pp. 14–18, 38–42.

Williams, L. (1990, March 25). "Using Self-Esteem to Fix Society's Ills." *New York Times.*

Firefighter Leadership and Professionalism in a Diverse Society

Key Terms

CARE Approach, p. 277   **management, p. 259**   **synergy, p. 277**
leadership, p. 262   **professionalism, p. 276**

Overview

This chapter begins by providing information in getting back to the basics of the mission, purpose, and values of the fire service. We discuss and explore management and the components of leadership in the rank and file. Next, we look at the styles of leadership, the importance of professionalism, and synergy that fuels the efforts of men and women in the fire service. The final section presents information and strategies to maximize our full potential in a multicultural organization and society by using the CARE Approach (compassionate, attentive, responsive, and eclectic) 24/7 in the firehouse, taking emergency calls, and when off-duty (Figure 7.1).

Objectives

After completing this chapter, participants should be able to:

- Describe the mission, sense of purpose, and values of the fire service.
- Identify any differences between management and leadership in the rank and file.
- Describe the components of leadership.
- Describe the different styles of leadership.
- Describe the importance of professionalism.
- Describe and model synergy behavior that promotes team spirit and cohesiveness in fire service agencies.
- Describe and use the components of the CARE Approach when in the firehouse setting and when on operational alarms.

FIGURE 7.1 ◆ A team of firefighters stand in silhouette against the flames of a burning building.

LEARNING TASKS

Professionalism and leader-directed behavior are ongoing characteristics of effective and efficient fire service organizations. These two behaviors are skills that can be learned and mastered in all ranks of the fire service. Learning about the different attributes of leadership and its application to the fire service in a multicultural society and workforce fosters professionalism. Talk with your agency and find out how it promotes the development and continued use of leadership in its external and internal operations. Be able to:

- Identify the methods of leadership the agency uses to instill its mission, sense of purpose, and values into firefighters and nonfirefighter personnel.
- Identify the methods of leadership the agency uses to demonstrate its mission, sense of purpose, and values to multicultural populations in its fire service district.
- Determine what existing management and leadership strategies the agency uses for operational and firehouse settings involving firefighter and nonfirefighter personnel.
- Determine what the organizational culture is for empowering all personnel at all levels (from the recruit firefighter to the fire chief) with leadership skills and abilities.
- Ascertain what ongoing strategies the agency has in place or is developing to improve leadership training and opportunities for all employees in the multicultural workforce.

PERCEPTIONS

The challenge of leadership and professionalism for firefighters and all first responders is more difficult than for other occupations. Soldiers at war, firefighters, emergency medical services (EMS) personnel, and police officers combat stress every day.

Soldiers are trained to kill the enemy and police are trained to use force, up to deadly force, if justified. Firefighters and EMS personnel differ from the military and the police in many facets but are trained to save lives and property. Firefighters and EMS personnel along with the police are expected to help people in stressful situations and do it nonstop every day, every week. No other civilian occupation is subjected to this extreme amount of stress as these three types of first responders, and no other occupation wants to deal with it either. An important observation is made by psychologist Mike McEvoy, whose background includes being a paramedic, firefighter, published writer, and EMS coordinator for Saratoga County, New York, when it applies to occupational stress versus job stress.

> Work as an emergency responder involves two distinctly different kinds of stress: occupational stress and job stress. Occupational stress pertains to the specific influences of emergency response work. . . . Job stress is stress related to the work environment, the boss, and relationships with coworkers. Emergency responders find enjoyment in occupational stressors, but few welcome job stress (McEvoy, 2004, p. 33) [Figure 7.2].

Occupational hazards injure and, unfortunately, kill first responders. The firehouse for firefighters is the work environment in which leadership must be proactive to reduce the job stress as it applies to coworkers and especially supervisors. We as multicultural educators and trainers have worked together with firefighters in training workshops to promote leadership strategies for dealing with the job stress issues in the firehouse. Leadership and professionalism go hand-in-hand in all ranks in the fire service and must be taught at all levels to benefit everyone. As writers we recognize the importance of the National Incident Management System (NIMS) for emergency incidents. We will not address NIMS because fire service personnel receive training on this topic in accordance with National Fire Protection Association standards.

FIGURE 7.2 ◆ Firefighters walk in a line up a hillside to stage a backfire.

INTRODUCTION

Fire service leaders in command positions recognize the value of leadership and professionalism. Kathy Saunders, the former fire chief of the Bloomington, Indiana, Fire Department and now the fire marshal for Knox County, Tennessee, is one of several professional fire service leaders who have a keen insight and proactive posture for leadership issues and challenges in the fire service.

In most fire departments, the amount of time normally spent on the fire ground or even on emergency medical responses is miniscule compared to the time spent in the firehouse, at training, or inspections, and on all the other myriad tasks during a firefighter's average shift. The type of leadership skills needed for incident command is vastly different than the leadership skills necessary to deal with personnel during the rest of the day. All too often midlevel fire officers are selected and promoted based on their ability at emergency incidents. As individuals move up the promotional ladder, they are expected to learn how to manage and lead firefighters through trial and error on the job. The ability to fight fire and to manage an emergency incident through strong working knowledge of the incident command system does not teach new fire officers at any level how to lead and manage people, except on the fireground. In order to be an effective leader in a fire department, the knowledge and skills necessary to manage personnel when NOT at emergency incidents must be learned. If the fire department does not provide this education, it is imperative that fire officers take the initiative to attain this information on their own (Saunders, 2003, p. 229).

These comments by Kathy Saunders are echoed throughout the fire service profession as more attention is being directed to the skills that need to be taught in leadership essentials for all firefighters. As the workforce of the fire service is becoming more diverse, it is sound judgment to focus more on the skills of leadership in all settings, those in the firehouse and also on the fire ground.

MISSION, PURPOSE, AND VALUES

Effective organizations must have a strong and durable structure that moves the organization forward to accomplishing its stated goals and objectives. A goal or objective is best described as striving to accomplish a measurable task by taking action. The intention to perform the task provides purpose to the goal and objective. The importance placed on performing the task is based on the value of the goal or objective and forms the organization's mission. Multiple goals that are deemed important make up the mission statement of organizations. Responsible organizations, especially in the case of the fire service, must write a mission statement and identify what the fire department will do in achieving its goals of protecting the people and property in the area they serve. Each fire department may vary, but overall, most fire department mission statements will state, "to save lives and to protect property." In order for mission statements to have meaning, each firefighter must personalize what his or her individual values, purpose, and goals are as they relate to the fire department (Figure 7.3). Once a firefighter takes on this ownership of responsibility, the firefighter and fire department will be more in sync in accomplishing the agency's mission statement.

FIGURE 7.3 ◆ Firefighters put out a blaze at a two story home while standing on ladders near a smoke-engulfed window and roof.

The mission statement provides an additional benefit of embedding the primary objectives and goals for the fire service organizations. We recommend that fire departments make their mission statements, vision statements, and core values visible in the firehouse on bulletin boards and even in the kitchen. Most fire departments have websites and post their mission statement with their primary objectives and goals on their home page. These objectives may include, but are not limited to (Bennett, 2004), the following suggestions:

- Establishing organizational identity
- Determining the abiding purpose
- Setting a philosophy
- Determining core values
- Establishing ethical standards

To illustrate examples of large metropolitan fire departments' mission statements by region throughout the United States, see Figure 7.4, and Figure 7.5 has examples of those of medium to smaller-sized fire departments. Some mission statements are lengthy and others are brief and concise. Value statements and vision statements are common among fire departments, and in Figure 7.6 we show three examples from large, medium, and small fire departments.

management
The performance of tasks that need to be accomplished to direct the continued operations of an organization. A manager's duties should align themselves to the organization's mission statement.

MANAGEMENT

Management is best described as the performance of tasks that need to be accomplished to direct the continued operations of an organization. A manager's duties should align themselves to the organization's mission statement. In the case of the fire service, the fire department's organizational chart outlining the chain of command ensures that the department's mission is carried out efficiently and effectively. This basic chain of command encompasses nonfire personnel and fire personnel positions consisting of firefighter, lieutenant, captain, battalion chief, assistant chief, and chief of

---

East: New York City Fire Department, New York City, New York, 2006

As first responders to fires, public safety and medical emergencies, disasters and terrorist acts, FDNY protects the lives and property of New York City residents and visitors. The Department advances public safety through its fire prevention, investigation and education programs. The timely delivery of these services enables the FDNY to make significant contributions to the safety of New York City and homeland security efforts.

West: Los Angeles Fire Department, Los Angeles, California, 2004

It is the mission of the Los Angeles Fire Department to preserve lives and property, promote public safety and foster economic growth through leadership, management and actions, as an all risk life safety response provider.

South: Memphis Fire Department, Memphis, Tennessee, 2005

Leadership, with responsible financial management, strategic planning, and customer service for both the employees and citizens.

North: Lansing Fire Department, Lansing, Michigan, 2004

The Lansing Fire Department is committed to serving the Lansing area community with the highest level of life and property protection. We will achieve this by providing excellent and compassionate service in an atmosphere that encourages innovation, professional development and diversity.

Midwest: Omaha Fire Department, Omaha, Nebraska, 2004

The mission of the Omaha Fire Department is to respond to community public safety needs minimizing risk to life, health, and property threatened by the hazards of fire, trauma, medical emergencies, acute illness, and other hazardous conditions by adopting a proactive philosophy through response, prevention and public education.

Northwest: Portland Fire Bureau, Portland, Oregon, 2006

The mission of Portland Fire & Rescue is to aggressively and safely protect life, property and the environment by providing excellence in emergency services, training and prevention.

FIGURE 7.4 ◆ Mission Statements of Large Fire Departments in the United States by Region
Source: Herbert Wong and Aaron Olson

the department. Management positions usually begin with the promotion of lieutenant. It is important and critical that all managers possess good leadership skills. These leadership skills are essential to the function of providing direction to all the people of the organization and also holding all employees accountable in accomplishing the fire department's mission. The chain of command provides legitimacy to managers who use leadership skills in leading their people. Unfortunately, not all managers are effective leaders. A person with good leadership skills may not be in a management position but, nonetheless, is respected by his or her peers and management and is called upon when decisions need to be made in any of the types of companies (Figure 7.7):

- Engine company
- Truck company

East: Castine Volunteer Fire Department, Castine, Maine
Castine Volunteer Fire Department is dedicated to providing the Town of Castine, Maine Maritime Academy and its surrounding communities with expedient and professional emergency response management, fire suppression, education and fire prevention services with a primary focus on the preservation of life and the protection of property. This is accomplished through a continuous program of training and the maintenance of an inventory of up to date equipment. CVFD continues to proudly serve as a vital, all volunteer organization made up of dedicated members of the community.
West: Carson City Fire Department, Carson City, Nevada, 2006
Carson City Fire Department's mission is to provide premium emergency services for the protection of life, property, and the environment of the citizens and visitors of Carson City, Nevada.
South: Arcola Pleasant Valley Volunteer Fire Department, Sterling, Virginia, 2004
The mission of the Arcola Pleasant Valley Volunteer Fire Department, as a component of the Loudoun County Fire and Rescue System, is to provide primary protection, suppression and emergency medical services within the service district. Principal focus is on efforts to mitigate or minimize the casual factors relating to fire and medical emergencies. In the event of an emergency, our focus is to respond expeditiously to save lives and minimize property and environmental damage.
North: Sugarcreek Township Fire Department, Sugarcreek Township, Ohio, 2006
Caring professionals protecting the community.
Midwest: Overland Park Fire Department, Overland Park, Kansas, 2006
Provide emergency and safety services to the community of the highest quality, efficiently and professionally.
Northwest: Spokane County Fire District 10, Spokane, Washington, 2006
Fire District 10 is dedicated to serving the West Plains by: protecting life, property and the environment; presenting education and prevention information; providing fire, medical and emergency service, safely; and performing professionally and politely.

FIGURE 7.5 ◆ Mission Statements of Medium to Smaller Sized Fire Departments in the United States by Region

Source: Herbert Wong and Aaron Olson

- ◆ Rescue company
- ◆ Wildland/brush company
- ◆ Hazardous materials company
- ◆ Emergency medical services (EMS) company

The optimum goal is that all managers possess effective and efficient leadership and management skills. The role of managers and leaders is to make sure that the resources of the fire department are used to accomplish the mission and objectives of the organization. People employed within the fire service are the most important

---

Large size: Los Angeles Fire Department, Los Angeles, California, 2004

To Residents:

We owe the residents of Los Angeles the highest quality of service possible, characterized by responsiveness, integrity and professionalism. We will continually strive for quality improvement.

To Fire Department:

We owe the Los Angeles Fire Department our full commitment and dedication. We will always look beyond the traditional scope of our individual positions to promote teamwork and organizational effectiveness.

To Each Other:

We owe each other a working environment characterized by trust and respect for the individual; fostering open and honest communication at all levels.

To Ourselves:

We owe ourselves personal and professional growth. We will seek new knowledge and greater challenges, and strive to remain at the leading edge of out profession.

Medium size: Spokane County Fire District 10, Spokane, Washington, 2006

As a firefighter I will:
- Respond quickly to aid those in need
- Do my utmost to save and protect life and property
- Act with compassion
- Rise beyond the call of duty
- Recognize that we are servants to the public
- Be a good steward of what the public has given us
- Thirst after knowledge that will improve our skills
- Put other crewmembers before myself
- Respect my family of fellow firefighters physically, emotionally and spiritually

---

Small size: Arcola Pleasant Valley Volunteer Fire Department, Sterling Virginia, 2004

Our vision is to be the leader of a volunteer run emergency organization that others look to emulate.

FIGURE 7.6 ◆ Examples of Value and Vision Statements from Large, Medium, and Small Fire Departments in the United States

Source: Herbert Wong and Aaron Olson

leadership

In a multicultural society and workforce, begins with what the leader must possess for values, attributes, and skills that shape a leader's character. These internal qualities are present while at work, alone, and with others.

resource and asset of the organization. Like all skills, leadership is a skill and it can be learned and mastered by management and nonmanagement in all the roles performed within the fire service.

LEADERSHIP

Leadership in a multicultural society and workforce begins with what the leader must possess for values, attributes, and skills that shape a leader's character. These internal qualities are present while at work, alone, and with others. They define the person's character and provide a solid foundation for future challenges. These values, attributes,

FIGURE 7.7 ◆ A fire service administrator prepares paperwork from a small file while sitting at a desk.

and skills are the same for all leaders, regardless of their position, although leaders refine their understanding of them as they receive more training and become more experienced in higher positions of greater responsibility. For example, a veteran firefighter or company officer with fire ground or emergency scene experience has a deeper understanding of selfless service and personal courage than a new firefighter does. However, experienced firefighters and managers must live and practice leadership if they are to be credible with their peers and the community they serve. "The ability to lead by example is one of the most necessary attributes in the fire service at this time" (Kolomay and Hoff, 2003, p. 16). To achieve leadership, we define leadership as the skill of influencing people by providing purpose, direction, and inspiration while modeling the behavior ourselves to accomplish the mission and improving the organization. This process begins with what businesses, government, schools, public safety agencies, and nonprofit organizations have learned over the centuries from the military in leadership for values, attributes, skills, and actions (U.S. Army, 1999).

VALUES

Values are important and worthy of measure. The ingredients that form values in leadership are loyalty, duty, respect, selfless service, honor, integrity, and personal courage (U.S. Army, 1999). Each one is espoused in the public safety professions of fire service, police, emergency medical services (EMS), 9-1-1, the military, and the criminal justice system. Each one will be described by proactive leadership behavior using interchangeably the words of first responders, firefighters, fire service personnel, and professionals, because the men and women in these roles are leaders in public safety.

LOYALTY

Leaders who demonstrate loyalty are faithful to the organization. This is demonstrated by observing the department's mission, vision, and value statements. They do not

FIGURE 7.8 ◆ Firefighters wash their fire truck with a hose, a brush, sponges, and rags in a parking lot of their fire station.

gossip or backstab the organization or other firefighters. Further, fire service personnel work within the department's policies without manipulating them for personal gain.

DUTY

Leaders who are devoted to duty meet professional, legal, moral, and ethical obligations. They perform department duties and tasks, meet professional standards, and lead by example (Figure 7.8). Firefighters follow the department's rules, policies, and procedures. Equally important, professionals always seek excellence in performance and do not settle for just meeting the minimum standards.

RESPECT

Leaders who demonstrate respect treat people as they personally desire to be treated. The climate of fairness and equal opportunity is embedded in their values. They are discreet, tactful and make performance corrections in private. Safety is never compromised in any task or assignment, and they look out for the well-being of others. Fire service personnel are polite and courteous and do not abuse their positions of authority.

SELFLESS SERVICE

Leaders who demonstrate selfless service put the welfare of the victim, fellow firefighters, subordinates, and other first responders before their own. They are attentive to crew and company morale. Fire service personnel share coworkers, subordinates, and other first responders' hardships (Figure 7.9). Also, professionals communicate credit for success to others and accept responsibility for their own shortcomings.

HONOR

Leaders who demonstrate honor pay homage to the code of ethics. Their actions are consistent with their rhetoric and they do not tolerate the unethical behavior of others.

FIGURE 7.9 ◆ A firefighter checks the dipstick on the tractor of a fire truck with a raised hood.

Firefighters respect tradition and show reverence to the flag of the United States of America and respect for patriotic ceremony. Fire service personnel value honesty, truthfulness, and courage.

INTEGRITY

Leaders who demonstrate integrity do what is morally, legally, and ethically right. They behave with high personal moral standards. Honesty in words and action is valued and should never be compromised. Professionals consistently show good moral judgment and behavior in the firehouse and on emergency and nonemergency calls. Because they are expected to show integrity, fire service personnel do what is right instead of what is politically correct.

PERSONAL COURAGE

Leaders who demonstrate personal courage take physical action that could possibly jeopardize their own safety (Figure 7.10). They make ethical decisions and take responsibility for their decisions and actions. Leaders support their subordinates who act within department guidelines and leaders accept responsibility for their own mistakes and shortcomings.

ATTRIBUTES

Desirable attributes associated with effective leadership are sound mental, physical, and emotional fitness (U.S. Army, 1999). All three need to be studied, learned, and practiced to become habits. As values, each one is espoused in the public safety professions of fire service, police, emergency medical services (EMS), 9-1-1, the military, and the criminal justice system. Each one will be described by proactive leadership behavior using interchangeably the words of first responders, firefighters, fire service

FIGURE 7.10 ◆ Firefighters at work on the roof of a burning house.

personnel, and professionals, because the men and women in these roles are leaders in public safety.

MENTAL ATTRIBUTES

Leaders who demonstrate desirable mental attributes possess and display a mind-set of self-determination of will, self-discipline, initiative, good judgment, self-confidence, intelligence, common sense, and cultural awareness. They think and act quickly and logically, even when there are no clear instructions or the original plan falls apart. These leaders analyze situations, combining complex ideas to generate feasible courses of action. In addition, they balance resolve and flexibility. They show a desire to succeed and do not quit in the face of adversity. First responders who possess these attributes do their fair share of work and are able to balance competing tasks. Last, they embrace and use the talents of all members to build team cohesion.

PHYSICAL ATTRIBUTES

Leaders who demonstrate desirable physical attributes maintain an appropriate level of physical fitness and stamina. This is shown by presenting a professional appearance and never compromising personal hygiene, grooming, and cleanliness. They meet fire service professional standards in demonstrating effective verbal and nonverbal communication skills. Enthusiasm is emanated and they use the appropriate coping skills with hardship. Firefighters perform physically demanding tasks and continue to function under adverse conditions (Figure 7.11). Last, professionals lead by example in performance, fitness, and appearance.

EMOTIONAL ATTRIBUTES

Leaders who demonstrate appropriate emotional attributes project self-confidence. They show the appropriate temperament during conditions of stress, chaos, and rapid change. Fire service personnel exercise self-control, balance, and stability. They

FIGURE 7.11 ◆ In needing to gain access to the compound, a firefighter cuts a chain in a chain link fence with a bolt cutter.

show a positive attitude and demonstrate mature and responsible behavior that inspires trust and earns respect. As responsible professionals, they understand and experience the impact of critical incident stress by supporting each other and using stress debriefings.

SKILLS

Effective and persuasive leaders must possess a certain level of knowledge to be competent. They must develop interpersonal skills and use them with the people they work with and supervise. They must also have conceptual skills and the ability to understand and apply principles and procedures. In addition, leaders need to possess the technical skills to perform and supervise complex tasks. Last, leaders must master their tactical skills, which are expected of the level of responsibility they have with their fire service organization (U.S. Army, 1999). As with values and attributes, each skill is espoused in the public safety professions of fire service, police, emergency medical services (EMS), 9-1-1, the military, and the criminal justice system. Each one will be described by proactive leadership behavior using interchangeably the words of first responders, firefighters, fire service personnel, and professionals, because the men and women in these roles are leaders in public safety.

INTERPERSONAL SKILLS

Leaders who demonstrate interpersonal skills coach, teach, counsel, motivate, and empower subordinates (Figure 7.12). They take the initiative to readily interact with others. Trust and respect are earned, and first responders actively contribute to problem-solving and decision making situations. To show how they are respected, leaders are sought out by peers for expertise and advice.

FIGURE 7.12 ◆ Four firefighters carry a basket full of tools in front of a crane at a quarry on a call for assistance.

CONCEPTUAL SKILLS

Leaders who demonstrate conceptual skills are able to reason critically and ethically. Firefighters think creatively, anticipate requirements and contingencies, and improvise in new situations. Fire service personnel use appropriate research material and pay attention to details.

TECHNICAL SKILLS

Leaders who demonstrate technical skills possess or develop the expertise necessary to accomplish all assigned tasks and functions (Figure 7.13). Knowing the standards for tasks to be accomplished is never a problem. As professionals they know the fundamentals of firefighting skills. These leaders know the National Incident Management System and are looked upon by the police and other first responders as invaluable resources. Last, they use technology, especially information technology, to enhance communication.

TACTICAL SKILLS

Leaders who demonstrate tactical skills apply the fundamentals of firefighter skills. Firefighters apply their professional knowledge, judgment, and supervisory skills at the appropriate leadership level. Further, these professionals combine and apply their skills with people, ideas, and things to accomplish their task or mission. And with precision, they apply their skills with people, ideas, and things to train for, plan, prepare, and execute on the fire ground or emergency scene (Figure 7.14).

ACTIONS

Earlier, we defined leadership as the skill of influencing people by providing purpose, direction, and inspiration while modeling the behavior to accomplish the mission and improving the organization. Three major components surfaced from this definition

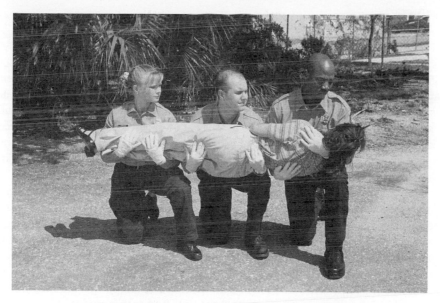

FIGURE 7.13 ◆ Three kneeling paramedics carefully lift a patient while turning her on her side.

and require action if the leader is to be effective. These components are influencing, modeling the behavior, and taking steps to improve the organization (U.S. Army, 1999). In a multicultural society and workforce, these actions are needed and important if leadership is to work. Each component and its subcomponents will be described by proactive leadership behavior using interchangeably the words of first responders, firefighters, fire service personnel, and professionals, because the men and women in these roles are leaders in public safety

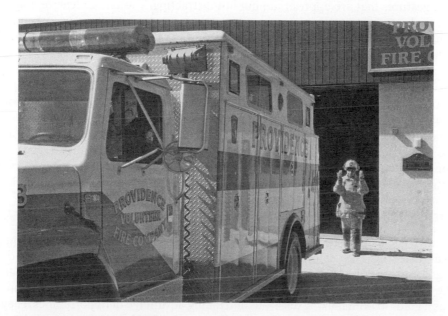

FIGURE 7.14 ◆ A volunteer firefighter carefully directs an ambulance backing up in front of a fire station.

INFLUENCING

The subcomponents of influencing are communication (oral and written), decision making, and inspiration (U.S. Army, 1999). Leaders who are effective in influencing use appropriate methods to reach goals while operating and improving. They motivate subordinates to accomplish assigned tasks and mission. These professionals set the example by demonstrating enthusiasm for and, if necessary, methods of accomplishing assigned tasks. They make themselves available to assist peers and subordinates and share information. Leaders encourage subordinates and peers to express candid opinions. Professionals actively listen to feedback and act appropriately based on it. They mediate peer conflicts and disagreements, and tactfully confront and correct others when necessary in private. Earning the respect and the willing cooperation of peers, subordinates, and superiors is a constant. Leaders challenge others to match their example and are compassionate in taking care of subordinates and their families, providing for their health, welfare, morale, and training. These leaders are persuasive in peer discussions and prudently rally peer pressure against peers when required. Last, they provide a vision for the future and shape the organizational climate by setting, sustaining, and ensuring a value-based environment.

COMMUNICATION

Leaders who communicate effectively display good oral, written, and listening skills. The ability to persuade others and to express thoughts and ideas clearly to individuals and groups characterizes quality communication.

Oral Communication Leaders who effectively communicate orally speak clearly and concisely. They speak enthusiastically and maintain listeners' interest and involvement. The skill of using culturally appropriate eye contact is practiced, and they utilize gestures that are appropriate but not distracting. Thoughts, ideas, feelings, sincerity, and conviction are communicated, and they express well-thought-out and well-organized ideas. These professionals use grammatically correct terms and phrases and use the communication process, as discussed in Chapter 4.

Written Communication Leaders who effectively communicate in writing are understood in a single reading by the intended reader. Fire service personnel use correct grammar, spelling, and punctuation, and their composition skills are good. When writing they use an appropriate format, a clear organization, and a simple style. The active voice is used and acronyms are essential and they write them out in first usage. Firefighters stay on topic and use correct facts and current data. Of equal importance, they are competent in using Word® documents and can print legibly when needed.

DECISION MAKING

Fire service leaders who make effective and timely decisions, use sound judgment and logical reasoning. They collect and analyze relevant information, changing situations to recognize and define emerging problems. In the absence of facts a professional will make deductions based on logical assumptions. These leaders uncover critical issues to use as a guide in both making decisions and taking advantage of opportunities. They are kept informed about developments and policy changes inside and outside the organization. By utilizing this information, leaders recognize and generate innovative solutions.

Decision making involves processing information from many different sources. A leader will formulate several different courses of action and choose the best one based upon analysis of costs and benefits, implications of the decision, and the need for action. The impact of a decision must be analyzed in order to determine how others will be affected and how the situation will change as a result of the action chosen. Once a decision is made, the effective leader will take charge and lead those affected through the change process.

To lead a group effectively through change, a leader must define the intent of goals and objectives. He or she remains resilient and decisive when discovering a mistake and acts in the absence of direction. A professional must function within the mission of the organization and keep the values and vision statements at the forefront in order to keep things fluid.

INSPIRATION

Inspiration is important to help subordinates to "catch the vision" and move toward change. A leader will motivate, encourage, and guide others toward the accomplishment of the mission statement objectives. Professionals do not show negative emotions when facing setbacks and articulate the reason for objectives, goals, and tasks. Anytime a leader can relate the change to the subordinate's Hierarchy of Needs, as defined by Maslow, there will be increased likelihood of successful change (Maslow, 1970). To do this, a leader articulates the reason for objectives, goals, and tasks.

Feedback is crucial to the change process. Accurate, timely, and positive feedback has a motivating effect on subordinates. Leaders actively listen for feedback from coworkers, subordinates, and supervisors. This feedback can be used to modify duties, tasks, requirements, and goals when justified. The process of providing feedback keeps subordinates, peers, and supervisors informed. This information sharing will enable all in the organization to be motivated by viewing the positive effects of change as they happen. Feedback also allows the leader the opportunity to recognize individual and company accomplishments with the intent of rewarding them appropriately.

A fire service leader who can identify poor performance and deal with it quickly and tactfully inspires subordinates by providing the subordinates with the information needed to make changes. Corrective action can be directed toward poor performance early. When a subordinate has clearly articulated expectations to aspire to, he or she works harder to meet those goals.

When subordinates have an expectation of success, they try that much harder to be successful. Before assigning tasks, a leader should consider duty positions, capabilities, and developmental needs. It is much better to provide advance notification when empowering a subordinate with a goal, objective, or task rather than surprising that person at the last minute with unrealistic expectations for task completion. Subordinates are inspired when they are credited with their own good performance. A leader takes responsibility for and corrects poor performance.

MODELING THE BEHAVIOR

The subcomponents of modeling the behavior are planning and preparing, executing, and assessing (U.S. Army, 1999). Leaders who effectively model the behavior accomplish the short-term and long term objectives, demonstrate tactical and technical competency appropriate to their rank and position, and complete individual and fire department tasks to standard and on time.

FIGURE 7.15 ◆ A team of paramedics lifts an injured man onto a gurney from the ground at an accident scene.

PLANNING AND PREPARING

Leaders effectively plan and develop feasible and acceptable plans for themselves and others. Professional planning and preparing accomplish the goals and objectives of the mission statement while expending minimum resources and looking beyond to the future. Forward planning is utilized to ensure each course of action will achieve the desired outcome. Reverse planning ensures that all tasks can be performed in the time available and that tasks depending on other tasks are performed in the correct sequence. Leaders conduct training rehearsals to be able to address likely contingencies. Flexibility is the key to successful training (Figure 7.15).

When training is to be conducted, scheduling conflicts should be recognized and resolved as soon as practical. Leaders need to notify peers and subordinates as far in advance as possible when their support is required. Professionals should always use a personal planning calendar because it is invaluable when organizing requirements.

EXECUTING

Leaders who effectively execute action use technical and tactical skills to meet mission standards, take care of people, and accomplish the mission with available resources. A professional ensures that individual and collective tasks are completed to standard and executes plans, adjusting them when necessary to accomplish the mission.

Leaders who successfully train and empower subordinates will encourage initiative, keep track of people and equipment, and make necessary on-the-spot corrections. Firefighters must remember that most work environments are fluid, so leaders must be able to adapt to changes. A leader will fight through obstacles, difficulties, and hardships to accomplish the mission (Figure 7.16).

Information is the key to success and leaders must keep the chain of command informed. Professionals should multitask by keeping track of assignments and deadlines. They should be prepared to adjust assignments, if necessary, and always follow up where needed.

FIGURE 7.16 ◆ First responders work together as a team to incapacitate an alcohol-impaired patient in front of a fire truck.

ASSESSING

Leaders who effectively assess use assessment techniques and evaluation tools to identify lessons learned and facilitate consistent improvement. They also establish and employ procedures for monitoring, coordinating, and regulating subordinates' actions and activities. A firefighter can do this through conducting initial assessments when beginning a new task or assuming a new position. It is important to assess progress on an ongoing basis, rather than waiting until a project is finished. A leader observes and assesses actions in progress without micromanaging.

Ultimately, activities must be analyzed to determine whether the desired objectives and outcomes have been achieved or affected. When the organization meets the standards, the leader puts procedures in place that will sustain the progress that has been made. Results are judged based on standards that should have been determined prior to beginning the change or project. To accomplish this, it is often helpful to conduct and facilitate a debriefing after the task is completed. A debriefing helps to determine causes, effects, and contributing factors for what worked well and what did not. Professionals recognize that a debriefing will help the organization to be more successful in future planning sessions.

IMPROVING

The subcomponents of improving the organization are developing, building, and learning (U.S. Army, 1999). Leaders who effectively improve the organization sustain skills and action that benefit themselves and each of their people for the future. A professional sustains and renews the organization for the future by managing change and exploiting individual and institutional learning capabilities. Last, a visionary leader creates and sustains an environment in which all leaders, peers, subordinates, and organizations can reach their full potential.

DEVELOPING

Leaders who effectively develop ongoing skills strive to improve themselves, subordinates, and the organization. This can be accomplished by the leader taking the role of mentor by investing adequate time and effort in counseling, coaching, and teaching individual subordinates and subordinate leaders. The firefighter sets the example by displaying high standards of duty performance, personal appearance, professional demeanor, and ethics (Figure 7.17).

Leaders create a climate that expects good performance, recognizes superior performance, and corrects poor performance. Tasks will be designed to provide practice in areas in which subordinates are weak. Subordinates are developed through a process of empowerment that starts with clearly articulated tasks and expectations. Realistic standards must apply. A leader guides subordinate leaders in thinking through problems for themselves and anticipates mistakes and freely offers assistance without being arrogant. As a subordinate leader progresses, opportunities will be provided for the subordinate to develop himself or herself. Training opportunities will be arranged that help subordinates achieve insight, self-awareness, self-esteem, and effectiveness. A leader encourages initiative, builds on successes, and improves weaknesses. This creates and contributes to a positive organizational climate.

Ultimately, a subordinate leader's goals must fit into the organization's goals. In other words, an employee's goals must be balanced by the organization's tasks, goals, and objectives that make up the mission statement.

FIGURE 7.17 ◆ A firefighter illustrates the proper wear of his protective gear, turnout gear, gloves, helmet, and PBI hood as he stands next to his fire engine.

BUILDING

A builder is a leader who effectively develops a strong working relationship with others and spends time and resources improving the organization. This fosters a healthy, ethical climate. A builder acts to improve the organization's collective performance and complies with and supports organizational goals. A builder encourages people to work effectively with each other and promotes teamwork and team achievement. This person usually has a reputation of being a team player and can remain positive when the situation becomes confused or changes.

A builder values and supports multicultural awareness as a tool to improve group dynamics in the fire service. This person also supports equal opportunity and prevents sexual harassment. Builders participate in organizational activities and functions and "step up to the plate" to participate in team tasks and missions without being requested to do so.

LEARNING

Effective leaders promote learning for themselves and others in the organization. This person seeks self-improvement in weak areas and acts to expand and enhance personal and organizational knowledge and capabilities. Learners typically apply lessons learned, ask incisive questions, envision ways to improve, design ways to practice, and endeavor to broaden their understanding. Firefighters who promote learning envision, adapt, and lead change.

Learners transform knowledge into experience and use it to improve future performance. This person typically makes knowledge accessible to the entire organization so his or her learning experience can be used to improve the organization too. The learner exhibits reasonable self-awareness and realizes the value of taking time off to grow and recreate. This enables the professional to embrace and manage change and adopt a future orientation.

LEADERSHIP STYLES

When we look at leadership styles, there are the extremes of autocratic and laissez-faire. The autocratic leader is a tyrant who barks out orders, uses fear to intimidate subordinates, and does not want to receive input or hear feedback. The laissez-faire nonleader is detached, provides no direction or interference, leaves the employees alone, and lets things run on their own and sometimes amuck, before he or she reacts to a major crisis (commonly an internal investigation complaint directed by a higher-ranking officer or a lawsuit initiated by a citizen or employee). These two extreme styles of leadership are morale killers and sabotage the effectiveness of the organization in accomplishing its mission. There are many levels between these two extremes and we will discuss three of them: controlling, participating, and empowering (U.S. Army, 1999). Keep in mind that adroit leaders mix elements of all of these styles in different situations, based on the nature of the circumstances and the skill level of the personnel available. If leaders do not adapt their style and use only one style of leadership, they become inflexible and have difficulty functioning in changing situations in which that style does not fit.

CONTROLLING

The controlling style of leadership is leader directed and consists of close supervision. Leaders using this style give detailed instructions on how, where, and when they want a task performed. The controlling style doesn't solicit input but does provide clarification to questions when asked by subordinates. Usually, the controlling style is used when time is short and leaders do not have time to explain the "why" aspects of the task. Also, this style is used when supervising inexperienced personnel.

PARTICIPATING

The participating style of leadership centers on both the leader and the team members. Once the assignment is received, the leader seeks input, information, and suggestions from his or her subordinates. This style is appropriate when leaders have time and experienced subordinates. It advances the team-building approach in allowing subordinates to take ownership in creating a plan and to work it. Subordinates will notice their supervisor trusts their input and is self-confident in their leadership skills to use this style. However, asking for input does not mean the leader is obligated to follow it. Ultimately, the leader is accountable for making the decision and putting together a plan of action.

EMPOWERING

Leaders who use the empowering style of leadership give employees the authority to make decisions and solve problems without having to go to the leader. In this process, supervisors need to encourage productive employees by giving them leadership roles and opportunities (Monroe, 2003). To have the empowerment emanate from the subordinate, leaders need to provide training for decision making and a clear understanding of the assignment or duties to be performed. Once this is understood, the subordinate is held accountable for his or her actions, but ultimately the leader is responsible for what does or does not happen. Empowerment is effective and is an excellent use of a fire department's most valuable asset, people.

PROFESSIONALISM

professionalism

Means approaching an activity, such as one's occupation or career, with a sense of dedication and expertise. A professional possesses integrity and demonstrates competence, regardless of rank, assignment, or tenure.

Professionalism means approaching an activity, such as one's occupation or career, with a sense of dedication and expertise. In comparison to an amateur, a professional is a committed high performer. A professional possesses integrity and demonstrates competence, regardless of rank, assignment, or tenure. The principal purpose of fire service education, whether in universities, colleges, or academies, is to improve performance and to increase professionalism among those in the fire service. What, then, seem to be the characteristics of the fire service professional? Some answers have been provided throughout this book as professionalism and leadership compliment each other in meaning. A professional in the fire service is one:

- ◆ Who performs an effective service at all times
- ◆ Who is developing and maintaining his or her career skill or competency level
- ◆ Who is properly educated and public service oriented
- ◆ Who exhibits ethical and sensitive behavior and ensures that other department members do the same as well
- ◆ Whose behavior and conduct on the job are appropriate and ethical, avoiding clear conflicts of interest
- ◆ Who capitalizes on diversity in people and organizations and seeks to develop human potential with regard to diversity

- Who is culturally competent to other coworkers and to their potential
- Who is aware of the impact of the fire service culture on the individual
- Who is aware of the latest developments in the fire service discipline

SYNERGY

Synergy implies cooperation and the integration of separate parts to function as a whole and to achieve a common goal. Synergy occurs through working together in combined action, attaining a greater total effect than a sum of the individual parts. Cultural synergy builds on the differences in people to promote mutual growth and accomplishment. Through such collaboration, similarities, strengths, and diverse talents are shared to enhance human activities and systems. In the fire service, synergistic leaders promote participation, empowerment, and negotiation within the organization and the community so that all fire personnel may work together to accomplish the mission of the fire department. Further, synergistic leaders demonstrate and promote the skills of facilitation, networking, coordination, and conflict resolution. Last, synergistic leaders are open-minded and effective cross-cultural communicators.

Leadership, professionalism, and synergy are three powerful, interconnected principles. Each one contributes to the success of the other and has application in all organizations. When combined within the public safety profession in general, and the fire service in particular, they enhance the effectiveness and efficiency of the organization in accomplishing its mission.

synergy
The benefit produced by the collaboration of two or more systems in excess of their individual contributions. Cultural synergy occurs when cultural differences are taken into account and used by a multicultural group.

CARE APPROACH

In presenting information and strategies to maximize our full potential in a multicultural organization and society we will discuss the use of the CARE Approach principle, which is an acronym for being compassionate, attentive, responsive, and eclectic (Olson, 2007). The **CARE Approach** is defined as a value-based concept and principle that promotes strategies to optimize harmony in individual, group, and organizational relationships. It can be used in, but is not limited to, approaches dealing with incidents, issues, or problems. It has universal application to business, communities, government, military, schools, nonprofit organizations, and all people. It is humanistic in design and maximizes the affective doman of human behavior for learning and performance. In the context of the affective domain, we mean values, beliefs, feelings, and attitudes. Some people have called the CARE Approach "the Golden Rule" in regards to "treat others the way you would want to be treated," and we agree with that comment. Unfortunately, not everyone knows or, by his or her actions, practices the Golden Rule in his or her public and private lives, and its meaning conjures up ambivalent reactions. In 2001 the CARE Approach was designed by coauthor Aaron T. Olson to provide awareness and strategies to improve upon workforce and community relationships needed in public safety. Since its development, thousands of people have received training on the CARE Approach, which include fire, police, corrections, 9-1-1, medical, business, and college students (Olson, 2007). The CARE Approach is taught as part of the curriculum with multicultural awareness, instructor development, and employee interpersonal relations training classes.

Fire service personnel will benefit from using these CARE Approach strategies at all times when on the scene of an emergency call; helping citizens; working with their coworkers and other public safety personnel; during training; and when at the fire-

CARE Approach
A value-based concept and principle that promotes strategies to optimize harmony in individual, group, and organizational relationships. CARE is an acronym for being compassionate, attentive, responsive, and eclectic. It can be used with, but is not limited to, approaches dealing with incidents, issues, or problems. It has universal application to business, communities, government, military, schools, nonprofit organizations, and all people. It is humanistic in design and maximizes the affective doman of human behavior for learning and performance.

house. There are four elements in the CARE Approach and each component will be discussed fully. All four elements align themselves in the triad of cognitive, personal, and environmental factors in human functioning, as explored in the social cognitive theory (Bandura, 1986; Schunk, 2000). Regardless of whether the circumstances are operational (formal) or personal (informal), the principles are useful and effective. The four elements of the CARE Approach are compassionate, attentive, responsive, and eclectic.

COMPASSIONATE

Compassionate behavior is motivated by an urge to help and is selfless. Firefighters, police officers, emergency medical service (EMS) providers, and 9-1-1 dispatchers and call takers are examples of individuals who made the choice to enter their professions because they want to help people. Recruiters representing public safety professions want people who possess compassion and are motivated to help people on emergency and nonemergency calls. These attributes are essential and make up the values of these public safety professions. Firefighters recognize the importance of showing compassion through their actions at emergency scenes; but after they remove their gloves, turnout coat, helmet, protective hood, bunker pants, and boots, they still need to take care of each other back at the firehouse. Stress related to the occupation, work, and other types of stress demand that firefighters be compassionate to each other because they are "family," and no one understands their situation better than each other. Personnel in no other public safety profession or occupation spend the amount of time with each other as firefighters do. The emphasis on being compassionate to each other must be reinforced by all fire service personnel, from the chief down to the probationary firefighter.

ATTENTIVE

Attentive behavior means being observant, considerate, and devoted. This second CARE Approach element helps firefighters interact with citizens and each other more effectively in the processing of information. Social, cultural, and social cognitive theories provide substance to this principle. Specifically, this observational learning involves the four subprocesses of attention, retention, production, and motivation (Bandura, 1986; Schunk, 2000). The payoff is that firefighters' awareness is enhanced, and they are better able to focus on the task they are attempting to accomplish.

RESPONSIVE

Responsive is the third element of the CARE Approach, and it means to take action in a situation or on a request. This enables firefighters to externalize their behavior in a proactive form based on their training, knowledge, and personal and organizational values. The ability to be responsive means one must practice one's listening and observational skills, and be able to ask inquisitive questions. In this process issues or pressing matters needing to be dealt with are handled in a timely and responsible manner. Further, the negative traits of inertia, avoidance, and the ignoring of problems are eliminated.

ECLECTIC

Eclectic is the fourth CARE Approach principle, which means to select a strategy from available systems, theories, or sources to deal with a problem or challenge. This provides several options and tailors one's choice to the specific need or issue. This is problem solving and involves assessment and analytical skills. In order to have these skills one must take advantage of formal and informal means of continuing education. Formal continuing education consists of attending professional courses offered by fire service training organizations or taking college classes at the undergraduate or graduate level. Informal continuing education includes reading books, journals, and professional magazines.

The CARE Approach is part of a term paper assignment in the Cultural Diversity in Criminal Justice Professions class, CJA (Criminal Justice Administration) 101, at Portland Community College, Portland, Oregon, in which students are asked to interview an adult in their community about multicultural issues from a checklist prepared by coauthor Aaron T. Olson, who designed the accredited college course and curriculum. One of the questions reads, "Ask them to identify any problem in our society and then you use each component of the CARE Approach in describing how it could help solve or reduce that problem. Be specific." Some students from the college fire science program have taken the course as well (Olson, 2007).

Students who receive instruction on the CARE Approach training sessions form small groups and then are presented with different scenarios to problem solve. They are asked to apply each component of the CARE Approach in an allotted time frame, to reach a group consensus in dealing with the scenario, and then to report their findings back to the entire class. A discussion follows each group presentation and the scenarios are customized to the specific profession. Figure 7.18 shows an example of CARE Approach scenarios presented to firefighters at in-service and academy training sessions, and Figure 7.19 shows another CARE Approach scenario presented to a coed group of firefighters at an academy training session (Olson, 2007).

The CARE Approach is an effective leadership strategy in all ranks of the fire service to improve coworker relationships in the firehouse, when on emergency and nonemergency calls, and when engaged with citizens. It touches on the affective domain of human behavior for learning and performance. It provides a practical and humanistic approach to dealing with incidents, issues, and problems that all too often go unnoticed or ignored. In using the CARE Approach, all fire service personnel will have another available "tool" from their "toolbox" to take actions in being compassionate, attentive, responsive, and eclectic to the need of their coworkers, citizens, other public safety professionals, and, equally important, their families and friends.

Group Activity: 10 Minutes
CARE Approach Application in the Firehouse

Instructions: Apply each component of the CARE Approach in your group exercise:

1. A firefighter returns to work after two weeks off after the death of a parent.
2. A firefighter is going through a painful divorce with a custody court battle.
3. A firefighter was unsuccessful in passing the test for lieutenant.
4. A firefighter returns to work after being on family leave for 12 weeks due to a new baby.

FIGURE 7.18 ◆ Example of CARE Approach Scenarios Presented to Firefighters at In-service and Academy Training Sessions

Source: A. T. Olson (2007). *The CARE Approach.* Accessed June 2, 2007, from http://atolson.com/careapproach.html

---

Group Activity
Group B: Female

Checklist for Each Group

1. Identify two challenges for this group in our society.

2. Are there any challenges for this group in the fire service? Why? If yes, identify them.

3. Use the CARE Approach© in helping to improve the challenges identified in #1 and #2. Be specific and make notes.

FIGURE 7.19 ◆ CARE Approach Scenario Presented to a Co-ed Group of Firefighters at an Academy Training Session

Source: A. T. Olson (2007), *The CARE Approach*, Accessed June 2, 2007, from http://atolson.com/careapproach.html

SUMMARY

To promote fire service leadership and professionalism in a diverse society, firefighters must be able to depend upon each other and their leaders in command for modeling respectful and responsible behavior to the men and women in the fire service profession and multicultural communities they serve. For a firefighter to reflect cultural awareness and understanding of all people, as well as to apply the principle of fair treatment to citizen and coworkers, he or she must have role models at all levels of the chain of command. Supervisors must master the art of leadership and practice these principles and skills at all times.

Professional fire service leaders must take the initiative to guide members of their department to become culturally competent. All fire personnel are leaders and are invaluable resources. Leaders in management positions must capitalize on the diversity of people within the fire department and community, establish synergy, and seek to develop human potential for the betterment of both. The challenges for the fire service, in particular, are to recognize and appreciate diversity within both the community and the workforce, while using such insights advantageously. Multicultural leadership and professionalism must become a source of renewal rather than tolerable legislated requirements within fire departments, communities, and society. To accomplish this all, firefighters must take on a leadership role, learn their mission statement, learn the components of leadership, and be empowered to maximize their full potential in a multicultural organization and society.

■ ■

Discussion Questions and Issues

1. **Mission Statement.** Search the Web for the fire department in your city or county, and ascertain whether it has a mission statement, value statement, and vision statement. Write down these statements and compare their content to the examples in this chapter. Discuss your findings in class.

2. **Values of Leadership.** There are seven values of leadership. Contrast each one in how it can maximize working fire-

fighter relationships between men and women in the firehouse and on emergency and nonemergency calls. Discuss your findings in class.

3. **Attributes of Leadership.** The desirable attributes of leadership are mental, physical, and emotional fitness. Identify which one is the most challenging for firefighters and why. Describe strategies to overcome the challenge and how these can benefit you and other firefighters.

4. **Skills of Leadership.** The skills of leadership are interpersonal, conceptual, technical, and tactical. Identify how each one would be used to help a team of firefighters who just returned to the firehouse after clearing from the scene of a police officer line-of-duty death. The police officer was well known and was killed in a motor vehicle crash when an intoxicated driver crossed the center line and struck the police officer's patrol car head on.

5. **Influencing.** You are a firefighter at the firehouse and a fellow firefighter, who is a 25-year veteran, confides to you, because you are still on probation, that he dislikes the new company lieutenant on your team. The veteran firefighter tells you that he is going to take his orders from the lieutenant but have nothing to do with her and ignore her at the firehouse. Discuss how you will use the influencing components of communicating, decision making, and inspiring in dealing with this situation.

6. **Improving.** You are the new company lieutenant who is experiencing the cold shoulder treatment from the 25-year veteran firefighter. He follows his orders from you but has nothing to do with you and you suspect it is because you are a woman. He has not said anything that would confirm your suspicions, but you know you must do something because this has been going on now for two weeks. Discuss how you use the improving components of developing, building, and learning to handle this situation.

7. **Leadership Style.** You are a veteran firefighter. Identify which style of leadership you hope your company lieutenant uses with you and your team when she discusses the required self-contained breathing apparatus (SCBA) inspection and maintenance at the firehouse. You are now the lieutenant. Which leadership style would you use in this situation?

8. **CARE Approach.** You and your team are at the scene of a single motor vehicle crash in which the driver ran into a tree, killing her four-year-old son. The mother has been transported to the hospital. The state police are investigating the fatal crash as a criminally negligent homicide because the mother was under the influence of methamphetamines. Three hours later you learn from your 9-1-1 dispatch that two of the state police officers who were on the crash scene had to attend the autopsy of the boy. Unfortunately, due to budget cuts this state police agency doesn't have a counselor available to assist in a critical incident stress debriefing. The dispatcher relays to you that the state police sergeant who is supervising the investigation asks you as a representative of your fire department for help in facilitating a debriefing. Use the CARE Approach and discuss specifically how you are going to handle this situation.

▪▪

Website Resources

Visit these websites for additional information about firefighter leadership and professionalism issues:

Asian Firefighters Association (AFA): http://www.asianfire.org
This website provides information about the Asian Firefighters Association, with the mission to create and maintain equality of opportunity for Asians and other minorities within the fire service throughout the San Francisco community.

Association for Women in Communications (AWC): http://www.awcdc.net
The Association for Women in Communications (AWC), Washington, DC, chapter is a professional organization that champions the advancement of women across all communications disciplines by recognizing excellence, promoting leadership, and positioning its members at the forefront of the evolving communications era. Founded in 1909, AWC has chapters throughout the United States.

Association of Public Safety Communications Officials, International, Inc. (APCO International): http://www.apcointl.com
The Association of Public Safety Communications Officials (APCO) International, Inc. is the world's oldest and largest not-for-profit professional organization dedicated to the enhancement of public safety communications. With more than 16,000 members around the world, APCO International exists to serve the people who manage, operate, maintain, and supply the communications systems used by police, fire, and emergency medical dispatch agencies throughout the world. APCO has APCO Institute, which is a not-for-profit educational institute that serves the unique needs of the public safety communications industry.

International Association of Black Professional Fire Fighters (IABPFF): http://www.iabpff.org
This website features fact sheets, news articles, and publications for cultivating and maintaining professional competence among firefighters, establishing unity, keeping alive the interest among retired members for the purpose of improving the social status of African Americans, and increasing professional efficiency.

International Association of Fire Chiefs (IAFC): http://www.iafc.org
It is an online resource for fire service executive and leadership issues and publications developed by the IAFC.

International Association of Fire Fighters (IAFF): http://www.iaff.org
It is a website for information about the range of issues confronting firefighters and their departments.

National Association of Hispanic Firefighters (NAHF): http://www.nahf.org
It is a website for information about the recruitment, retention, and advancement of the Hispanic firefighter by developing and conducting national unbiased culture awareness programs in these areas.

National Career Academy Coalition (NCAC): http://www.ncacinc.org
It is a website for information concerning career academies and career development of professions.

National Coalition Building Institute (NCBI): http://www.ncbi.org
A nonprofit leadership training organization based in Washington, DC. Since 1984, NCBI has been working to eliminate prejudice and intergroup conflict in communities throughout the world. It is a website for information about training programs, accomplishments, chapters, discussion groups, and campus and new programs. Leadership teams embody all sectors of the community, including elected officials, emergency responders, government workers, educators, students, business executives, labor union leaders, community activists, and religious leaders.

National Communication Association: http://www.natcom.org
NCA is a nonprofit organization of approximately 7,700 educators, practitioners, and students who work and reside in every state and more than 20 foreign countries. The purpose of the association is to promote study, criticism, research, teaching, and application of the artistic, humanistic, and scientific principles of communication. Founded in 1914, NCA is the oldest and largest national organization to promote communication scholarship and education. Its national headquarters are based out of Washington, DC.

National Fire Academy: http://www.usfa.fema.gov/nfa
It is a website for information about training, careers, jobs, publications, and research for the fire service.

National Volunteer Fire Council: http://www.nvfc.org
It is a website in partnership with the U.S. Fire Administration that offers different

training programs to state and EMS associations nationwide. It is also a resource of information for career and volunteer fire service.

U.S. Fire Administration: http://www.usfa.fema.gov
A resource of the fire service resources, programs, and statistics collected and published by the Federal Emergency Management Administration (FEMA).

Wildland Fire Leadership Development Program: http://www.fireleadership.gov
It is a website designed to provide information regarding the implementation of the Wildland Fire Leadership Development Program. In addition, it provides a resource to allow individuals to strive for a higher performance level as a leader through self-directed learning opportunities.

Women Chief Fire Officers Association (WCFOA): http://www.womenfireofficers.org
It is a website for information to provide a proactive network that supports, mentors, and educates current and future women chief officers.

Women in the Fire Service, Inc. (WFS): http://www.wfsi.org
It is a website for information about training conferences, careers, jobs, publications, and research pertinent to networking women in today's firefighting world.

References

Arcola Pleasant Valley Volunteer Fire Department, Sterling, Virginia. (2004). *Mission/Vision Statement.* Retrieved June 25, 2006, from http://www.arcolavfd.org/mission

Bandura, A. (1986). *Social Foundations of Thought and Action: A Social Cognitive Theory.* Englewood Cliffs, NJ: Prentice-Hall.

Bennett, R. (2004). *Fundamentals of Fire Protection.* Quincy, MA: National Fire Protection Association.

Carson City, Nevada, Carson City Fire Department. (2006). *Mission Statement.* Retrieved June 25, 2006, from http://www.carson-city.nv.us/CCFD

Castine Volunteer Fire Department, Castine, Maine. (2006). *Main Page (Mission Statement).* Accessed June 25, 2006, from http://www.castinefire.org/FireDept/MainPage.htm

Kolomay, R., and R. Hoff. (2003). *Firefighter Rescue & Survival.* Tulsa, OK: PennWell.

Lansing, Michigan, Fire Department. (2004). *Mission Statement.* Retrieved June 24, 2006, from http://www.lansingfire.com

Los Angeles, California, Fire Department. (2004). *Mission Statement.* Retrieved June 24, 2006, from http://www.lafd.org/mission.htm

Maslow, A. H. (1970). *Motivation and Personality,* 2nd ed. New York: Harper & Row.

McEvoy, M. (2004). *Straight Talk About Stress: A Guide for Emergency Responders.* Quincy, MA: National Fire Protection Association.

Memphis, Tennessee, Fire Department. (2005) *Mission Statement.* Retrieved June 24, 2006, from http://www.cityofmemphis.org/framework.aspx?page=28

Monroe, L. (2003). *The Monroe Doctrine: An ABC Guide to What Great Bosses Do.* New York: Public Affairs.

New York City, New York, Fire Department. (2006). *Mission Statement.* Retrieved June 24, 2006, from http://home2.nyc.gov/html/fdny/html/general/mission.shtml

Olson, A. T. (2007). *The CARE Approach.* Retrieved June 2, 2007, from http://atolson.com/careapproach.html

Omaha, Nebraska, Fire Department. (2004). *Mission Statement.* Retrieved June 24, 2006, from http://omaha-fire.org/staff/index.php

Overland Park, Kansas, Fire Department (2006). *Mission Statement.* Retrieved June 25, 2006, from http://www.opfd.com/index.html

Portland, Oregon, Fire & Rescue. (2006). *Mission Statement.* Retrieved

June 24, 2006, from http://www. portlandonline. com/fire/index. cfm? a=7737&c=26324

Saunders, K. (2003). *The Fire Chief's Handbook,* 6th ed. Tulsa, OK: PennWell.

Schunk, D. H. (2000). *Learning Theories: An Educational Perspective,* 3rd ed. Upper Saddle River, NJ: Prentice Hall.

Spokane, Washington, Spokane County Fire District 10. (2006). *Mission Statement.* Retrieved June 25, 2006, from http://www. scfd10.org/mission.html

Sugarcreek Township, Ohio, Sugarcreek Township Fire Department. (2006). *Mission Statement.* Retrieved June 25, 2006, from http://www.sugarcreektownship.com/fw/ main/Fire_Department-10.html

U.S. Army. (1999). *Military Leadership.* Field Manual 22-100. Washington, DC: Headquarters, Department of the Army.

Firefighters and Homeland Security and Disaster Preparedness

8 CHAPTER

Key Terms

community
assessment, p. 302
**Department of Homeland
Security (DHS), p. 288**

domestic terrorism, p. 292
international terrorism, p. 299
**terrorist attack pre-incidence
indicators (TAPIs), p. 301**

terrorism, p. 288
**weapons of mass destruction
(WMD), p. 309**

Overview

In this final chapter, we highlight the application of the multicultural and diversity strategies to the emerging roles of firefighters in homeland security and disaster preparedness. The president of the United States created the Department of Homeland Security (DHS) as a cabinet-level department on November 25, 2002, with the main objective of protecting the United States from terrorism and ensuring homeland security (Figure 8.1). With the development of the DHS, fire service officers and agencies now face many new roles and responsibilities. These include detecting threats of terrorism, developing and analyzing information regarding threats and vulnerabilities, using information and intelligence that has been provided from the DHS, and coordinating locally the homeland security efforts against terrorism. The six major goals of the Department of Homeland Security focus upon:

1. Preventing attacks of terrorism
2. Reducing vulnerabilities to terrorist attacks
3. Analyzing threats of terrorism and issuing warnings accordingly
4. Providing security for our transportation systems and for the borders of the United States
5. Preparing for emergency readiness and response
6. Minimizing the damage from terrorism and facilitating recovery from terrorist attacks

FIGURE 8.1 ◆ Lower Manhattan and the harbor of New York before the skyline of the city was changed forever with the terrorism on the United States on 9/11/2001.

This chapter highlights the roles and functions of fire service officers and agencies in their work in addressing the six goals of homeland security in multicultural and diverse communities, as well as the fire service participation in natural disasters preparation and efforts. Clearly, first responder practices and procedures already define much of the homeland security work and activities within fire service agencies. The emphasis in this chapter is on the specific aspects of multicultural fire service that are applicable to homeland security and disaster preparedness issues in the context of our diverse communities.

Objectives

After completing this chapter, participants should be able to:

- Describe the role that the fire service has in homeland security and disaster preparedness within multicultural communities.
- Address the six goals of homeland security in multicultural and diverse communities.
- Highlight the complexities of homeland security and disaster preparedness involving multijurisdictional efforts.
- Provide specific response strategies for homeland security and disaster preparedness in terms of collaborative work within local, state, and federal agencies involving multicultural and diverse communities.
- Delineate recommendations for fire service officers and other emergency services personnel for protecting homeland security and for assistance in natural disasters within multicultural communities.

LEARNING TASKS

Knowing the key roles for the fire service in homeland security and disaster preparedness for the ethnic, racial, and cultural groups in one's service area is critical toward addressing the multicultural challenges of your area. Talk with your agency and find out how it is involved in fire service to diverse communities. Be able to:

- ◆ Find out the roles for the fire service in homeland security and disaster preparedness for your area.
- ◆ Identify the action plans related to the six homeland security goals for your area.
- ◆ Determine the existing homeland security and disaster preparedness challenges that your fire service agency has in dealing with citizens in multicultural communities.
- ◆ Find out the leadership expectations and agency standards for multijurisdictional efforts in your area.
- ◆ Ascertain how greater diversity in fire departments can be translated into more effective efforts for protecting homeland security and for assistance in natural disasters in your county and city fire departments.

PERCEPTIONS

Americans' sense of homeland security was "shattered" following the September 11, 2001, attacks on American soil. In one concerted and orchestrated set of terrorist attacks (Figure 8.2), it was clear to all Americans that the security of our homeland was vulnerable.

> . . . with the grace of Allah, the battle has moved inside America. We will strive to keep it going—with Allah's permission—until victory is attained or until we meet Allah through martyrdom. (Osama bin Laden interview, October 21, 2001, quoted in Robbins, 2002).

FIGURE 8.2 ◆ Flames erupt from the South Tower of the World Trade Center in New York City as United Airlines flight 175 crashes into the side of the building during the terrorist attacks of September 11, 2001.

FIGURE 8.3 ◆ Firefighters and other rescue workers stand next to dust-enshrouded rubble of the destroyed World Trade Center resulting from the terrorism on the United States on 9/11/2001.

Department of Homeland Security (DHS)

The cabinet-level federal agency responsible for preserving the security of the United States against terrorist attacks. This department was created as a response to the 9/11 terrorist attacks.

terrorism

A violent act or an act dangerous to human life, in violation of the criminal laws of the United States or any segment to intimidate or coerce a government, the civilian population, or any segment thereof, in furtherance of political or social objectives.

Today as our nation works to ensure the general welfare of our citizens, the fire service has become an integral part of the nation's response system—in fact, we are America's first responders that provide for the common defense of this country within the geographic borders of the United States. The events of September 11, 2001 clearly demonstrated that in time of disaster or crisis, whatever the cause, *the local fire service is the critical link, the first line of defense and the primary response to any such event.* The fire service can no longer be looked upon as a local resource, but must be seen as a national asset (Randy R. Bruegman, President, International Association of Fire Chiefs, in an Open Letter to Congress, 2003) [Figure 8.3].

The United States rallied quickly to institute measures for overcoming the fear and shock of the terrorists' attacks. As noted by Donohue (2002), over 97 percent of the business conducted by Congress between September 11, 2001, and January 11, 2002, was related to terrorism and more than 450 bills, amendments, and resolutions relating to counterterrorism were proposed. Clearly our reaction to the fear and vulnerability generated by terrorism was to move forward.

The nation advanced and created the **Department of Homeland Security (DHS)**, which incorporated the roles and functions of 22 federal agencies with 170,000 employees. The DHS resulted in the largest reorganization of the federal government since the creation of the Department of Defense in 1947. The DHS, in turn, extended, created, and developed new roles and functions for the fire service in providing homeland security and for emergency preparedness for natural disasters, which never before were within the central realm of work for the fire service, public safety, and legal systems.

These emerging roles and functions for firefighters presented new challenges for the day-to-day work within the fire service system and for the Federal Response Plan (FRP) against **terrorism** and for homeland security. Additionally, they stretched the resources and knowledge base necessary to implement these areas of work, especially within multicultural communities. This chapter focuses upon the challenges and opportunities presented to multicultural fire service agencies and officers in detecting and preventing terrorism and the actions necessary to address the goals of homeland security and disaster preparedness within our diverse communities.

Homeland security, fostered through the gathering, processing, and application of intelligence, provides our nation with a blanket of protection against attacks. As a result of the geographic isolation of the United States, most Americans in the past have felt secure about their "homeland" and did not perceive that attacks could be launched on U.S. shores. Both Mexico and Canada were seen as friendly neighbors and not as sources of threat or danger to Americans' sense of homeland security.

Prior to September 11, 2001, many Americans (the person on the street to governmental leaders) perceived that enemies would need to surmount the tremendous burden of first having to cross vast expanses of oceans just to get their troops in our country for an attack. Pearl Harbor was seen as an "exception," and no troops had actually landed on American soil (Figure 8.4). Moreover, extensive logistics for replenishing supplies and troops were thought to be necessary to sustain any kind of attack effort even if enemy troops could land on our shores. As a result of the perceived advantage of the intervening oceans to the United States, much of the efforts and activities for homeland security, in the past, were located within the functions of the federal government (e.g., FBI, CIA, military, Department of State) to be implemented outside of the United States.

FIGURE 8.4 ◆ Pearl Harbor attack showing American battleships (USS *West Virginia,* USS *Tennessee)* burn after the Japanese attack in 1941 in Hawaii.

Local and state emergency and public safety agencies such as the fire service, EMS, fire marshal, police, sheriff, and prison departments had limited roles in homeland security. Based upon the homeland security consulting by one of the coauthors (Wong) with government agencies, the lesser involvement of local fire service in homeland security was unlike most countries around the world in which local and regional fire service agencies work closely with national departments on issues of homeland security. The roles of local and regional fire service agencies in the United States with respect to homeland security and disaster preparedness have increased greatly in recent times, and will most likely begin to resemble many of their global counterparts in cities and nations around the world.

Throughout the "Cold War" era with Russia and the communist bloc countries, and also in the Vietnam War era, there were many indications that homeland security encompassed much more than merely keeping enemy troops from marching on U.S. soil. Some examples for the need for greater homeland security efforts in the United States during this period were (1) spies infiltrating into agencies of the United States government, (2) Soviet missiles based in Cuba that would lead to the Cuban Missile Crisis and the Bay of Pigs invasion of Cuba, (3) assassinations of President John F. Kennedy and Rev. Martin Luther King, Jr., and (4) violent riots and demonstrations against the Vietnam War. However, the notion of the security of our country continued to be viewed as less the work of local fire and public safety agencies, and more the work of the national agencies (e.g., CIA, National Security Agency [NSA], FBI, and Defense Intelligence Agency [DIA]). Even for many of the college student demonstrations and riots related to the protest of the war in Vietnam, security efforts were conducted via the mobilization of national troops into the local regions (e.g., the use of the National Guard at Kent State University [Figure 8.5] and at the University of California, Berkeley). In part, as a result of the student riots on campus and surrounding areas, public safety agencies began to develop procedures and acquire specialized tools for gathering intelligence and for securing local areas (e.g., fire department bomb squads, hazardous materials teams, special weapons and tactics [SWAT] teams). The lines of operation continued to be clearly

FIGURE 8.5 ◆ Police officers in riot gear arrest demonstrators during a sit-in demonstration on a New York street.

drawn: (1) Issues affecting the security of our nation (even though local in nature) continued to be handled by federal agencies such as the Federal Bureau of Investigation (FBI) and Federal Emergency Management Agency (FEMA), and (2) issues related to the well-being and public safety of the local community remained the jurisdiction of the fire service and public safety agencies of the respective areas.

Until after the attacks of September 11, 2001, almost all of the homeland security efforts were seen as national in direction, leadership, and scope. The United States policies for counterterrorism were still centered on dealing with enemies and events primarily outside of the United States. For example, the United States policy regarding terrorism included:

1. Make no concessions and strike no deals.
2. Bring terrorists to justice for their crimes.
3. Isolate and bring pressure on countries that sponsor terrorism to force them to change their behaviors.
4. Develop the counterterrorism capabilities of those countries that work with the United States and that request assistance (Pillar, 2001; PDD-39 in FEMA, 1997).

Starting around the mid-1990s, national policies and procedures for dealing with terrorism and homeland security were enacted; however, the focus was still very much on leadership and direction on a national level with implementation by federal agencies (PDD-39 in FEMA, 1997). Involvement of local and regional fire service agencies was minimal in the planning, policy development, and implementation of the federal response to terrorism and homeland security. Subsequent inquiries following the terrorist attacks on September 11, 2001, made it very clear that the detection and prevention of terrorism, and the response to both foreign or domestic terrorists, would require the coordinated and collaborative efforts of local fire service agencies and officers with federal departments and organizations. The role of the Department of Homeland Security (DHS) is one of facilitating and developing the avenues of communication between these entities to ensure viable and effective leadership, policies, and procedures for homeland security in the United States.

Today, local and state fire and public safety leadership involvement in homeland security and disaster preparedness is a given:

MacMillan and 130 other senior US security officials attending a counterterrorism conference in Jerusalem found themselves caught up in an unfolding manhunt for a suspected Palestinian suicide bomber. An hour later, after a dramatic helicopter and motorcycle chase through police roadblocks on the main Jerusalem-Tel Aviv highway, Israeli police told the Americans that the would-be bomber had been captured, with 15 pounds of explosives packed with nails and shrapnel. The blow-by-blow account of the operation provided a real-time introduction to Israeli security pressures for the participants in the four-day conference, designed to encourage information sharing and expertise between Israeli and US counterterrorism officials. The gathering boasted the largest group of US law enforcement, emergency services, and homeland security officials ever to assemble in Israel. . . . The largest delegation came from California, including Los Angeles Police Chief William J. Bratton, the former police commissioner in Boston, and Joanne M. Hayes-White, chief of the San Francisco Fire Department. . . . White, San Francisco's fire chief, said she found the conference "truly impressive." "Our whole world changed after 9/11. It is something that we plan for now, but we have very little expertise in it. To come to a place like this, really you're learning from the true experts," she said (Kalman, 2006).

domestic terrorism

Involves groups or individuals whose terrorist activities are directed at elements of government, organizations, or population without foreign direction.

This chapter provides information on working with multicultural communities in the emerging areas of **domestic terrorism** and homeland security. Fire service efforts of homeland security within our local, state, regional, national, and global multicultural communities are addressed. Research and documented findings indicate that acts of terrorism and efforts toward homeland security start with key elements that are "local" in prevention, response, and implementation (Howard and Sawyer, 2004; Nance, 2003). Generally speaking, acts of terrorism in the United States usually involve local fire service agencies and other public safety personnel as first responders (Paradise, 2003). Thus, specific strategies and practices regarding the role of fire service personnel as first responders are provided as critical background information. This chapter highlights fire service community assessment, prevention, response, control, and reporting strategies related to the war on terrorism, homeland security, and disaster preparedness within multicultural communities. The chapter ends with key concerns relevant to fire service officers, and specific challenges involved in emerging roles and practices in dealing with terrorism, homeland security, and disaster preparedness within multicultural communities.

HISTORICAL INFORMATION AND BACKGROUND

Terrorism and homeland security have always been a part of organized society (Ahmad and Barsamian, 2001, Hoffman, 1998). From the earliest history of establishing order and government, extremist groups and individuals have used property damage and violence against people to generate fear and compel "change" in society and organizations (Figure 8.6). The Department of Justice, Federal Emergency Management Agency, U.S. Fire Administration, and National Fire Academy (1997) training manual, *Emergency Response to Terrorism: Basic Concepts*, highlights some of the following historical examples of terrorism affecting homeland security over the past 300 years, and current events provide additional contemporary examples.

Eighteenth Century

+ Infected corpses were used by the Russians in areas held by Sweden.
+ Organized violence against government taxation included Shay's Rebellion in 1786 and the Whiskey Rebellion in 1791.
+ British officials provided blankets from smallpox patients to Native Americans.

Nineteenth Century

+ Ku Klux Klan began acts of violence against African Americans.
+ An unknown person threw a bomb into a peaceful labor rally at Haymarket Square in Chicago, killing seven and injuring many others.
+ Catholic churches were burned in Boston, Philadelphia, and other cities.
+ Labor activists used 3,000 pounds of dynamite to blow up the Hill and Sullivan Company mine in Idaho, along with other housing units.

Twentieth Century to Current Times

+ In 1954, Puerto Rican nationalists wounded five members of Congress by gunfire.
+ In 1975, a bombing at New York City's LaGuardia Airport by Croatian nationalists killed 11 and injured 75.
+ In 1983, two left-wing radicals detonated a bomb in the cloakroom of the U.S. Senate in the Capitol Building.

FIGURE 8.6 ◆ The Democratic party platform satirized in a scathing political cartoon from the 1856 presidential campaign. The cartoon depicts scenes of violence (Preston Brooks beating Senator Sumner with a cane) and slavery, belying the high-minded language of the platform.

- Numerous airline hijackings and bomb threats occurred in the latter half of the twentieth century.
- The 1993 World Trade Center bombing in New York City killed six people.
- The 1995 Alfred P. Murrah Federal Building bombing in Oklahoma City killed 168 people.
- In 1996–1997, multiple bombing incidents occurred in Atlanta, including one at the 1996 Summer Olympic Games.
- On September 11, 2001, 3,047 died and many more injuries occurred with the attacks on the World Trade Center buildings, the Pentagon, and a hijacked plane in Pennsylvania (Figure 8.7).
- During the morning commute hours on March 11, 2004, a coordinated series of terrorists' bombs hit the Madrid, Spain, train system, resulting in 191 people killed and 2,050 people injured.
- On July 7, 2005, a series of terrorist bombs exploded in London's public transportation system during the morning rush hour killing 52 commuters (and four terrorists) and injuring over 700 people.

The targets and tactics of terrorists have changed over time (Hoffman, 1998), and the strategies for homeland security and protection have equally changed over time (Nance, 2003). In the past, targets of terrorists were often more individually directed. The death of a unique, single individual, such as a head of state, president, and/or prime minister, would produce the major disruption that the terrorist desired. In modern times, governments and organizations have become far more bureaucratic and decentralized. The targets of terrorists have included unique individuals, their surrounding networks, affiliated organizations and institutions, and any functions or processes associated with any targeted individual or groups. Today, terrorists attack

FIGURE 8.7 ◆ The rubble and debris at the Pentagon in Arlington County, Virginia, in the aftermath of the terrorist attacks of September 11, 2001.

not only prominent individuals and their organizations but also a wider range of targets that have been considered immune historically. For example, prior to modern times, terrorists have granted certain categories of people immunity from attack (e.g., women, children, elderly, disabled, and doctors). By not recognizing any category of people as excluded from attack, terrorists today have an unlimited number of targets for attack. The apparent randomness and unpredictability of terrorist attacks make the work of homeland security for the fire service officers extremely challenging, and solutions require a high level of sophistication.

As a result of the greater range of potential victims and the demographic background of terrorists, multicultural knowledge, skills, and resources constitute critical elements in the (1) preparation of local communities for safety and homeland security with regard to terrorism, (2) prevention of possible terrorists' incidents and enhancement of disaster preparedness, (3) participation in emergency response to terrorism, (4) observation and information gathering involving terrorists, and (5) follow-up actions and assistance in the prosecution of crimes involving and/or resulting from terrorism.

MYTHS AND STEREOTYPES ABOUT TERRORISTS AND HOMELAND SECURITY

Knowledge of the diversity and historical background of constituents within various multicultural communities, as well as their concerns and life experiences, will facilitate the homeland security and public safety mission of fire service officers. It is important to have an understanding about some of the myths and stereotypes that are held of groups associated with terrorism and homeland security, and how these stereotypes might contribute to prejudice, discrimination, and biased encounters with members of these populations. Stereotypic views of multicultural groups who might be potential terrorists reduce individuals within this group to simplistic, one-dimensional caricatures as either "incompetent cowards" or "suicidal bogey persons" who cannot be stopped. It is important for fire service officers to be aware of the different stereotypes of potential terrorists. From the training experiences with fire departments and other public safety agencies by one of the coauthors (Wong), the key to effectiveness in multicultural public safety and homeland security with any ethnic or racial group is not that we completely eliminate myths and stereotypes about these groups, but that we are aware of these stereotypes and can monitor our thinking and our behaviors when the stereotypes do not apply to those persons with whom we are interacting.

Some of the current stereotypes that might affect fire service officers' perceptions of terrorists include the following:

1. **Arab and Middle Eastern Nationality or Ethnic/Cultural Background.**   As a result of the September 11, 2001, attacks by al-Qaida and our images of the terrorists involved, it is easy to stereotype terrorists as having Arab and Middle Eastern backgrounds (Figure 8.8). Terrorists can be foreign or domestic, and clearly, the majority of the domestic terrorist incidents and attacks in the United States have not come from groups or individuals of Arab and Middle Eastern backgrounds. For example, a review of the FBI bomb incidents report with the most current data available (FBI, 1999) shows that the vast majority of the terrorist incidents in the United States are conducted by Americans against other Americans. Moreover, sources of potential terrorist threats include a variety of groups of diverse backgrounds.

Foreign terrorist groups that might affect the homeland security of the United States include (examples from Howard and Sawyer, 2004):

- ✦ Abu Nidal Organization (Libya)
- ✦ Aum Shinrikyo, Aleph (Japan)
- ✦ Harakat ul-Mujahidin (Pakistan)
- ✦ Sinn Fein (Ireland)
- ✦ Euzkadi Ta Askatasuna (Spain)

FIGURE 8.8 ◆ A surveillance camera image shows suspected terrorists Mohammed Atta and Abdulaziz Alomari going through security at a Portland, Maine, airport in the early morning of September 11, 2001

- Red Army Faction (Germany)
- Red Brigade (Italy)
- Popular Front for the Liberation of Palestine (PLO)
- Sendero Luminoso (Peru)
- Black Tigers (Sri Lanka)
- November-17 (Greece)
- Interahamwe (Africa)
- Abu Sayyaf (Philippines)
- AUC (Colombia)
- People's War (Nepal)
- Free Aceh (Indonesia)
- Cosican Army (France)

Any of the current terrorist organizations around the world could combine forces or utilize operatives from other organizations to "look different" than expected or stereotyped.

2. Insane and/or behave like automatons. Unlike the movie or video industry's portrayal of terrorists, these individuals are usually not insane, although their actions may appear insane or not rational (Dershowitz, 2002). According to Malcolm W. Nance, a 20-year veteran of the U.S. intelligence community's Combating Terrorism Program and author of *The Terrorist Recognition Handbook* (2003), most terrorists are generally intelligent, rational, decisive, and clear thinking. In the heat of an attack, terrorists may focus and harness their energies to accomplish their mission with dedication, motivation, ruthlessness, and commitment that may appear to outsiders as "insane."

3. Not as skillful or professional as U.S. public safety and fire service officers and personnel. As would be obvious, terrorists are not all similar in skills, training, and background. They vary in terrorist-training background from foreign-government-trained professionals who make up the foreign intelligence agencies of countries such as Cuba, Iran, and North Korea to novice, untrained civilian militia, vigilante, and criminals. The training and experience of *some* of the terrorists render them as skillful in their craft as any public safety professional in the United States. On the other hand, others have skill levels similar to petty criminals on the street. The use of intelligence information (i.e., not assumptions or stereotypes) to assess the skill and ability levels of terrorists or terrorist groups is central to homeland security.

Throughout this text, the authors have emphasized attitudes and skills required of fire service officials in a multicultural society: (1) respecting cultural behaviors that may be different from one's own, (2) observing and understanding behaviors important to diverse communities, and (3) analyzing and interpreting diverse behaviors for application within multicultural communities. Similarly, multicultural skills and knowledge are also elements that can contribute to detecting and predicting terrorist actions and activities for homeland security when coupled with an intelligence-based approach. Just as it is important to understand one's own biases and stereotypes about multicultural communities before one can effectively serve those communities, fire service officers need to change their perceptions of who the terrorists are before they can effectively detect terrorist activities.

1. Anyone could be a terrorist and threaten our homeland security. Fire service officers who start to look at specific groups may be blinded and overlook a group or people who may actually be the real terrorists sought. For example, if one were going by Arab descent or nationality, one would have missed Richard C. Reid, the shoe bomber (i.e., who was of British citizenship), and Jose Padilla, the alleged plotter for releasing a "dirty bomb" in the United States (born in Chicago of U.S. citizenship and Puerto Rican descent). If one thought that terrorists were inferior and not intelligent, one would have missed apprehending Theodore Kaczynski, the Unibomber, who had an Ivy League college education and taught mathematics at several major universities. Moreover, as noted by Stern (2004):

> The official profile of a typical terrorist—developed by the Department of Homeland Security to scrutinize visa applicants and resident aliens—applies only to men. That profile was developed before the advent of Islamist chat rooms recruiting operatives for a global jihad, before the war in Iraq increased anti-American sentiment world-wide, and before women started serving as suicide bombers for Islamist terrorist organizations. . . . Although women represent a fraction of terrorists worldwide, it is naïve to assume they're not recruited to violent extremist groups.

2.   **Learn to acknowledge the terrorist's motivation, skills, and capabilities.**   Because of the actions of terrorists, fire service officers may view terrorists with contempt along with other negative stereotypic descriptors such as "crazies," "camel jockeys," "suicidal," "scum of the earth," and "rag heads." Such contempt and disdain for terrorists may indeed be the perspective that blinds fire service officers and others to the fact that terrorists must be acknowledged for what they are: motivated, ruthless human beings who use destruction, death, and deceit to meet their lethal goals. For example, prior to the bombing of the Alfred P. Murrah Federal Building in Oklahoma City, one homeland security stereotype held that U.S. domestic terrorists were not capable of mass destruction, but were merely a criminal nuisance element of society (Heymann, 2001).

3.   **Observe and interpret street-level behaviors with proper intelligence information.**   Terrorists are not invisible "ghosts" (i.e., terrorists' behaviors and actions are usually visible to the trained observer); however, stereotypes based on race, ethnicity, gender, age, nationality, and other demographic dimensions may blind a fire service officer from observing and correctly interpreting these behaviors. Moreover, racial profiling would be one of the actions that could blind fire service efforts in detecting a terrorist's intent (Howard and Sawyer, 2004).

Without proper intelligence, it is almost impossible to interpret effectively behaviors that one observes. For example, three of the September 11, 2001, terrorists were involved in traffic stops by the public safety officers for speeding on three separate occasions prior to the terrorist attacks:

- Mohammed Atta, the al-Qaida 9/11 skyjacker who piloted the American Airlines Flight 77 into the North Tower of the World Trade Center, was stopped in July 2001 by the Florida State Police for driving with an invalid license. A bench warrant was issued for his arrest because he had ignored the issued ticket. Atta was stopped a few weeks later in Delray Beach, Florida, for speeding, but the officer was unaware of the bench warrant and let him go with a warning (Kleinberg and Davies, 2001, p. A1).
- Zaid al-Jarrah, the al-Qaida 9/11 skyjacker who investigators believe was a pilot on United Airlines Flight 93, which crashed in Pennsylvania, was stopped and cited by the Maryland State Police on September 9, 2001, for driving 90 miles per hour in a 65-mph zone near the Delaware state line (a $270 fine). Trooper Catalano, who made the stop, reported that he looked over the car several times, videotaped the entire transaction, and said that the stop was a "regular, routine traffic stop" ("A Nation Challenged," 2002, p. A12).
- Hani Hanjour, one of the al-Qaida 9/11 skyjackers aboard the plane that crashed into the Pentagon, was stopped for speeding within a few miles of the military headquarters six weeks before the attack and was ticketed by the Arlington, Virginia, police for going 50 miles per hour in a 35-mph zone. Hanjour surrendered his Florida driver's license to the police officer who stopped him, but was allowed to go on his way (Roig-Franzia and Davis, 2002, p. A13).

4. Analyze and utilize intelligence and other source information. For homeland security, key sources of information include not only local, state, and federal intelligence sources but also knowledge, relationships, and networks developed within the local multicultural communities. As noted by Nance (2003), "Terrorism against America can only be defeated through careful intelligence collection, surveillance, cooperative efforts" among fire service, public safety, and intelligence agencies, and "resolving the root complaints of the terrorist-supporting population."

DEVELOPING AND ANALYZING INFORMATION REGARDING THREATS AND VULNERABILITIES

Since the establishment of the Department of Homeland Security, fire service agencies have been receiving a variety of intelligence information, advisories, warnings, and other pertinent communications regarding the efforts and activities to be implemented in local communities. Local fire service agencies and officers, because of their knowledge of the existing multicultural communities and networks, have been called upon to aid in the data gathering and development of useful intelligence regarding possible terrorists. Such efforts have often sparked a mixed reception by fire service agencies. Some have regarded such requests for cooperative efforts as putting a strain on the established relationships of community services within multicultural communities. As we have noted in prior sections, public confidence and trust in public safety agencies are essential for the effective prevention of and response to terrorism and for homeland security. Residents in diverse communities representing different races, ethnic backgrounds, religions, and other aspects of diversity must be able to trust that they will be treated fairly and protected as part of homeland security. This is of particular importance when there might be some perceptions, assumptions, and/or stereotypes among community residents as to who might be a "terrorist among us." When local fire service agencies and officers are called upon to provide assistance, the tasks of information gathering and interviewing possible terrorist suspects within multicultural communities involve the following critical steps:

1. Contact with community leaders. Fire service officers need to work closely with community leaders to establish a cooperative plan for gathering the needed information. Often, announcements made by the community leaders to inform the multicultural communities of the reasons for information gathering and the processes involved can be very effective. Community leaders could, for example, work with fire service offices to develop a communication plan for the effective dissemination of information.

2. Utilize a communication plan. It is essential to have a well-developed and pilot-tested communication plan to ensure that the proper messages are provided to the multicultural communities involved. This is particularly important when language translations and interpreting are necessary. A carefully pretested "core message" regarding data-gathering procedures and interviews will usually work better than on-the-spot messages subject to the momentary interpretation of those who first hear such information.

3. Clearly define and spell out the implications for participation in the data-gathering process. When it comes to possible accusations of terrorism and/or fears related to homeland security, members from multicultural communities who might be stereotyped as terrorists or have experienced prejudice by others in this regard would naturally be quite cautious of any contact with fire service agencies and officers. Providing clearly defined procedures and information regarding how the data gathering and interviews would be used would be most important. Additionally, if there are federal agency consequences (e.g., being reported to the ICE for a visa violation), these elements should be clearly stated to ensure an ongoing relationship of trust with members of the multicultural communities involved.

4. Utilize fire service personnel and translators/interpreters from the same or similar multicultural communities. To the extent that it is possible, use of fire service personnel who are from the same or similar ethnic/cultural communities would add to sustaining the building of relationship and trust building necessary in such information-gathering processes. Clearly, the availability of bilingual personnel will make a significant difference to speakers of other languages. Fire service agencies need to be concerned about how they participate in homeland security data-gathering efforts in order to avoid possible negative effects on their relationships with multicultural communities.

LOCAL COMMUNITY ISSUES AND GLOBAL/NATIONAL/REGIONAL ISSUES FOR HOMELAND SECURITY

The targets and methods of terrorists have become more diverse and more difficult to predict. The terrorist attacks on September 11, 2001, clearly demonstrate that the global issues involving the United States in other parts of the world can result in terrorism within our own borders and affect victims in our local cities and towns. As highlighted by the Department of Justice:

> Terrorists in the United States continued a general trend in which fewer attacks are occurring in the United States, but individual attacks are becoming more deadly. Extremists in the United States continued a chilling trend by demonstrating interest in—and experimenting with—unconventional weapons. Over the past ten years, a pattern of interest in biological agents by criminals and extremists has developed. America and Americans have also been a favorite choice of target for terrorists. Reprisals for U.S. legal actions against domestic and international terrorists increase the likelihood that America will be the target of terrorist attacks either in the United States or overseas (Terrorist Research and Analytical Center, 1995).

Often, clues surrounding the terrorist incident and/or attack will reveal whether domestic or **international terrorism** was involved and will point toward possible motives and the perpetrators behind the incident. Such clues can be used for homeland security preparedness by local cities and communities, as well as to allay the fears and concerns of some of the members from multicultural communities who might be stereotyped. Clues that might be helpful surrounding a terrorist event include the following:

international terrorism
Involves groups or individuals whose terrorist activities are foreign-based and/or directed by countries or groups outside the United States or whose activities transcend national boundaries.

Timing of the Event For many years to come, September 11 will be a day for which all United States facilities around the world will operate at a heightened state of security and awareness because of the al-Qaida simultaneous terrorist attacks. Within the United States, April 19 will continue for some time to be a day of heightened alert because it is the anniversary of both the bombing of the Alfred P. Murrah Federal Building in Oklahoma City and the fire at the Branch Davidian compound in Waco, Texas (Figure 8.9). The more that the fire service works to understand the makeup of a multicultural community, the greater the likelihood that officers will have the knowledge to predict potential terrorist targets within communities, as well as to be alerted to relationships between targets and possible timings of terrorist events.

FIGURE 8.9 ◆ Flames engulf the Branch Davidian compound April 19, 1993, in Waco, Texas, where 81 Davidians, including leader David Koresh, perished as federal agents tried to drive them out of the compound.

Occupancy, Location, and/or Purpose Related to the Target
These include the following types of elements:

• *Controversial businesses* are those that have a history of inviting the protest and dislike of recognized groups, which include one or more components of extremist elements. For example, controversial businesses would include oil refineries, abortion clinics, logging mills, nuclear facilities, and tuna fishing companies. Controversial businesses have been ideal targets for arson and bomb threats requiring the assistance of the fire service.

• *Public buildings and venues with large numbers of people* are seen by terrorists as opportunities for attention getting with mass casualties and victims. For some terrorists, causing massive destruction and casualties in targeted public buildings or in venues containing large numbers of people would be linked to the identity of the operators/owners or to the events associated with the building or venue. Examples of these targets would include the World Trade Center, International Monetary Fund/World Bank, entertainment venues, athletic events, tourist destinations, shopping malls, and convention centers (Figure 8.10).

• *Symbolic and historical agencies, organizations, or programs* are targets that involve a relationship between the target and the organization, event, and/or services of the organization that specifically offend extremists. Examples of symbolic and historical targets would include the offices of the Internal Revenue Service (IRS) for tax resisters, the offices of the Bureau of Alcohol, Tobacco and Firearms (ATF) for those who oppose any form of gun control, and African American churches and Jewish synagogues for those who are members of white supremacist groups.

• *Infrastructure systems and services* include those structures and operations that are vital for the continued functioning of our country. Throughout the United States, these targets would include communication companies, power grids, water treatment facilities, mass transit, telecommunication towers, and transportation hubs. Terrorists' attacks on any of these targets have the potential for disabling and disrupting massive areas and regions, resulting in chaos, especially with respect to huge numbers of injuries and fatalities across large geographical areas.

FIGURE 8.10 ◆ Asian parents and children at International Peace March.

Fire service officers' and agencies' knowledge of possible targets linked to occupancy, location, and/or purpose of an organization within the local, multicultural communities will enhance the ability to prevent and to prepare for a terrorist incident and to be better prepared for homeland security and disaster efforts. For any of the **terrorist attack pre-incidence indicators (TAPIs)**, fire service officers would nearly always be among the first responders to the scene (Walter, Edgar, and Rutledge, 2003). Clearly, fire service officers and leaders see the expanding roles of firefighting professionals in homeland security:

> "This conference reflects the growing roles firefighters are playing in not only fire suppression and prevention but also national security," said Dennis Brodigan, the Matanuska-Susitna Borough emergency services director. About 280 participants representing more than 30 fire departments around the state signed up, according to Tara Mellon, the Central Mat-Su Fire Department staffer who coordinated the conference. The Alaska State Firefighters and Alaska Fire Chiefs associations present the event each year. Some sessions centered on more traditional duties: extricating victims from mangled cars, firehose nozzles, firefighting foam, thermal cameras and arson detection. But a few focused on the kind of emergency response that has only really surfaced since the terrorist attacks of 2001: medical response to weapons of mass destruction or coordinating the command hierarchy during a terrorist event. Tom Burgess, Alaska's former office of homeland security director and now a federal official, planned to attend the weapons of mass destruction course, Brodigan said. About 20 responders participated in a different three-day course on the National Incident Management System, the command hierarchy developed by the U.S. Department of Homeland Security to coordinate numerous different agencies responding to domestic incidents (Hollander, 2005).

terrorist attack pre-incidence indicators (TAPIs)

A term used by the intelligence community to describe actions and behaviors taken by terrorists before they carry out an attack.

COMMUNITY ASSESSMENT

community assessment
Reviewing and analyzing all of the localities that are (1) likely targets for terrorism and (2) problems during community disasters, and working to improve the security and safety of these locations.

Preventing, limiting, and reducing a community's vulnerability to attack requires a careful and ongoing **community assessment** to locate and measure the risks involved in the particular community. Such assessments would involve reviewing and analyzing all of the localities that are likely targets for terrorism and working to improve the security of these locations. Some fire service agencies have developed a rating system and have cataloged their possible problem areas into different priority levels. The following four priority-level assessment categories could be used within a community assessment or with respect to a particular fire service department or agency (Department of Justice and Federal Emergency Management Agency, 1997):

1. **First priority level—fatal.** These are processes and functions that if they fail would result in death, severe financial loss, massive legal liabilities, or catastrophic consequences. Such processes and functions would include all essential mission-critical elements such as electrical power (Figure 8.11), communications, information systems, and command and leadership functions.

2. **Second priority level—critical.** These would be processes and functional units of the agency, department, or organization that are critical to its operations and would be difficult to do without for any long period of time. Some examples of "critical" functional units might include computer terminals, elevators, perimeter security lights, and heat/air conditioning in the building. Examples of critical processes might include building entry and exit security and two-way communication radio network.

3. **Third priority level—important.** These would include those processes and functional units that are not critical to the department, agency, or organization such as facsimile, copier, videotape player, video camera, and tape recorders. Some examples of important processes would include files and record access, administrative training, and new-hire orientation.

4. **Fourth priority level—routine.** These would include processes and functional units that are not strategically mission-important (although they may cause some inconvenience) to the

FIGURE 8.11 ◆ The Fanny Bridge in Tahoe City, California, overlooks the water flowing over the dam from Lake Tahoe into the Truckee River.

agency, department, or organization such as coffeemakers, microwave ovens, and room thermostats. Nonessential and routine processes might include providing parking lot stickers and a duty list for staff personnel who offer information at the reception area.

Community assessments of functional units and processes should be ongoing and scheduled routinely so that information of possible terrorist targets could be updated on a regular basis. Clearly the work of the fire professional organizations has functional roles from this perspective:

> "Most, if not all, of our codes and standards can be seen as directly or indirectly touching the concept of homeland security by either addressing emergency preparedness and response or improving the design of the built environment," says Casey Grant, assistant chief engineer for NFPA and secretary of the Standards Council. Grant says NFPA's codes and standards address the nation's emergency-preparedness community on two levels. "They're focused on 'big-picture' subjects, such as disaster response, and on more specific but just as critical operations-level matters, like training and equipment," he notes (Paradise, 2004).

BUILDING MULTICULTURAL COMMUNITY NETWORKS AND RESOURCES

In order to detect and to deter terrorist incidents and attacks, the development of strong community networks, relationships, and resources is critical. As a result of many fire service efforts within multicultural communities on a regular basis, much of the day-to-day, natural elements for building networks and resources are developed as seen in the following example:

> Van Eyll said Chaska's Spanish-speaking population is growing, and the largest concentration of Spanish-speakers lives in Riverview Terrace. While several of Chaska's police officers speak Spanish, few on the volunteer firefighting force do, though Van Eyll said he hoped more will start learning. So the department recruited students from Chaska High School teacher Maria Ochoa's advanced Spanish class to help translate. . . . The team of 12 students split up into groups of two to three students and went door-to-door with firefighters to visit the Riverview Terrace neighborhood of 242 homes. At the first two stops for Jaspers and Korpi, there were English-speakers at home. But at the third house, the girls got to put their language skills to work. They introduced Krause and Lt. Alan Trebiatowski as "bomberos" (firefighters) and explained they were replacing smoke detector batteries for the residents. Despite having to work their way through the "wire" quandary—they managed to make the residents understand what they meant by referring to wiring as the path that the electricity followed—the girls were able to translate most of what the firefighters needed to say. "Oh, snap! That was so cool!" exclaimed Jaspers as they were leaving the first home. "We did it!" she said as the two high-fived each other. "They were able to understand us completely," Korpi said. She said doing the community service project was a more valuable experience than hours of class time (Mathur, 2006).

As we have noted earlier, the ability to obtain critical information from a community assessment depends on ongoing, positive, and trusted relationships within the

FIGURE 8.12 ◆ Group of Mesa Fire Department and EMS people with equipment going through training scenarios involving homeland security and natural disasters.

multicultural communities. It is equally important to develop the internal networks and relationships within the different agencies and departments of the local and statewide public safety community (Figure 8.12), as illustrated by this example involving several agencies from Providence, Rhode Island:

> More than 100 emergency personnel hurried to a bomb drill shortly after 7:30 a.m. at Providence Station. The Providence Emergency Management Agency & Office of Homeland Security conducted the exercise to test city, state and federal readiness to respond to a crisis. Information of a possible terrorist attack on a train had all on high alert, said Leo Messier, director of the emergency management and homeland security agency. . . . Fire Department personnel checked the area for chemicals and assisted the "victims." Federal, state and local officers worked to secure the scene. Personnel from the state medical examiner's office responded to deal with the "dead." And employees of Family Services of Rhode Island arrived to counsel the living. Agencies that played roles included the state Emergency Management Agency, the state fire marshal's office, the FBI, the federal Transportation Security Administration, Rhode Island Hospital and the city's Community Emergency Response Team. Messier said the most critical, and successful, aspect of the drill was establishing a unified command post, to ensure the efficient exchange of information (Pina, 2006).

USING ETHNIC AND MULTICULTURAL MEDIA AS A RESOURCE

For any terrorist, homeland security, or natural disaster incident, fire service agencies must help to provide a constant flow of credible information to the local community. Clearly, if positive relationships have been formed with the diversity of residents and the multicultural community leaders, the fire service agency's credibility would not be suspect and/or in question. Media sources used for the release of information should include the ones used by most populations, as well as those used by specific multicultural groups, especially that broadcast or print in the language of the key multicultural groups of the community. Some public safety agencies may be concerned about not having their information already prepared and translated into the language of the diverse ethnic and cultural groups; however, given the nature and importance of the information, most ethnic media will do the translation in order to provide critical information to their "audience."

Policies, procedures, and mechanisms need to be in place for releasing information to enable the safety and security of all community residents. This means that the involvement of ethnic and multicultural media is critical; the readiness of translations and interpretations are key in ensuring appropriate responses, alleviating any unnecessary fears or concerns, and providing assurances that public safety agencies and officers are in control of the situation. The media have been shown to be the most efficient and reliable means to disseminate information rapidly to the community regarding a terrorist, homeland security, or natural disaster incident and to notify community residents of the subsequent response by the fire service and public safety agencies.

Strategic media events should be held to inform the public continuously that the incident is well under control, the personnel are well prepared and trained to handle the incident, and the community's recovery is fully expected. Whether one is dealing with a terrorist incident or an everyday disaster event, the basis for effective communications with the community through the media is a strong partnership and an established positive relationship with the media. Fire service agencies need to understand that whether in crisis management or in daily community services, the media have a dual role: to obtain information, stories, and perspectives on the incident for news reporting objectives, and to inform the community efficiently of impending dangers and threats stemming from the incident. Good media relations allow for the strategic communication of critical incident information to the communities served by the fire service agencies. To reach multicultural communities, fire service officers must go beyond the conventional and mainstream media sources, and must become familiar with all media, in English and in other languages, that will reach the broadest populations in the shortest amount of time.

WORKING WITH MULTICULTURAL COMMUNITIES ON PREVENTION AND RESPONSE

In an earlier section, we quoted Nance (2003), who highlighted a key element for homeland security and for winning the fight against terrorism that needs to include cooperative efforts among fire service, public safety, and intelligence agencies, and "resolving the root complaints of the terrorist-supporting population." Clearly, fire service agencies are not in a position to resolve the "root complaints" involved with any terrorist-supporting nation or community. Fire service agencies are, however, in an excellent position to prevent the erosion of positive community relationships with the multicultural communities that might include community members who hold concerns of "root complaints" and other issues of unfairness, bias, and discrimination. Moreover, when it comes to local coordination efforts, such as in natural disaster relief such as those involved for Hurricane Katrina (Figure 8.13), coordination of

FIGURE 8.13 ◆ A girl and her father is at a shelter after being rescued by a firefighter from New Orleans Fire Department after being trapped in her home in high water in Orleans parish during the aftermath of Hurricane Katrina, August 30, 2005, in which the hurricane made landfall as a Category 4 storm with sustained winds in excess of 135 mph.

these efforts with community responses is vital to working within multicultural communities as seen in the following Indian reservation example:

> The Fort McDermitt Paiute-Shoshone Tribe has criticized the federal response after an August wildfire that scorched a portion of its reservation along the Oregon border. As Hurricane Katrina ravaged the Gulf Coast, the blaze destroyed one home and 16 sheds while blackening 3,000 acres of the reservation along U.S. 95 about 70 miles north of Winnemucca. Members of the impoverished tribe also lost winter firewood, pasture land for horses, tools, equipment and other belongings. "It wasn't much, but it was everything to them," said Richard Harjo, chairman of the Nevada Indian Commission. "We've never had a catastrophic event like this." Tribal members said they sought help from the federal government after the Aug. 29 fire, only to find themselves caught in a bureaucratic maze without answers. "It's almost the same situation as the Katrina situation, with no plan in place," said Sherry Rupert, executive director of the Nevada Indian Commission ("Tribe Criticizes Federal Response After Nevada Wildfire," 2005).

TRAINING FIRE SERVICE AGENCIES IN MULTICULTURAL COMMUNITY HOMELAND SECURITY ISSUES

Fire service professionals have recognized that prejudices and stereotypes left unchecked and acted upon can result in not only unfairness, humiliation, and citizen complaints but also missed opportunities for building long-term fire service–community relationships important for homeland security and disaster preparedness. Social scientists have noted that stereotypes and prejudices become more pronounced when fear, danger, and personal threat enter the picture, as may happen in possible terrorism and/or disaster incidents. Training in understanding multicultural communities and in understanding some of the multicultural groups in the community might have prevented the following situation for two recent European American immigrants:

> Pavel Lachko, a Russian student at the University of Texas at Arlington, was in America for about a month when he received a rude cultural lesson he won't soon forget: In these post–September 11 times, young foreign men aren't allowed innocent mistakes. On September 6, Lachko and his roommate, fellow Russian student Boris Avdeev, wandered on their bicycles into the employee parking lot of the Arlington police station looking for someone to give them directions. Rather than giving them answers, police locked them in handcuffs, held them for questioning by an agent from the Department of Homeland Security and charged them with criminal trespassing. . . . "I heard the male officer say into his radio that we are Pakistanis," Lachko says. "We said, 'No, no, no. We are not Pakistanis. We're Russians.'" . . . Johnson said he did not know whether the two young men were initially taken to be Middle Easterners, and he insisted the arrests had nothing to do with their nationality (Korosec, 2003).

Cultural awareness and understanding of the different multicultural populations of the community would have helped to prevent misidentifying the nationalities of the students. Moreover, training in recognizing cultural and language differences would have allowed the fire service officers in the example above to correctly recognize and identify some of the specific behaviors and cues of possible terrorists and discern the behaviors of likely nonterrorists.

EDUCATING MULTICULTURAL COMMUNITIES ON HOMELAND SECURITY AND DISASTER PREPAREDNESS

Most cities and states currently have websites and online information on homeland security and disaster preparedness for their communities. However, information and education regarding homeland security–specific actions for the prevention of terrorism are not available for most communities in the United States. This lack of information is even more pronounced for some of the multicultural communities with recent immigrants and refugees (including those from South and Central America, Asia, Eastern Europe, and the Middle East). The role of fire service and public safety officers in providing such information offers another viable avenue for developing positive relationships with multicultural communities and for community-policing efforts. Clearly, all communities are interested in acquiring specific information regarding activities of homeland security, as the following example illustrates:

> The East Side college has entered into a partnership with the American Institute of Homeland Defense to offer both continuing education courses and college-credit courses in homeland security and anti-terrorism preparation. Also, AIHD has designated St. Philip's as their Homeland Security Regional

Center of Excellence for the San Antonio area. . . ."There's a great need for homeland security training for first responders," said John Carnes, dean of applied sciences and technology at St. Philip's. "That might mean that the first person on the scene could be someone from the fire department or the police department or a hotel employee." "Our (program) will be focused on first responders," added Ray Boryczka, director of continuing education for St. Philip's. "We're actually going to be training the people who will be the first ones on the scene." Boryczka said that since passing the word along about this new program, the college has received requests from organizations and individuals interested in what the program will offer. "Our phones have been ringing off the hook," he said. "It's a much bigger response than what we first expected, and these weren't fire departments, police departments or EMTs (emergency medical technicians). These were calls from the general public and from school districts and from businesses. This is a hot issue. The sensitivity is still there" (Conchas, 2003).

Fire service agencies may also want to include into their citizens' fire academies information regarding homeland security and disaster preparedness (Figure 8.14) and ways that the multicultural communities could participate and provide assistance to the fire service departments. Additionally, fire service agencies might utilize their community volunteer personnel to provide such information in a bilingual mode to ensure proper information to communities with large populations of immigrants and refugees. Community-based presentations at familiar neighborhood centers, associations, churches, mosques, and temples may facilitate greater multicultural population attendance and participation. Likewise, collaborative programs

FIGURE 8.14 ◆ Firefighters check for radiation during a mock dirty bomb attack in Seattle with the members of the Seattle Fire Department wearing protective suits. This is part of a 36-hour terrorism response exercise involving a joint effort by the U.S. Department of Homeland Security and the U.S. Department of State, in partnership with state and local agencies and organizations.

of fire service agencies and community-based organizations have resulted in positive dissemination of information and ongoing building of trust with multicultural communities:

> Tens of thousands of people are taking emergency response courses at fire houses, volunteering at police departments or joining local reserve medical corps as part of a growing citizen brigade. . . . The idea of developing citizen response teams began more than 20 years ago in Los Angeles, where the fire department wanted to train neighbors to help neighbors, especially when a major disaster, such as an earthquake, delays the department's response. Since Sept. 11, 2001, spurred by federal grants, CERT has taken off nationwide, going from fewer than 170 local programs to more than 1,500, said Karen Marsh, an official with Citizen Corps. The training courses have drawn thousands in Florida, where they played an important supporting role during the hurricanes last year (Malone, 2005).

THE FIRST-RESPONSE CHALLENGE FOR THE FIRE SERVICE IN MULTICULTURAL COMMUNITIES

Fire service officers as "first responders" to a terrorism attack confront tremendous challenges, risks, and responsibilities (see *Protecting Our Nation: The American Fire Service Position Paper on the Department of Homeland Security*, 2002). The terrorist attack or disaster scene is complicated by the confusion, panic, and casualties of the attack, as well as any residual effects if **weapons of mass destruction (WMD)** have been used. Moreover, it has been quite typical for terrorists to deliberately target responders and rescue personnel at the disaster scene. Terrorists have utilized "secondary devices" to target fire service and other public safety personnel responding to a terrorist's attack. For example, in the 1997 bomb attack at an Atlanta abortion clinic, a second bomb went off approximately one hour after the initial explosion and was very close to the command post established for the first bomb attack.

As in all hazardous fire service situations, *officer safety* and *self-protection* are top priorities. However, in some terrorists' attacks, as in the September 11, 2001, multiple-site scenario, no amount of self-protection at the scene would have prevented the deaths of the 69 fire, police, and port officers from New York and New Jersey who were killed in the attack. Neither could the U.S. Secret Service agent, the FBI agent, and the U.S. Fish and Wildlife Service agent aboard Flight 93 that crashed in rural Pennsylvania have protected themselves from becoming victims.

Fire service personnel, as first responders to terrorists' acts, need to know that the forms of self-protection against WMD can be defined in terms of the principles of time, distance, and shielding. Fire service personnel need to have sufficient cross-cultural language skills to communicate the importance of these principles for effective action and response within multicultural communities during a WMD incident:

1. *Time* is used as a tool in a terrorist disaster scene. Spend the shortest amount of time possible in the affected area or exposed to the hazard. The less time one spends in the hazard area, the less likely one will become injured. Minimizing time spent in the hazard area also reduces the chances of contaminating evidence within the disaster scene.

weapons of mass destruction (WMD)

Three types of weapons are most commonly categorized as WMD: (1) nuclear weapons, (2) biological weapons, and (3) chemical weapons. The subcategories of WMD include activities that may be labeled as "agro terrorism," which is to harm our food supply chain, and "cyber terrorism," which is to harm our telecommunication, Internet, and computerized processes and transactions.

2. *Distance* from the affected terrorist area or hazardous situation should be maximized. The greater the distance from the affected area while performing one's functions, the less the exposure to the hazard. Maintaining adequate distance from the hazard areas also eases the evacuation of the injured, allows for entry by other emergency personnel requiring immediate access, and facilitates crowd control by fire service officers.

3. *Shielding* can be used to address specific types of hazards. Shielding can be achieved through buildings, walls, vehicles, and personnel protective equipment including chemical protective clothing, fire protective clothing, and self-contained breathing apparatuses.

Fire service officers need to be able to understand the various types of danger and harm that may result from a terrorist situation not only for their own self-protection but also in order to understand the reactions of affected victims and to be able to provide effective assistance to those affected by terrorist activities as well. Fire service personnel need to be aware that the many people who form multicultural communities may have had prior experiences (or have heard of prior experiences) with WMD incidents within their home countries (e.g., Vietnamese Americans' experience with thermal harm during the napalm bombing of their villages, or Iraqi Americans' experience with chemical harm from mustard gas attacks within their communities).

FIRE PROTECTION AND PUBLIC SAFETY CONSIDERATIONS

The extent that fire service and other first responders are able to act quickly and accurately in conducting a hazard and risk assessment of the affected population of a terrorist, homeland security incident, or disaster event will determine the approach to public safety and protection. The following three options are available:

1. Evacuation of all threatened and affected population
2. Protection-in-place for all
3. Combination of evacuation and protect-in-place by evacuating some population areas and protecting others in-place

Effective communication of information and explanation of fire protection procedures to members of multicultural communities may be key challenges for many fire service agencies (Figure 8.15). During a terrorist incident or natural disaster, there would not be time to locate bilingual personnel or interpreters for assistance in communicating with speakers of other languages, and the customary resources such as the AT&T language bank would be ineffective or unavailable. Explaining why multicultural community residents have to evacuate their homes and familiar surroundings, even in the event of a terrorist incident, will require personnel with language skills.

Evacuating the public from the affected area is based upon a decision that indicates that the public is at greater risk by remaining in or near the incident area. The decision to evacuate is determined in part by the following elements:

1. *Degree or severity* of public dangers, harm, and threats as estimated by the hazards and risk assessment.

2. *Number* of individuals or the magnitude of the population area affected by the danger or threat.

3. *Resources* available to evacuate the affected population to include fire, emergency, public safety, and other personnel; school buses; privately owned vehicles; and/or public mass transit.

FIGURE 8.15 ◆ Evacuation becomes a gridlock as drivers and passengers wait outside their vehicles on the interstate highway leaving downtown New Orleans August 28, 2005. Authorities in New Orleans ordered hundreds of thousands of residents to flee because Hurricane Katrina strengthened into a rare top-ranked storm and barreled toward the vulnerable U.S. Gulf Coast city.

4. *Notification* and instructional resources to provide information to the public before and during the evacuation. For multicultural communities, these may include the use of local community leaders as well as ethnic-group media such as radio/television, mobile public address systems, and door-to-door contact. For communities with residents who are primarily speakers of other languages, the need for interpreters and translators would be very important in ensuring timely communication and notification.

5. *Route security* is a key factor that fire service officers and agencies need to ascertain and to maintain so that evacuees are not subjected to further terrorist attacks.

6. *Opportunity* to implement the evacuation requires assessing ongoing risks and hazards such as airborne chemical contamination, unexploded bombs, and other unexpected terrorist activities.

7. *Special needs* of the evacuees are taken care of to include accurate information regarding the terrorist event and updated summaries of the recovery effort.

Protection-in-place allows the affected populations to remain within the confines of their own dwellings. Like evacuations, the decision for protection-in-place depends on the risk and hazard assessment of the terrorist incident. Basically, if the anticipated dangers and the presenting hazards to the public are made less by having the affected population remain in-place, the protection-in-place option is the better solution. Consideration needs to be made for multicultural communities in which there are large, multifamily residents within one household involving family members who are frail, elderly, and/or very young children. For some within multicultural communities, the preference for the entire extended family to be together may be paramount. In their home countries, if family members became separated or moved, this often meant having members "disappear" and never be seen again. Protection-in-place, if safe and possible, may be much more easily implemented than attempts to evacuate such families, especially in multicultural communities where there may be some language issues involved in receiving and providing specific instructions for an evacuation.

FIRE SERVICE RESPONSE STRATEGIES INVOLVING MULTICULTURAL POPULATIONS

The Incident Commander has responsibility for the overall entry and exit routes from the incident area. The fire departments are usually the first responders to most catastrophes, including a terrorist attack. However, depending on the nature, severity, and circumstances of the terrorist attack, the on-scene Incident Commander who will oversee the Incident Command System may or may not be from the fire service. The overall structure and line-of-authority for all responders to a terrorist attack should operate under an Incident Command System (as referenced in 29 CFR 1910.120), which is described in the forthcoming section (and detailed in a Federal Emergency Management Agency (1992) document entitled, *The Federal Response Plan for Public Law 93-288, as amended*). Clear communication within the command structure is central to the effective work of the on-scene incident management system, which may consist of personnel from many local fire service departments, as well as state, regional, and federal agencies and personnel. Effective language skills supported by prior multicultural communications training would be a positive benefit in these multijurisdictional, diverse leadership style, crisis command systems. For example, fire service officers and managers who are experienced in knowing that different multicultural groups have different norms, values, communication styles, languages, and practices could transfer those experiences to understanding and working with the differences found within multijurisdictional and multi-agency personnel involved in a terrorist incident. The key role of the Incident Commander is to ensure coordination and collaboration of the responders to the terrorist incident (Figure 8.16).

The responsibilities for responding to a terrorist incident or attack involving nuclear, biological, and chemical WMD materials are outlined in Presidential Decision Directive-39 (PDD-39)—details provided in FEMA (1997). PDD-39 identifies

FIGURE 8.16 ◆ The fire chief, as Incident Commander, provides key leadership role in fire suppression, homeland security, and natural disasters.

the Federal Bureau of Investigation (FBI) as the lead agency for crisis management during terrorist attacks and incidents involving nuclear, biological, and chemical materials. PDD-39 also identifies the Federal Emergency Management Agency (FEMA) as the lead agency for consequence and recovery management during terrorist attacks and incidents involving nuclear, biological, and chemical materials. The Federal Response Plan (FRP) is used as the outline and vehicle for coordinating consequence management efforts under PDD-39. FRP directs *other* federal agencies to support the FBI and FEMA, as needed.

WORKING WITH IMMIGRATION AND CUSTOMS ENFORCEMENT (ICE)

One of the key situations for a fire service agency involves decision making when the local public safety agencies and departments become involved in enforcing and working with the Immigration and Customs Enforcement (ICE) section, now a part of the Department of Homeland Security (Figure 8.17). The ramifications for multicultural community relationships and network building are significant. The decision making and strategic responses to terrorism within these collaborative and possible joint efforts are not clear-cut, nor "givens," for any fire service or public safety agency.

Immigration and Customs Enforcement cooperation with local fire service agencies is of particular concern for multicultural communities where there has been a history of legal as well as illegal immigration (e.g., Latino/Hispanic American, Asian/Pacific American, Arab and Middle Eastern American). Indeed, some multicultural community members will be prosecuted and deported. However, where no wrong has been committed, then fire service must keep a check on potential stereotyping and the prejudicial act of singling out individuals based on stereotypes.

FIGURE 8.17 ◆ Fire departments and other public safety agencies are asked to cooperate with Immigration and Customs Enforcement (ICE), a part of the U.S. Department of Homeland Security. This is to enforce immigration violations as illustrated by the law enforcement officials hoisting contraband in a net on a boat during a drug bust and search for immigration violations.

SUMMARY

The experience of fire service agencies involved in homeland security and the war on terrorism within multicultural communities is evolving. Likewise, the emerging roles and functions for fire service in homeland security and disaster preparedness present many challenges for the day-to-day work within the community services system. Homeland security challenges for fire service agencies include the following: (1) detecting and preventing attacks of terrorism; these can be made more complex within multicultural communities because of the past histories of those communities with public safety and immigration services in the United States, as well as in their native homelands; (2) avoiding stereotypes and possible biased perceptions based on ethnicity, culture, race, and religion that may be evoked within multicultural communities; and (3) working with multicultural community leaders regarding their perceptions of fire service actions and public safety efforts in homeland security. Fire service officers should realize that some citizens may have been victims of terrorism, as well as being harmed by the antiterrorism efforts within their native homeland (while innocent of any involvement with terrorism). These citizens carry with them stereotypes of "people in uniform" such as those in fire service as something to be feared and avoided. Fire service officials need to go out of their way to work with multicultural communities to establish trust, to provide outreach efforts, and to win cooperation in order to effectively accomplish their goals for homeland security and disaster preparedness efforts. Building partnerships that are focused upon community collaboration in the fight against terrorism is important both locally and nationally.

Attitudes and skills that enhance multicultural fire service include (1) respecting cultural behaviors that may be different from one's own, (2) observing and understanding behaviors important to diverse communities, and (3) analyzing and interpreting diverse behaviors for services application within multicultural communities. These same multicultural attitudes and skills form the core elements, when coupled with an intelligence-based approach, in detecting and predicting terrorist actions and activities for homeland security.

Since the establishment of the Department of Homeland Security, fire service agencies have been receiving a variety of intelligence information, advisories, warnings, and other pertinent communications regarding the efforts and activities to be implemented within local communities. Local fire service agencies and officers, because of their knowledge of the existing multicultural communities and networks, have been called upon to aid in the data gathering and development of useful intelligence regarding possible terrorists. Such efforts have often produced a mixed reception by fire service agencies. When local fire service agencies and officers are requested to provide assistance for homeland security information gathering of possible terrorist suspects, some of the following steps are recommended with the multicultural communities involved:

1. **Make contact with community leaders.** Fire service officers need to work closely with community leaders to establish a cooperative plan for gathering the needed information. Community leaders could work with the fire service agencies' fire prevention office to develop a communication plan for the effective dissemination of information.

2. **Utilize a communication plan.** It is critical to have a well-developed communication plan to ensure that the proper messages are provided to the multicultural communities involved. This is particularly important when language translations and interpreting are necessary.

3. **Define explicitly and spell out the implications for participation in the data-gathering process.** Providing clearly defined procedures and information regarding how the data gathering and interviews would be used is most important. Additionally, if there are possible consequences (e.g., with the ICE), these elements should be delineated to ensure an ongoing relationship of "trust" with members of the multicultural communities involved.

4. **Utilize fire service personnel and translators/interpreters from the same or similar multicultural communities.** Use of fire service personnel who are from the same or similar ethnic or cultural communities would contribute to maintaining relationships, building the trust necessary in such information-gathering processes.

Some multicultural community members, because of their past experiences with public safety, may be reluctant to participate in homeland security efforts. It is important for fire service departments and officials to build relationships and working partnerships with multicultural communities to ensure effective participation in combating terrorism and implementing homeland security.

A key challenge for fire service efforts in homeland security and disaster preparedness includes sustaining the community services within multicultural neighborhoods. In times of scarce funding resources, community service efforts might take a backseat to homeland security and the fight on terrorism. Moreover, because contact with multicultural communities will increase because of the emerging roles for fire service in homeland security, the need for additional bilingual firefighting officers and personnel will be paramount for these communities. As part of emergency preparation, the need to provide clear communications and to ensure understanding of emergency instructions will require additional bilingual personnel and translators for these diverse communities.

■■■

Discussion Questions and Issues

1. **Homeland Security Roles for the Fire Service in Your Local Community.** The chapter provided the six major goals of the Department of Homeland Security (DHS). Review the homeland security roles of the fire service in your community using the six major DHS goals. List the special challenges and opportunities involved in implementing the top two or three DHS goals for fire service agencies in your community. What are some of the unique fire service challenges related to implementation of these six goals within multicultural communities?

2. **Terrorist Targets in Your Local Community.** The chapter provided a four priority-level approach to implementing a community assessment of possible terrorist and homeland security targets. Review the possible terrorist targets in your community using the four priority-level categorization. List the special fire service challenges involved in protecting the top two priority levels of possible terrorist targets in your community. List some of the unique fire service challenges in working with multicultural communities in this regard (either within your community or in a nearby community with multicultural neighborhoods).

3. **Working with Multicultural Groups in Your Community on Homeland Security.** The authors recommended some guidelines and approaches for collaborative efforts in information gathering and in implementing homeland security efforts within multicultural communities. Select one or more multicultural groups that are part of your community (or part of a nearby community). How would you apply the suggested guidelines and approaches to working with these multicultural groups?

4. **Recruiting and Using Multicultural and Language-Expert Fire Service Personnel.** The authors noted the importance of having multicultural and language-expert personnel in working with multicultural communities. How would you go about recruiting and using such personnel in your local community? Do you see roles for the use of volunteers who have such backgrounds and language expertise in homeland security efforts? What are the advantages and disadvantages in using community volunteers in homeland security and disaster preparedness activities?

■■■

Website Resources

Visit these websites for additional information about multicultural fire service issues in responding to homeland security and national disaster efforts.

Association of Former Intelligence Officers (AFIO): http://www.afio.com
This website provides the *Weekly Intelligence News,* which summarizes some of the homeland security issues from the perspectives of the AFIO.

Central Intelligence Agency—*Factbook*: http://www.odci.gov/cia/publications/factbook/index.html
This website provides background and current information about groups and countries affecting the homeland security of the United States.

Department of Homeland Security (DHS): http://www.dhs.gov/
This website provides extensive information regarding the DHS, as well as the threat level for the United States at any point in time. Suggestions and tips are provided for local communities in their preparation for homeland security.

Department of State—Travel Warnings and Consular Information: http://travel.state.gov/travel_warnings.html
This website highlights countries and areas around the world that might be of a terrorist threat to citizens of the United States and its homeland security.

Federal Bureau of Investigation: http://www.fbi.gov/
This website provides the latest annual report of terrorism incidents in the United States.

Federal Emergency Management Agency: http://www.fema.gov/
This website provides a range of useful information related to the management of community, state, and national emergencies and homeland security issues.

U.S. Army Chemical and Biological Defense Command (CBDCOM): http://www.cbdcom.apgea.army.mil/cbdcom/
This website provides background and technical information on biological and chemical weapons that might be encountered by homeland security in a terrorist attack.

U.S. Coast Guard (USCG): http://www.uscg.mil/USCG.shtm
The U.S. Coast Guard is responsible for the protection of all of our ports and waterways within the Department of Homeland Security. The website highlights many of the specific aspects of homeland security for a fire service agency.

References

Ahmad, E., and D. Barsamian. (2001). *Terrorism: Theirs and Ours.* New York: Seven Stories Press.

Bruegman, R. R. (2003). "An Open Letter to Congress from the President of the International Association of Fire Chiefs." Fairfax, VA: IAFC.

Conchas, E. (2003, October 22). "College Gets Homeland Security Nod; St. Philip's to Offer Anti-Terrorism Preparation Classes." *San Antonio Express-News,* p. H1.

Department of Justice, Office of Justice Programs and Federal Emergency Management Agency, U.S. Fire Administration, and National Fire Academy. (1997). *Emergency Response to Terrorism: Basic Concepts.* Washington, DC: U.S. Department of Justice.

Dershowitz, A. M. (2002). *Why Terrorism Works: Understanding the Threat, Responding to the Challenge.* New Haven CT: Yale University Press.

Donohue, L. K. (2002, March). "Fear Itself: Counterterrorism, Individual Rights, and U.S. Foreign Relations Post 9-11." Paper presented at the International Studies Association Convention, New Orleans (and parts published in "Bias, National Security, and Military Tribunals." [2002, July]. *Criminology and Public Policy).*

FBI Bomb Data Center. (1999). *1999 Bombing Incidents.* General Information Bulletin 99-1. Washington, DC: Federal Bureau of Investigation.

Federal Bureau of Investigation. (1996). *Terrorism in the United States 1995.* Washington, DC: Terrorist Research and Analytical Center, National Security Division, Federal Bureau of Investigation (FBI).

Federal Emergency Management Agency. (FEMA). (1992). *The Federal Response Plan for Public Law 93-288, as amended.* Washington, DC: Author.

Federal Emergency Management Agency (FEMA). (1997). *The Federal Response Plan Notice of Change: Terrorism Incident Annex.* Washington, DC: Author.

Heymann, P. B. (2001). *Terrorism and America: A Commonsense Strategy for a Democratic Society.* Cambridge, MA: MIT Press.

Hoffman, B. (1998). *Inside Terrorism.* New York: Columbia University Press.

Hollander, Z. (2005, October 5). "Conference Reflects New Firefighter Roles; Security: A Few Exercises Focused on Weapons of Mass Destruction or Terrorist Attacks." *Anchorage Daily News,* p. G9.

Howard, D. R., and R. L. Sawyer. (Eds.). (2004). *Terrorism and Counterterrorism: Understanding the New Security Environment.* Guilford, CT: McGraw-Hill/Dushkin.

Kalman, M. (2006, March 24). "A First Hand Lesson in Fighting Terror." *Boston Globe,* p. A10.

Kleinberg, E., and D. Davies. (2001, October 19). "Delray Police Stopped Speeding Terror Suspect." *Palm Beach Post (Florida)*, p. A1.

Korosec, T. (2003, September 25). "A Texas Welcome: Two UTA Students Make a Wrong Turn, End Up Jailed in Arlington." *Dallas Observer*.

Malone, J. (2005, March 18). "Volunteers Turn Out to Keep Us Safe; Emergency Response Courses Teach Citizens How to Help the Pros." *Austin American-Statesman (Texas)*, p. A17.

Mathur, S. L. (2006, May 24). "Spanish Students Help Firefighters Translate." *Star Tribune (Minneapolis, MN)*, p. 1W.

Nance, M. W. (2003). *The Terrorist Recognition Handbook*. Guilford, CT: Lyons Press.

"A Nation Challenged: The Terrorists; Hijacker Got a Speeding Ticket." (2002, January 9). *New York Times*, p. A12.

National Commission on Terrorism. (1999). *Countering the Changing Threat of International Terrorism*. Washington, DC: Author. Available at http://www.fas.org/irp/threat/commission.html

National Volunteer Fire Council (2002). *Protecting Our Nation: The American Fire Service Position Paper on the Department of Homeland Security*. Washington, DC: Author

Paradise, J. R. (2004, January–February). "21st Century Civil Defense." *NFPA Journal*.

Pillar, P. R. (2001). "The Dimensions of Terrorism and Counterterrorism." In *Terrorism and U.S. Foreign Policy*. Washington, DC: Brookings Institution Press.

Pina, A. A. (2006, June 5). "City Stages Bomb Drill at Providence Station." *Providence Journal*, p. C1.

Robbins, J. S. (2002). "Bin Laden's War." In D. R. Howard and R. L. Sawyer (Eds.), *Terrorism and Counterterrorism: Understanding the New Security Environment*. Guilford, CT: McGraw-Hill/Dushkin.

Roig-Franzia, M., and P. Davis. (2002, January 9). "For Want of a Crystal Ball; Police Stopped Two Hijackers in Days Before Attacks." *Washington Post*, p. A13.

Stern, J. (2004, January 4). "Women Slip by Radar with Jihad Agendas." *Contra Costa Times*.

Terrorist Research and Analytic Center (1995). *Terrorism in the United States*. Washington DC: Department of Justice.

"Tribe Criticizes Federal Response After Nevada Wildfire." (2005, October 23). *Associated Press*.

U.S. Department of State. (1995). *Patterns of Global Terrorism*. Washington, DC: U.S. Department of State, Publication No. 10239.

U.S. Department of State. (1997). *Patterns of Global Terrorism. 1996*. Washington, DC: Office of the Coordinator for Counterterrorism, U.S. Department of State, Publication No. 10433.

Walter, A., C. Edgar, and M. Rutledge. (2003). *First Responder Handbook: Fire Service Edition*. Clifton Park, NY: Thomson Delmar Learning.

Self-Assessment of Communication Skills in the Fire Service

INSTRUCTIONS

For each item, check the box for the response in the column that best describes your approach to the communication process.

	Seldom	Occasionally	Often	Always
1. In communicating, I project a positive image of myself (e.g., voice, approach, tone).	❑	❑	❑	❑
2. When appropriate, I try to show my "receiver" (the person with whom I am communicating—for example, fire victims, citizens asking for information, and coworkers) that I understand what is being communicated from his or her point of view. I do this by restating this point of view and by showing empathy and concern.	❑	❑	❑	❑
3. I am sensitive to culturally different usages of eye contact and I establish eye contact where appropriate, but avoid intense eye contact with people for whom less eye contact is more comfortable.	❑	❑	❑	❑
4. I am aware of when my own emotions and state of mind affect my communication with others. (For example, I know my own needs, motives, biases, prejudices, and stereotypes.)	❑	❑	❑	❑
5. I refrain from using insensitive and unprofessional language while on the job (including language used in written and computer communications).	❑	❑	❑	❑
6. I try not to let the person with whom I am communicating push my "hot buttons," which would negatively affect my communication (e.g., cause me to go out of control verbally).	❑	❑	❑	❑
7. When speaking with individuals from groups that speak English differently from the way I do, I try not to imitate their manner of speech in order to be "one of them."	❑	❑	❑	❑

	Seldom	Occasionally	Often	Always
8. With nonnative speakers of English, I try not to speak in an excessively loud voice or use incorrect English (e.g., "You no understand me?") in an attempt to make myself clear.	❑	❑	❑	❑
9. I am aware that many immigrants and refugees do not understand fire service procedures, and I make special efforts to explain these procedures (including their rights).	❑	❑	❑	❑
10. I check in a supportive manner to see whether people have understood my message and directions, and I encourage people to show me that they have understood me.	❑	❑	❑	❑
11. With nonnative speakers of English, I make a special point to simplify my vocabulary, eliminate the use of slang and idioms, and try to use phrases that are not confusing.	❑	❑	❑	❑
12. I make extra efforts to establish rapport (e.g., show increased patience, give more explanations, show professionalism and respect) with individuals from groups that have typically and historically considered the fire service and public safety agencies unresponsive.	❑	❑	❑	❑
13. I am sensitive to cultural or gender differences between me and the receiver of my communication.	❑	❑	❑	❑
14. I convey respect to all citizens while on duty regardless of their race, color, gender, or other difference from me.	❑	❑	❑	❑
15. When using agency communication channels or media, I communicate professionally, avoiding inappropriate or derogatory remarks.	❑	❑	❑	❑

Impact of Diversity on Fire Service Behaviors and Practices Survey

INSTRUCTIONS

There has been a great deal of discussion in recent years about whether the job of fire service officer has been changing. Some of the discussion revolves around issues related to contact with people from different cultural, racial, or ethnic groups. Please check or enter one answer for each question.

1. **Comparing the job of fire service officer today with that of fire service officer a few years ago, I think that today the job is:**
   - ❑ A lot more difficult
   - ❑ Somewhat more difficult
   - ❑ About the same in difficulty
   - ❑ Somewhat easier
   - ❑ A lot easier

2. **When I am on fire service duty in a neighborhood of a different racial or ethnic group than myself, I must admit that I am more concerned about my safety than I would be if I were on fire service duty in a white neighborhood.**
   - ❑ Strongly agree
   - ❑ Agree
   - ❑ Neither Agree nor Disagree
   - ❑ Disagree
   - ❑ Strongly disagree

3. **If a fire service officer notices a group of young people gathering in a public place and the young people aren't known to the officer, they should be watched very closely for possible trouble.**
   - ❑ Strongly agree
   - ❑ Agree
   - ❑ Neither Agree nor Disagree
   - ❑ Disagree
   - ❑ Strongly disagree

4. **If a fire service officer notices a group of young people from another racial or ethnic group gathered in a public place, the officer should plan on watching them very closely for possible trouble.**
   - ❑ Strongly agree
   - ❑ Agree
   - ❑ Neither Agree nor Disagree
   - ❑ Disagree
   - ❑ Strongly disagree

5. **How often do you think it is justifiable to use derogatory labels such as "scumbag" and "dirtbag" when dealing with possible victims and community residents?**
   ❑ Very frequently
   ❑ Frequently
   ❑ Occasionally
   ❑ Once or twice in a year
   ❑ Never

6. **When I interact on fire service duty with civilians who are of a different race, ethnicity, or culture, my view is that:**
   ❑ They should be responded to *very firmly* to make sure that they understand the powers of the fire service
   ❑ They should be responded to *a little more firmly* to make sure that they understand the powers of the fire service
   ❑ They should be responded to *the same* as anyone else
   ❑ They should be responded to *somewhat differently*, taking into account their different backgrounds
   ❑ They should be responded to *very differently*, taking into account their different backgrounds

7. **When I encounter citizens of a different race, ethnicity, or culture who have committed a violation of the fire regulations or the law, my view is that:**
   ❑ They should be responded to *very firmly* to make sure that they understand the powers of the fire service
   ❑ They should be responded to *a little more firmly* to make sure that they understand the powers of the fire service
   ❑ They should be responded to *the same* as anyone else
   ❑ They should be responded to *somewhat differently*, taking into account their different backgrounds
   ❑ They should be responded to *very differently*, taking into account their different backgrounds

8. **When interacting on fire service duty with civilians who have a complaint or a question and who are of a different race, ethnicity, or culture, I try to be very aware of the fact that my usual gestures may frighten or offend them.**
   ❑ Strongly agree
   ❑ Agree
   ❑ Neither Agree nor Disagree
   ❑ Disagree
   ❑ Strongly disagree

9. **When interacting on fire service duty with victims who are of a different race, ethnicity, or culture, I try to be very aware of the fact that my usual behavior may frighten or offend them.**
   ❑ Strongly agree
   ❑ Agree
   ❑ Neither Agree nor Disagree
   ❑ Disagree
   ❑ Strongly disagree

10. **How often have you run into a difficulty in understanding what a civilian was talking about because of language barriers or accents?**
    ❑ Very frequently
    ❑ Frequently
    ❑ Occasionally
    ❑ Once or twice in a year
    ❑ Never

11. **How often have you run into a difficulty in understanding what a fire or emergency victim was talking about because of language barriers or accents?**
    - ❑ Very frequently
    - ❑ Frequently
    - ❑ Occasionally
    - ❑ Once or twice in a year
    - ❑ Never

12. **How often have you run into some difficulty in making yourself clear while talking to a civilian because of language barriers or accents?**
    - ❑ Very frequently
    - ❑ Frequently
    - ❑ Occasionally
    - ❑ Once or twice in a year
    - ❑ Never

13. **How often have you run into some difficulty in making yourself clear while talking to a fire or emergency victim because of language barriers or accents?**
    - ❑ Very frequently
    - ❑ Frequently
    - ❑ Occasionally
    - ❑ Once or twice in a year
    - ❑ Never

14. **How important is it that the fire department provide training to make its members aware of the differences in culture, religion, race, or ethnicity?**
    - ❑ Very important
    - ❑ Important
    - ❑ Neither important nor unimportant
    - ❑ Unimportant
    - ❑ Very unimportant

15. **Personally, I believe that the training I have received on group differences is:**
    - ❑ Far too much
    - ❑ Somewhat too much
    - ❑ About the right amount
    - ❑ Somewhat too little
    - ❑ Far too little

16. **The training in the area of group differences has been:**
    - ❑ Extremely helpful
    - ❑ Very helpful
    - ❑ Somewhat helpful
    - ❑ Not too helpful
    - ❑ Not helpful at all

17. **My own view is that our department's quality of service could be improved by:**
    - ❑ Placing greater emphasis on hiring on the basis of the highest score obtained on the entrance exam, making no attempt to diversify by race, ethnicity, or gender
    - ❑ Placing greater emphasis on diversity by race, ethnicity, or gender and somewhat less emphasis on the numerical rank obtained on the entrance examination
    - ❑ Giving equal weight to both the score obtained on the entrance examination and diversification by race, ethnicity, or gender

18. **What percentage of civilian or internal complaints against fire or emergency service employees are adjudicated equitably?**
    - ❑ Over 80 percent
    - ❑ Between 60 and 80 percent
    - ❑ Between 40 and 60 percent
    - ❑ Between 20 and 40 percent
    - ❑ Less than 20 percent

19. **Some civilian or internal complaints are adjudicated more favorably toward people from diverse groups rather than toward the majority population.**
    ❑ Strongly agree
    ❑ Agree
    ❑ Neither Agree nor Disagree
    ❑ Disagree
    ❑ Strongly disagree

20. **I think that fire or emergency service employees of a different race or ethnicity receive preferential treatment on the job.**
    ❑ Strongly agree
    ❑ Agree
    ❑ Neither Agree nor Disagree
    ❑ Disagree
    ❑ Strongly disagree

21. **The racial diversity of my coworkers has made it easier for me to see issues and incidents from another perspective.**
    ❑ Strongly agree
    ❑ Agree
    ❑ Neither Agree nor Disagree
    ❑ Disagree
    ❑ Strongly disagree

22. **I think that employees of a different race or ethnicity than myself receive preferential treatment on this job.**
    ❑ Bothers me because I do not think it is justified
    ❑ Does not bother me because I think it is justified
    ❑ Is fair only because it makes up for past discrimination
    ❑ I do not believe minorities get preferential treatment

23. **In certain situations, having a partner of a different race or ethnicity than myself is more advantageous than having a partner of my same race or ethnicity.**
    ❑ Strongly agree
    ❑ Agree
    ❑ Neither Agree nor Disagree
    ❑ Disagree
    ❑ Strongly disagree

24. **I have received negative feedback from members of the community regarding the conduct of other fire service officers.**
    ❑ Strongly agree
    ❑ Agree
    ❑ Neither Agree nor Disagree
    ❑ Disagree
    ❑ Strongly disagree

25. **I have received negative feedback from members of the community regarding the conduct of officers who are of a different race or ethnicity in particular.**
    ❑ Strongly agree
    ❑ Agree
    ❑ Neither Agree nor Disagree
    ❑ Disagree
    ❑ Strongly disagree

26. **I have received more negative feedback from members of the community regarding the conduct of officers from different races and ethnic backgrounds than about the conduct of white officers.**
    ❑ Strongly agree
    ❑ Agree
    ❑ Neither Agree nor Disagree
    ❑ Disagree
    ❑ Strongly disagree

27. **In terms of being supervised:**
    ❑ I would much rather be supervised by a man
    ❑ I would somewhat rather be supervised by a man
    ❑ I would much rather be supervised by a woman
    ❑ I would somewhat rather be supervised by a woman
    ❑ It does not make a difference whether a man or a woman supervises me

28. **In terms of being supervised by a man:**
    ❑ I would much rather be supervised by a nonminority
    ❑ I would somewhat rather be supervised by a nonminority
    ❑ I would much rather be supervised by a minority
    ❑ I would somewhat rather be supervised by a minority
    ❑ It does not make a difference to which group my supervisor belongs

29. **In terms of being supervised by a woman:**
    ❑ I would much rather be supervised by a nonminority
    ❑ I would somewhat rather be supervised by a nonminority
    ❑ I would much rather be supervised by a minority
    ❑ I would somewhat rather be supervised by a minority
    ❑ It does not make a difference to which group my supervisor belongs

If this questionnaire is being used for a training class, please check the one answer in the following questions that best applies to you.

30. **What is your gender?**
    ❑ male
    ❑ female

31. **What is your ethnicity or race?**
    ❑ African American or black
    ❑ Asian/Pacific American
    ❑ Latino/Hispanic American
    ❑ Native American, American Indian, Alaskan Native
    ❑ White
    ❑ other _____

32. **How many years have you been employed by the fire department?**
    ❑ 0 to 5 years
    ❑ 6 to 10 years
    ❑ 11 to 20 years
    ❑ more than 20 years

33. **What is your current rank?**
    ❑ Officer
    ❑ Engineer
    ❑ Sergeant
    ❑ Lieutenant
    ❑ Captain
    ❑ Chief
    ❑ Chief of department
    ❑ Other _____

34. **What is the highest academic degree that you hold?**
    ❑ High school diploma
    ❑ Associate's degree
    ❑ Bachelor's degree
    ❑ Master's degree
    ❑ Other _____

Listing of Consultants and Resources

The following is a partial list of professional consultants and resources that may be useful to fire services (current as of March 2008). Please note that we are not endorsing the services of these consultants but are making our readers aware of their specialized multicultural and cross-cultural training and consultation. We recommend searching the Internet for a more detailed listing.

CROSS-CULTURAL DIVERSITY CONSULTANTS AND RESOURCES

Aaron T. Olson, Organization and Training Services Consultant. P.O. Box 345, Oregon City, OR 97045. Telephone: (971) 409-8135. Coauthor of *Multicultural Law Enforcement: Strategies for Peacekeeping in a Diverse Society,* 4th edition; *Concepts and Tools in Practice for Multicultural Law Enforcement,* 2nd edition, and *Instructor's Manual,* 4th edition; and *Multicultural and Diversity Strategies for the Fire Service.* U.S. Army veteran and retired Oregon State Police patrol sergeant and supervisor, and Criminal Justice Instructor at Portland Community College, Portland, Oregon. As a consultant, he provides communications, leadership, instructor development, and multicultural training to business, government, and public safety organizations (i.e., firefighter recruit and in-service academies). Website: http://www.atolson.com. E-mail: information@atolson.com

Change Works Consulting (CWC). 28 South Main Street, #113, Randolph, MA 02368. Telephone: (781) 986-6150. CWC staff use an innovative philosophy and framework to address diversity, inclusion, antiracism, and organizational change. From education and training, to strategic planning, from assessments to diversity councils, from planning diversity initiatives to writing insightful articles, CWC works with clients to provide services that help create more equitable and respectful environments. Website: http://changeworksconsulting.org

Deena Levine & Associates. P.O. Box 582, Alamo, CA 94507. Telephone: (925) 947-5627. Coauthor of *Multicultural Law Enforcement: Strategies for Peacekeeping in a Diverse Society,* 4th edition. She provides cross-cultural consulting for businesses and organizations. Website: http://www.dlevineassoc.com. E-mail: deena_levine@astound.net

Elsie Y. Cross Associates, Inc. 7627 Germantown Avenue, Philadelphia, PA 19118. Telephone: (215) 248-8100. Provides organization development in planning and implementing diversity initiatives. Offers a quarterly e-journal on a broad range of issues that affect anyone interested or engaged in diversity initiatives and programs. Website: http://www.eyca.com

George Simons International. 236 Plateau Avenue, Santa Cruz, CA 95060. Telephone: (831) 531-4706. Provides consulting, coaching, training, e-learning, and customized tools to empower organizations and their people to work, communicate, and negotiate effectively across cultures. Website: http://www.diversophy.com

The GilDeane Group, Inc. 13751 Lake City Way, NE, Suite 210, Seattle, WA 98125. Telephone: (206) 362-0336. Is a consulting and training firm that helps people and their organizations work effectively across cultural borders. Believes "cultural intelligence" is key to this effectiveness. Website: http://www.gildeane.com

Guardian Quest. 900 Lakewood Place, Aurora, IL 60506. Telephone: (630) 449-0958. Provides training and consulting on organizational assessments, diversity, ethics, leadership, and train the trainer to businesses, government, and public safety agencies. Retired Deputy Chief Ondra Berry from the Reno Police Department is cofounder of the organization. Website: http://guardianquest.com

Herbert Z. Wong & Associates. 111 Deerwood Road, Suite 200, San Ramon, CA 94583. Telephone. (925) 837-3595. Dr. Herbert Z. Wong is couuthor of *Multicultural and Diversity Strategies for the Fire Service; Multicultural Law Enforcement: Strategies for Peacekeeping in a Diverse Society,* 4th edition; and *Concepts and Tools in Practice for Multicultural Law Enforcement,* 2nd edition. Herbert Z. Wong & Associates consultants engage in organizational surveys, cultural assessments, and workforce diversity training. The consulting firm specializes in diversity training and the assessment of organizational culture for implementing the strategic advantages of diversity for the fire service. Local, county, state, and federal public safety and regulatory agency diversity assessment and training are also provided.

Loden Associates, Inc. 140 Hacienda Drive, Tiburon, CA 94920. Telephone: (415) 435-8507. Loden Associates takes a comprehensive approach to the design and implementation of diversity initiatives that maximizes synergy with strategic business goals and support for organizational culture change. All client consultation and program development are based on proven behavioral science principles that facilitate broad acceptance of change as they minimize resistance. Website: http://www.loden.com

ODT Inc. P.O. Box 134, Amherst, MA 01004. Telephone: (800) 736-1293. A web store source of alternative information on maps, corporate diversity materials, and books. Offers an array of fascinating and thought-provoking resources to expand your view of the world. Website: http://www.odt.org

Simulations Training Systems. P. O. Box 910, Del Mar, CA 92014. Telephone: (800) 942-2900 and (858) 755-0272. Produces simulation training programs for businesses, schools, government agencies, and nonprofit organizations dealing with the areas of cross-cultural relations, diversity, empowerment, team building, the use and abuse of power, ethics, and sexual harassment. Website: http://www.simulationtrainingsystems.com

Sounder, Betances and Associates, Inc. 5448 North Kimball Avenue, Chicago, IL 60625. Telephone: (773)463-6374. Works with public and private organizations of all sizes in providing consultation and training services on diversity awareness, diversity competency, diversity mentoring, leadership, and many more areas. Has a specialized program customized for public safety agencies. Website: http://betances.com

ANTI-BIAS ORGANIZATIONS AND GOVERNMENT AGENCIES

African American Firefighter Museum
1401 South Central Avenue
Los Angeles, CA 90021
Telephone: (213) 744-1730
Fax: (213) 744-1731
Website: http://www.aaffmuseum.org

American-Arab Anti-Discrimination Committee (ADC)
4201 Connecticut Avenue, NW, Suite 500
Washington, DC 20008
(202) 244-2990
Website: http://www.adc.org

Anti-Defamation League (ADL)
442 Park Avenue South
New York, NY 10016
(212) 684-6950
Website: http://www.adl.org

Asian American Justice Center (Formerly the National Asian Pacific American Legal Consortium)
1140 Connecticut Avenue, NW, Suite 1200
Washington, DC 20036
(202) 296-2300
Website: http://www.advancingequality.org

Center for Democratic Renewal National Office (CDR)
P.O. Box 50469
Atlanta, GA 30302-0469
(404) 221-0025
Website: http://www.thecdr.org

Coalition Against Anti-Asian Violence
c/o Asian American Legal Defense and Education Fund
99 Hudson Street, 12th Floor
New York, NY 10013
(212) 966-5932
Website: http://www.aaldef.org/

Congress on Racial Equality (CORE)
30 Cooper Square
New York, NY 10003
(212) 598-4000
Website: http://www.core-online.org

Gay & Lesbian Alliance Against Defamation (GLAAD)
150 West 26th Street, Suite 503
New York, NY 10001
(212) 807-1700
Website: www.glaad.org

Jewish Council for Public Affairs
443 Park Avenue South, 11th Floor
New York, NY 10016
(212) 684-6950
Website: http://www.jewishpublicaffairs.org

Museum of Tolerance
9786 W. Pico Boulevard
Los Angeles, California 90035
(310) 553-8403
Website: http://www.museumoftolerance.com

National Association for the Advancement of Colored People (NAACP)
Washington Bureau
1025 Vermont Avenue, NW
Washington, DC 20009
(202) 638-2269
Website: http://www.naacp.org

National Conference for Community and Justice (NCCJ)
328 Flatbush Avenue
Box 402
Brooklyn, NY 11217
(718) 783-0044
Website: http://www.nccj.org

National Congress of American Indians (NCAI)
1301 Connecticut Avenue, NW, Suite 200
Washington, DC 20036
 (202) 466-7767
Website: http://www.ncai.org

National Council of La Raza (NCLR)
Raul Yzaguirre Building
1126 16th Street, NW
Washington, DC 20036
(202) 785-1670
Website. http://www.nclr.org

National Gay and Lesbian Task Force
1325 Massachusetts Avenue, NW, Suite 600
Washington, DC 20005
(202) 393-2241
Website: http://www.thetaskforce.org

National Organization for Women (NOW)
1000 16th Street, NW, Suite 700
Washington, DC 20036
(202) 331-0066
Website: http://www.now.org

The Prejudice Institute
2743 Maryland Avenue
Baltimore, MD 21218
Telephone: (410) 243-6987
Fax: (410) 243-6987
Website: http://www.prejudiceinstitute.org

Southeast Asia Resource Action Center (SEARAC)
1628 16th Street, NW, 3rd Floor
Washington, DC 20009
(202) 667-4690
Website: http://www.searac.org

Southern Poverty Law Center
400 Washington Avenue
Montgomery, AL 36104
(334) 956-8200
Website: http://www.splcenter.org

U.S. Commission on Civil Rights
624 9th Street, NW
Washington, DC 20425
(202) 376-8312
Website: http://www.usccr.gov

The U.S. Commission on Civil Rights (1) investigates complaints alleging that citizens have been deprived of their right to vote resulting from discrimination; (2) studies and collects information relating to discrimination on the basis of race, color, religion, sex, age, disability, or national origin; (3) reviews federal laws, regulations, and policies with respect to discrimination and equal protection under the law; and (4) submits reports to the president and Congress on civil rights issues.

U.S. Department of Justice
Bureau of Justice Statistics
810 Seventh Street, NW
Washington, DC 20531
(800) 732-3277
Website: http://www.ojp.usdoj.gov/bjs

U.S. Department of Justice Community Relations Service Headquarters
810 Seventh Street, NW
Washington, DC 20531
(202) 305-2935
Website: http://www.usdoj.gov/crs

U.S. Equal Employment Opportunity Commission (EEOC)
1801 L Street, NW
Washington, DC 20507
(202) 663-4400
Website: http://www.eeoc.gov

U.S. Fire Administration
16825 South Seton Avenue
Emmitsburg, MD 21727
Telephone: (301) 447-1000
Fax: (301) 447-1346
Website: http://www.usfa.fema.gov

Women in the Fire Service, Inc.
P.O. Box 5446
Madison, WI 53705
Telephone: (608) 233-4768
Fax: (608) 233-4879
Website: http://www.wfsi.org

Multicultural Community and Fire Service Workforce Survey

The first set of questions asks for your opinions about how certain segments of the community view fire service. Using the response sheets (Attitude Assessment Survey Response Sheet) on pages 335-338, put the number of the response that you think best describes each group's perception. Remember, give the response based on how you feel each group would answer the statements.

1. In your opinion, how would this group rate the job this fire department does?

	Very good	Good	Fair	Poor	Very Poor
Business community	❑	❑	❑	❑	❑
Minority residents	❑	❑	❑	❑	❑
Community leaders	❑	❑	❑	❑	❑
Most residents	❑	❑	❑	❑	❑
Community Volunteers	❑	❑	❑	❑	❑

2. This group generally cooperates with fire service.

	All of the Time	Most of the Time	Sometimes	Rarely	Never
Business community	❑	❑	❑	❑	❑
Minority residents	❑	❑	❑	❑	❑
Community leaders	❑	❑	❑	❑	❑
Most residents	❑	❑	❑	❑	❑
Community Volunteers	❑	❑	❑	❑	❑

3. Overall, this group thinks that fire department acts to protect the individuals' communities.

	Strongly Agree	Agree	Neither Agree nor Disagree	Disagree	Strongly Disagree
Business community	❑	❑	❑	❑	❑
Minority residents	❑	❑	❑	❑	❑
Community leaders	❑	❑	❑	❑	❑
Most residents	❑	❑	❑	❑	❑
Community Volunteers	❑	❑	❑	❑	❑

4. This group feels that the current relationship between the fire department and the community is described by which of the following?

	Very good	Good	Fair	Poor	Very Poor
Business community	❑	❑	❑	❑	❑
Minority residents	❑	❑	❑	❑	❑
Community leaders	❑	❑	❑	❑	❑
Most residents	❑	❑	❑	❑	❑
Community Volunteers	❑	❑	❑	❑	❑

5. **Overall, this group feels this department responds to citizen complaints about fire service officers in an objective and fair manner.**

	Strongly Agree	Agree	Neither Agree nor Disagree	Disagree	Strongly Disagree
Business community	❏	❏	❏	❏	❏
Minority residents	❏	❏	❏	❏	❏
Community leaders	❏	❏	❏	❏	❏
Most residents	❏	❏	❏	❏	❏
Community Volunteers	❏	❏	❏	❏	❏

6. **This group thinks most contacts with fire service are negative. The next questions ask for your opinions about procedures and practices within the fire department.**

	Very good	Good	Fair	Poor	Very Poor
Business community	❏	❏	❏	❏	❏
Minority residents	❏	❏	❏	❏	❏
Community leaders	❏	❏	❏	❏	❏
Most residents	❏	❏	❏	❏	❏
Community Volunteers	❏	❏	❏	❏	❏

7. **Overall, fire service supervisors in this department respond to citizens' complaints about employees in an objective and fair manner.**
   - ❏ Strongly agree
   - ❏ Agree
   - ❏ Neither Agree nor Disagree
   - ❏ Disagree
   - ❏ Strongly disagree

8a. **Most fire service officers in this department are sensitive to cultural and community differences.**
   - ❏ Strongly agree
   - ❏ Agree
   - ❏ Neither Agree nor Disagree
   - ❏ Disagree
   - ❏ Strongly disagree

8b. **Most civilian employees in this department are sensitive to cultural and community differences.**
   - ❏ Strongly agree
   - ❏ Agree
   - ❏ Neither Agree nor Disagree
   - ❏ Disagree
   - ❏ Strongly disagree

9a. **This fire department adequately prepares officers to work with members of the community who are of a different race or ethnicity than the majority of the population.**
   - ❏ Strongly agree
   - ❏ Agree
   - ❏ Neither Agree nor Disagree
   - ❏ Disagree
   - ❏ Strongly disagree

9b. **This fire department adequately prepares civilian employees to work with members of the community who are of a different race or ethnicity than the majority of the population.**
   - ❏ Strongly agree
   - ❏ Agree
   - ❏ Neither Agree nor Disagree
   - ❏ Disagree
   - ❏ Strongly disagree

10. The fire service administration is more concerned about fire service–community relations than it should be.
❑ Strongly agree
❑ Agree
❑ Neither Agree nor Disagree
❑ Disagree
❑ Strongly disagree

11a. Special training should be given to fire service officers who work with community members who are of a different race or ethnicity than the majority population.
❑ Strongly agree
❑ Agree
❑ Neither Agree nor Disagree
❑ Disagree
❑ Strongly disagree

11b. Special training should be given to civilian employees who work with community members who are of a different race or ethnicity than the majority population.
❑ Strongly agree
❑ Agree
❑ Neither Agree nor Disagree
❑ Disagree
❑ Strongly disagree

12. Special training should be given to assist fire officers in working with which of the following segments of the community?

	Yes	No
African American/Black (includes Caribbean, Haitian, etc.)	❑	❑
Asian/Pacific	❑	❑
Latino/Hispanic	❑	❑
Gay Lesbian, Bisexual & Transgener	❑	❑
Middle Eastern/Arab American	❑	❑
American Indian/Native American	❑	❑

13. How often are racial slurs and negative comments about persons of a different race or ethnicity expressed by personnel in this department?
❑ Often
❑ Sometimes
❑ Rarely
❑ Never

14a. Persons of a different race or ethnicity in this city are subject to unfair treatment by some fire service officers in this department.
❑ Strongly agree
❑ Agree
❑ Neither Agree nor Disagree
❑ Disagree
❑ Strongly disagree

14b. Persons of a different race or ethnicity in this city are subject to unfair treatment by some civilian employees in this department.
❑ Strongly agree
❑ Agree
❑ Neither Agree nor Disagree
❑ Disagree
❑ Strongly disagree

15a. Prejudicial remarks and discriminatory behavior by fire service officers are not tolerated by line supervisors in this department.
❑ Strongly agree
❑ Agree
❑ Neither Agree nor Disagree
❑ Disagree
❑ Strongly disagree

15b. Prejudicial remarks and discriminatory behavior by civilian employees are not tolerated by line supervisors in this department.
❑ Strongly agree
❑ Agree
❑ Neither Agree nor Disagree
❑ Disagree
❑ Strongly disagree

16. Transfer policies in this department have a negative effect on fire service–community affairs.
❑ Strongly agree
❑ Agree
❑ Neither Agree nor Disagree
❑ Disagree
❑ Strongly disagree

17. Citizen complaint procedures in this department operate in favor of the citizen, not the employee.
❑ Strongly agree
❑ Agree
❑ Neither Agree nor Disagree
❑ Disagree
❑ Strongly disagree

18. Internal discipline procedures for employee misconduct are generally appropriate.
❑ Strongly agree
❑ Agree
❑ Neither Agree nor Disagree
❑ Disagree
❑ Strongly disagree

19. The procedure for a citizen to file a complaint against a department employee should be which of the following?
❑ Citizen sends complaint in writing to the department
❑ Citizen telephones complaint to the department
❑ Citizen comes to the department
❑ Any of the above are acceptable means
❑ None of the above are acceptable means
Explain. _____
_____
_____

20. With regard to discipline for misconduct, all employees in this department are treated the same in similar situations, regardless of race or ethnicity.
❑ Strongly agree
❑ Agree
❑ Neither Agree nor Disagree
❑ Disagree
❑ Strongly disagree

21. **What kind of discipline do you think is appropriate for the first incident of the following types of misconduct? (Assume intentional.)**

Type of Misconduct	Verbal Warning	Training/ Counseling	Oral Reprimand	Formal Reprimand	Suspension	Termination
Discrimination	❏	❏	❏	❏	❏	❏
Use of racial slurs	❏	❏	❏	❏	❏	❏
Criminal conduct	❏	❏	❏	❏	❏	❏
Poor service	❏	❏	❏	❏	❏	❏
Discourtesy to citizen	❏	❏	❏	❏	❏	❏
Improper procedure	❏	❏	❏	❏	❏	❏

This section examines your views about fire service–community relations training and community participation.

Please circle the response that best describes your opinion.

22. **Do you think training in fire service–community relations was adequate to prepare you to work with all segments of the community?**
❏ Yes
❏ No
❏ Don't know

Explain. If no, please describe why the training was not satisfactory. _____
_____
_____
_____

23. **How often do you have opportunities to participate in positive contacts with community groups?**
❏ Frequently
❏ Sometimes
❏ Rarely
❏ Never

24. **Do you think this fire department has an adequate community relations program?**
❏ Yes
❏ No
❏ Don't know

Please explain. _____
_____
_____
_____

25. **What subject areas related to community relations would be helpful on an in-service training basis?**
_____
_____
_____

26. **What do you think is the most important thing that citizens need to understand about the fire department?**

_____

_____

_____

27. **How can the fire department best educate the public about fire service policies and practices?**
    ❏ Through fire service officer contacts with citizens
    ❏ Through public meetings
    ❏ Through the media
    ❏ Selected combinations of the responses above
    ❏ Other _____
    ❏ Don't know
    Explain. _____

    _____

    _____

28. **Listed are steps that fire departments can take to improve fire services as they relate to community relations.**

    Please indicate how important you think each of the following should be to this administration by placing the number that best describes your response next to the appropriate question.

    1 = Very Important   *2 = Important*   *3 = Not at all important*

    _____ Hire more fire service officers
    _____ Focus on more serious emergencies and disasters
    _____ Improve response time
    _____ Increase salaries
    _____ Provide more training
    _____ Raise qualifications for potential applicants
    _____ Be more courteous to public
    _____ Increase community fire service
    _____ Reduce discrimination
    _____ Provide dedicated time for community involvement
    _____ Other _____

Glossary

acculturation The process of becoming familiar with and comfortable in another culture. The ability to function within a different culture or environment, while retaining one's own cultural identity.

affirmative action Legally mandated programs with the objective to increase the employment or educational opportunities of groups that have been disadvantaged in the past.

alien Any person who is not a citizen or national of the country in which he or she lives.

anti-Semitism Latent or overt hostility toward Jews, often expressed through social, economic, institutional, religious, cultural, or political discrimination and through acts of individual or group violence.

applicant screening The methods, processes, and procedures used by fire service agencies to assess applicants along a range of dimensions to perform fire service roles, tasks, and functions.

assimilation The process by which ethnic groups that have emigrated to another society begin to lose their separate identity and culture, becoming absorbed into the larger community.

awareness Bringing to one's conscious mind that which is only unconsciously perceived.

bias Preference or inclination to make certain choices that may be positive (bias toward excellence) or negative (bias against people), often resulting in unfairness.

bigot A person who steadfastly holds to bias and prejudice, convinced of the truth of his or her own opinion and intolerant of the opinions of others.

CARE Approach A value-based concept and principle that promotes strategies to optimize harmony in individual, group, and organizational relationships. CARE is an acronym for being compassionate, attentive, responsive, and eclectic. It can be used with, but is not limited to, approaches dealing with incidents, issues, or problems. It has universal application to business, communities, government, military, schools, nonprofit organizations, and all people. It is humanistic in design and maximizes the affective domain of human behavior for learning and performance.

community assessment Reviewing and analyzing all of the localities that are (1) likely targets for terrorism and (2) problems during community disasters, and working to improve the security and safety of these locations.

consent decree An out-of-court settlement whereby the accused party agrees to modify or change behavior rather than plead guilty or go through a hearing on charges brought to court. (Source: CSIS Project Glossary, http://csisweb.aers.psu.edu/glossary/c.htm)

cross-cultural Involving or mediating between two cultures.

cultural competence A developmental process that evolves over an extended period of time; both individuals and organizations are at various levels of awareness, knowledge, and skills along a continuum of cultural competence. Culturally competent organizations place a high value on the following (1) developing a set of principles, attitudes, policies, and structures that will enable all individuals in the organization to work effectively and equitably across all cultures; and (2) developing the capacity to acquire and apply cross-cultural knowledge and respond to and communicate

337

effectively within the cultural contexts that the organization serves. (Adapted from Cross, T., Bazron, B. J., Dennis, K. W., and Isaacs, M. R. [1989]. *Toward a culturally competent system of care. Volume 1: Monograph on effective services for minority children who are severely emotionally disturbed.* Washington, DC CASSP Technical Assistance Center, Georgetown University Child Development Center.)

culture A way of life developed and communicated by a group of people, consciously or unconsciously, to subsequent generations. It consists of ideas, habits, attitudes, customs, and traditions that help to create standards for a group of people to coexist, making a group of people unique. In its most basic sense, culture is a set of patterns for survival and success that a particular group of people has developed.

Department of Homeland Security (DHS) The cabinet-level federal agency responsible for preserving the security of the United States against terrorist attacks. This department was created as a response to the 9/11 terrorist attacks.

discrimination The denial of equal treatment to groups because of their racial, ethnic, gender, religious, or other form of cultural identity.

diversity The term used to describe a vast range of cultural differences that have become factors needing attention in living and working together. Often applied to the organizational and training interventions in an organization that seek to deal with the interface of people who are different from each other. Diversity has come to include race, ethnicity, gender, disability, and sexual orientation.

domestic terrorism Involves groups or individuals whose terrorist activities are directed at elements of government, organizations, or population without foreign direction.

dominant culture Refers to the value system that characterizes a particular group of people that dominates the value systems of other groups or cultures. See also *macro culture/majority or dominant group.*

émigré An individual forced, usually by political circumstances, to move from his or native country and who deliberately resides as a foreigner in the host country.

ethnic group Group of people who conceive of themselves, and who are regarded by others, as alike because of their common ancestry, language, and physical characterisics.

ethnicity Refers to the background of a group with unique language, ancestral, often religious, and physical characteristics. Broadly characterizes a religious, racial, national, or cultural group.

ethnocentrism Using the culture of one's own group as a standard for the judgment of others, or thinking of it as superior to other cultures that are merely different.

fire agency Use synonymously as fire department and firehouse.

fire executive Current term in the fire service for fire chief and other fire service senior officers.

glass ceiling An invisible but often perceived barrier that prevents some ethnic or racial groups and women from becoming promoted or hired.

heterogeneity Dissimilar; composed of unrelated or unlike elements.

immigrant Any individual who moves from one country, place, or locality to another. An alien admitted to the United States as a lawful permanent resident.

international terrorism Involves groups or individuals whose terrorist activities are foreign-based and/or directed by countries or groups outside the United States or whose activities transcend national boundaries.

leadership In a multicultural society and workforce, begins with what the leader must possess for values, attributes, and skills that shape a leader's character. These internal qualities are present while at work, alone, and with others.

macro culture/majority or dominant group
The group within a society that is largest and/or most powerful. This power usually extends to setting cultural norms for the society as a whole. The term *majority* (also *minority,* see below) is falling into disuse because its connotations of group size may be inaccurate.

management The performance of tasks that need to be accomplished to direct the continued operations of an organization. A manager's duties should align themselves to the organization's mission statement.

micro culture or minority Any group or person who differs from the dominant culture. Any group or individual, including second- and third-generation foreigners, who is born in a country different from his or her origin and has adopted or embraced the values and culture of the dominant culture.

multiculturalism The existence within one society of diverse groups that maintain their unique cultural identity while accepting and participating in the larger society's legal and political system.

paradigm shift What occurs when an entire cultural group begins to experience a change that involves the acceptance of new conceptual models or ways of thinking and results in major societal transitions (e.g., the shift from agricultural to industrial society).

parity The state or condition of being the same in power, value, rank, and so forth; equality.

pluralistic The existence within a nation or society of groups distinctive in ethnic origin, cultural patterns, religion, or the like. A policy of favoring the preservation of such groups within a given nation or society.

prejudice The inclination to take a stand for one side (as in a conflict) or to cast a group of people in favorable or unfavorable light, usually without just grounds or sufficient information.

professionalism Means approaching an activity, such as one's occupation or career, with a sense of dedication and expertise.

A professional possesses integrity and demonstrates competence, regardless of rank, assignment, or tenure.

race A group of persons of (or regarded as of) common ancestry. Physical characteristics are often used to identify people of different races. These characteristics should not be used to identify ethnic groups, which can cross racial lines.

racism Total rejection of others by reason of race, color, or, sometimes more broadly, culture.

racist One with a closed mind toward accepting one or more groups different from one's own origin in race or color.

recruiting incentives Financial and other incentives to recruit women and minorities where such programs are lawful (monetary incentive programs may be adversely affected by Fair Labor Standards Act considerations).

recruitment strategies Approaches used by organizations to attract and bring in job applicants and candidates to enable a diverse workforce.

refugee A person who flees for safety and seeks asylum in another country. In addition to those persecuted for political, religious, and racial reasons, "economic refugees" flee conditions of poverty for better opportunities elsewhere.

role identification The possibilities to identify with roles in which many family members were firefighters or had belonged to a group of peers in which many peers became firefighters; such role identifications are facilitative of individuals joining the fire service.

role modeling Leadership and positive examples seen among members senior in the fire services to facilitate positive actions and behaviors of all fire service personnel.

stereotype To believe or feel that people and groups are considered to typify or conform to a pattern or manner, lacking any individuality. Thus a person may categorize behavior of a total group on the basis of limited experience with one or a few representatives of that group. Negative stereotyping classifies many people in a group by slurs, innuendoes, names, or slang expressions that depreciate the group as a whole and the individuals in it.

subculture A group with distinct, discernible, and consistent cultural traits existing within and participating in a larger cultural grouping.

synergy The benefit produced by the collaboration of two or more systems in excess of their individual contributions. Cultural synergy occurs when cultural differences are taken into account and used by a multicultural group.

terrorism A violent act or an act dangerous to human life, in violation of the criminal laws of the United States or any segment to intimidate or coerce a government, the civilian population, or any segment thereof, in furtherance of political or social objectives.

terrorist attack pre-incidence indicators (TAPIs) A term used by the intelligence community to describe actions and behaviors taken by terrorists before they carry out an attack.

transgender The term *transgender* covers a range of people, including heterosexual cross-dressers, homosexual drag queens, and transsexuals, who believe they were born in the wrong body. This term includes those who consider themselves to be both male and female, or intersexed, as well as those who take hormones to complete their gender identity without a sex change.

weapons of mass destruction (WMD) Three types of weapons are most commonly categorized as WMD: (1) nuclear weapons, (2) biological weapons, and (3) chemical weapons. The subcategories of WMD include activities that may be labeled as "agro terrorism," which is to harm our food supply chain, and "cyber terrorism," which is to harm our telecommunication, Internet, and computerized processes and transactions.

Index

Note: Italicized page numbers refer to the Glossary section. Page numbers followed by *f* indicate figures.